DYNAMICS
OF SNOW
AND ICE MASSES

CONTRIBUTORS

GEORGE D. ASHTON

W. D. HIBLER III

D. H. MALE

W. S. B. PATERSON

R. I. PERLA

C. F. RAYMOND

R. Q. ROBE

DYNAMICS
OF SNOW
AND ICE MASSES.

Edited by

SAMUEL C. COLBECK

U.S. Army Cold Regions Research
and Engineering Laboratory
Hanover, New Hampshire

 1980

ACADEMIC PRESS
A Subsidiary of Harcourt Brace Jovanovich, Publishers

New York London Toronto Sydney San Francisco

ACADEMIC PRESS, INC.
111 Fifth Avenue, New York, New York 10003

United Kingdom Edition published by
ACADEMIC PRESS, INC. (LONDON) LTD.
24/28 Oval Road, London NW1   7DX

Library of Congress Cataloging in Publication Data

Main entry under title:

Dynamics of snow and ice masses.

    Includes bibliographies and index.
    1. Ice. 2. Snow. I. Colbeck, Samuel C.
GB2403.2.D95   551.3'1   79–27949
ISBN 0–12–179450–4

PRINTED IN THE UNITED STATES OF AMERICA

80 81 82 83   9 8 7 6 5 4 3 2 1

CONTENTS

LIST OF CONTRIBUTORS

Numbers in parentheses indicate the pages on which the authors' contributions begin.

GEORGE D. ASHTON (261) U.S. Army Cold Regions Research and Engineering Laboratory, Hanover, New Hampshire 03755

W. D. HIBLER III (141) U.S. Army Cold Regions Research and Engineering Laboratory, Hanover, New Hampshire 03755

D. H. MALE (305) Division of Hydrology, University of Saskatchewan, Saskatoon, Saskatchewan, Canada

W. S. B. PATERSON (1) Polar Continental Shelf Project, Department of Energy, Mines and Resources, Ottawa, Ontario K1A 0E4, Canada

R. I. PERLA (397) Environment Canada, Canmore, Alberta T0L 0M0, Canada

C. F. RAYMOND (79) Geophysics Program AK-50, University of Washington, Seattle, Washington 98195

R. Q. ROBE (211) U.S. Coast Guard Research and Development Center, Avery Point, Groton, Connecticut 06340

PREFACE

The original studies of snow and ice masses were made in the high mountains and near the remote poles by people with a scientific curiosity. In addition to their well-established academic appeal, these studies now assume tremendous practical importance. Petroleum-related transportation in the Arctic Basin, winter navigation on northern lakes and rivers, competing demands for snowmelt and glacial runoff, exploitation of mineral deposits in glacial valleys, damage caused by avalanches, use of icebergs as a freshwater source, and disposal of radioactive wastes on large ice sheets are a few examples of causes of the increasing interest in snow and ice masses. In view of these academic and engineering concerns, it is necessary to have a book to serve as an introduction and reference to the growth, motion, and decay of snow and ice masses.

This book treats the dynamical aspects of snow and ice masses on the geophysical scale. It is divided into seven chapters, each of which describes the basic features of a particular snow or ice mass. In each chapter a conceptual framework is established on a physical basis and a mathematical description is provided with as many references to the technical literature as space allows. No attempt is made to address particular applications of the information, but the physical and mathematical descriptions of the properties and processes provide for both an understanding of snow and ice masses and a basis through which particular problems can be addressed.

This book provides an outline of the subject for study at the graduate-student level, a review for our colleagues in snow and ice studies, and an introduction for other people interested in snow and ice. For example, it serves as an introduction to ice-sheet dynamics for climatologists, to sea ice for oceanographers, and to snowmelt for hydrologists. For them we have tried to provide broad coverage of the subject while emphasizing the most essential properties and processes.

This book would not have been possible without the complete cooperation of many people at the Cold Regions Research and Engineering Laboratory. In particular, I acknowledge the cooperation of Wesley Pietkiewicz and Edward Perkins of the Technical Information Branch. Mark Ray was valuable to me as an assistant editor and Pam Sirois as a typist. Dr. George Ashton and Dr. Kay Sterrett provided encouragement as well as financial support through the work unit, Properties of Snow and Ice.

1 ICE SHEETS AND ICE SHELVES

W. S. B. Paterson

Polar Continental Shelf Project
Department of Energy, Mines and Resources
Ottawa, Ontario, Canada

DYNAMICS OF SNOW AND ICE MASSES
Copyright © 1980 by Academic Press, Inc.
All rights of reproduction in any form reserved.
ISBN 0-12-179450-4

LIST OF SYMBOLS

a Ablation rate measured as thickness of ice per unit time

A_0 Flow law constant [Eq. (7)]

A_1 Flow law constant for perfect plasticity [Eq. (8)]

A Flow law parameter [Eqs. (6) and (7)]

b Mass balance; thickness of ice added by accumulation (positive) or removed by ablation (negative) in unit time

B Constant in hyperbolic sine flow law [Eq. (77)]

c Accumulation rate measured as thickness of ice per unit time

c' Specific heat

c_b Accumulation rate at base of ice shelf

c_s Accumulation rate at surface

C Constant in sliding law [Eq. (19)]

C_1 Constant (Section VI.F)

C_2 Constant (Section VI.F)

D_0 Constant [Eq. (103)]

$D(L)$ Water depth at ice-sheet–ice-shelf junction

F Rate of internal heat production per unit volume

g Acceleration due to gravity

G Longitudinal gradient of stress deviator [Eq. (48)]

h Ice thickness

h_0 Ice thickness at origin (Section III.D)

h_1 Height of ice shelf surface above sea level

h' Transition depth in Dansgaard–Johnsen flow model [Eqs. (59) and (61)]

h_b y coordinate of bed

h_s y coordinate of surface

H Thickness at center of ice sheet

H_1 Surface elevation, at center of ice sheet, above horizontal plane [Eq. (39)]

I Dawson's integral

J Mechanical equivalent of heat (unity in S.I. units)

k Thermal diffusivity

K Thermal conductivity

K_0 Constant [Eq. (28)]

K_1 Constant [Eq. (88)]

K_2 Constant [Eq. (105)]

K' Constant [Eq. (59)]

l Characteristic length in temperature distribution [Eqs. (92) and (96)]

L Half-width of ice sheet

L_0 Initial half-width of growing or shrinking ice sheet

L_1 Distance from center of ice sheet to equilibrium line

n Flow law parameter [Eq. (6)]

q, q_x Ice flux in x direction,
$$\int_0^h u(y)\,dy$$

q_z Ice flux in z direction,
$$\int_0^h w(y)\,dy$$

Q Activation energy for creep

r Constant in Dansgaard–Johnsen flow model (Section III.E)

R Gas constant (8.314 J mole^{-1} deg^{-1})

s Shape factor for glacier cross section

S Bed smoothness parameter

t Time

T Temperature

T_0 Amplitude of surface temperature variation (Section VI.E)

T_1 Constant temperature (Section VI.F)

T_2 Constant temperature (Section VI.F)

T' Shear stress defined by Eq. (47)

T_b Temperature at base of ice sheet

T_s Temperature at surface of ice sheet

T_c Climatic warming rate [Eq. (95)]

u, v, w Velocity components in x, y, z directions

u_b, w_b Components of sliding velocity

U Velocity component averaged over ice thickness

W Constant surface warming rate [Eq. (94)]

x, y, z Space coordinates; except where stated otherwise, the origin is on the bed at the center of the ice sheet, the x axis is horizontal, the y axis vertical positive upward, the z axis chosen to

make the system right handed

Z Width of glacier or ice shelf

α Surface slope

α_0 Surface slope as defined in Section III.D

β Bed slope

β_0 Bed slope as defined in Section III.D

δ Phase angle of surface temperature variation Section VI.E)

$\dot{\varepsilon}$ Effective strain rate [Eq. (13)]

$\dot{\varepsilon}_x, \dot{\varepsilon}_y, \dot{\varepsilon}_z$ Normal strain rate components

$\dot{\varepsilon}_{xy}, \dot{\varepsilon}_{yz}, \dot{\varepsilon}_{xz}$ Shear strain rate components

λ Increase in air temperature per unit decrease in surface elevation

μ Flow law parameter [Eq. (10)]

ρ, ρ_i Density of ice

ρ_r Density of rock

μ_w Density of seawater

σ_0 Constant in hyperbolic sine flow law [Eq. (77)]

$\sigma_x, \sigma_y, \sigma_z$ Normal stress components

$\sigma_x', \sigma_y', \sigma_z'$ Stress deviator components

τ Effective shear stress [Eq. (14)]

τ_0 Yield stress

τ_b Basal shear stress

τ_s Shear stress at side of ice shelf

$\tau_{xy}, \tau_{yz}, \tau_{xz}$ Shear stress components

ϕ Geothermal heat flux

ω Angular frequency of surface temperature variation (Section VI.E)

I. INTRODUCTION

A. Reasons for Studying Ice-Sheet Dynamics

A glacier may be defined as a large mass of perennial ice that originates on land by the densification and recrystallization of snow and shows evidence of past or present flow [Meier (1974)]. In the upper part of a glacier, the amount of snow falling on the surface each year exceeds that lost by melting and evaporation. At lower elevations, all the previous winter's snow and some ice are removed by melting and runoff during the summer. Calving of icebergs into the sea or a lake may also result in significant loss of mass. However, the profile of the glacier does not change much from year to year, because ice flows under gravity from the *accumulation area* to the *ablation area*.

Numerous ways of classifying glaciers have been proposed. The simple scheme of Meier (1974) is adequate for our purposes. Meier divides glaciers into three main groups: (1) glaciers that are continuous sheets, moving outward in all directions, called *ice sheets* if they are the size of Antarctica or Greenland and *ice caps* if they are smaller; (2) glaciers whose paths are confined by the surrounding bedrock, called *mountain* or *valley glaciers* according to their location; and (3) glaciers whose lower parts spread out on level ground or float on the ocean, called *piedmont glaciers* and *ice shelves,* respectively. This chapter deals with ice sheets, ice caps, and ice shelves; the other types are treated in Chapter 2. The distinction between ice sheets and ice caps, based solely on size, seems arbitrary and serves no useful purpose; we will not adhere strictly to it. The physical principles governing the behavior of all types of glaciers are the same. However, the interest and importance of various aspects of the behavior may differ for the different types.

There are several reasons for studying the dynamics of glaciers.

(1) It is well known that glaciers advance and retreat in response to changes in climate. However, the relationship is much more complex than is usually assumed. Detailed deductions about past climates from geological evidence of the former extent of glaciers and predictions of their possible future behavior are impossible without detailed knowledge of glacier dynamics [Meier (1965); Paterson (1969, Chapter 11); Untersteiner and Nye (1968)]. Attempts to predict glacier advances may have considerable practical importance because of hydroelectric installations, mines, highways, and pipelines currently being built near glaciers. Moreover, ice avalanches from glaciers have caused catastrophic damage and loss of life [Lliboutry (1971, 1975); Röthlisberger (1974)] and the sudden drainage of glacier-dammed lakes has resulted in disastrous floods [Björnsson (1974)] and mud slides [Lliboutry *et al.* (1977)]. Also, major changes in the Greenland and Antarctic ice sheets would have significant effects on world climate and sea level.

(2) The higher parts of polar ice sheets contain a detailed and continuous record of past climate over a period of the order of 100,000 yr. This is because the ratios of the heavy to light atoms of oxygen and hydrogen

$(O^{18}/O^{16}, D/H)$ in precipitation depend on the temperature [Dansgaard (1961)]. Thus the variation of oxygen isotope ratio with depth in a core drilled through an ice sheet can be interpreted as the variation of temperature in past time [Johnsen *et al.* (1972); Dansgaard *et al.* (1973)]. In addition, the ice contains small amounts of microparticles [Thompson (1974)], volcanic ash [Gow and Williamson (1971)], and trace elements resulting both from natural causes and pollution [Herron *et al.* (1977); Cragin *et al.* (1977)] and from fallout from nuclear bomb tests [Crozaz (1969); Lorius (1973)]. Air bubbles in the ice contain samples of the atmosphere at the time the ice was formed [Raynaud and Delmas (1977)]. The variation of all these quantities with time can be studied if the age of the ice at each depth is known. Although methods of dating ice based on radioactivity [Clausen (1973); Oeschger *et al.* (1976)] and other techniques [Hammer *et al.* (1978)] are being developed, ages of existing deep cores have been calculated on the basis of simplified flow models [Dansgaard and Johnsen (1969a)]. Choice of a suitable model requires knowledge of ice sheet dynamics.

(3)   Recent theoretical studies have shown that ice sheets possess several potential sources of instability (see Section VII). These dynamical instabilities, which are unrelated to climatic changes, could result in a major spreading of the ice sheet. Indeed Hughes (1973) has suggested that the ice sheet in West Antarctica may be disintegrating at present. Rapid spreading of a major portion of the Antarctic Ice Sheet would increase the amount of ice on the surrounding ocean and thus reduce the amount of solar radiation absorbed by the earth. This could produce a worldwide climatic change. Only further theoretical analyses and field observations will resolve the question of whether such instabilities can occur.

(4)   In spite of numerous theories, the cause of ice ages is still unknown [Lliboutry (1965, pp. 897–924); Flint (1971, Chapter 30)]. Any explanation of the growth and decay of continental ice sheets must take account of their dynamics; this in turn requires studying the dynamics of existing ice sheets. An understanding of the mechanisms of erosion and deposition by existing glaciers is also essential for proper interpretation of the deposits left by ice-age ice sheets.

(5)   Folds and other features produced in glaciers by flow closely resemble structures observed in metamorphic rocks. This is not surprising because ice and rock are both polycrystalline solids and so they should deform in similar ways. However, rock deformation takes place in the earth's interior and may require millions of years, whereas ice deformation is much more rapid and can be observed in glaciers. Studying the origin of structures in glaciers can thus shed some light on fundamental geological processes.

(6)   Flow in the earth's interior is the process that builds mountains and causes the continents to drift. Mathematical models of these processes are being developed, but the theory is less well-developed than that of glacier flow. This is because conditions in the earth's interior are not as well known as those in glaciers and the flow of rock is more complex than that of ice. Thus application of some of the methods and results of glacier flow analysis to problems of flow in the earth might be an improvement over current treatments. Weertman (1962) and Lliboutry (1969) have done this. Thus recent theories of glacier flow may well have applications beyond the immediate purpose for which they were developed.

B.   Features of Existing Ice Sheets

Today, glacier ice covers about 10% of the world's land surface and stores about three-quarters of all the fresh water on earth. The Antarctic Ice Sheet contains about 90% of this with most of the remainder in Greenland. The Antarctic Ice Sheet and the ice shelves that surround much of it contain about 24×10^6 km^3 of ice and cover an area of about 14×10^6 km^2 or almost one and a half times that of Europe [Suyetova (1966)]. The Ice Sheet (Fig. 1) can be divided into the East and West Antarctic Ice Sheets on the basis of differences in surface and bedrock topography. The West Antarctic Ice Sheet is the part lying south of South America and lies between the meridians of 45° W and 165° E approximately. The Transantarctic Mountains form the physical division between the two ice sheets. The East Antarctic Ice Sheet, which contains over 80% of all the ice [Bardin and Suyetova (1967)], is an elliptical dome with a maximum elevation of over 4000 m. Although the greatest ice thickness is about 4500 m, this is in a small area and most of the bedrock surface is above sea level and has low relief. The West Antarctic Ice Sheet has a very rugged bedrock floor; much of it lies well below sea level, and would remain so if the ice were removed and isostatic rebound occurred. Ice thicknesses are comparable with those in East Antarctica, but the ice surface is lower and much more irregular.

Annual precipitation in water equivalent varies from about 30 mm in the central part of East Antarctica to about 600 mm near the coast [Lorius (1973)]. The average is about 170 mm, a value comparable to that in desert areas [Dolgushin *et al.* (1962)]. These figures exclude the Antarctic Peninsula, which receives more precipitation. Almost the entire surface of the ice sheet, including the ice shelves, is an accumulation area; iceberg calving accounts for almost all the loss in mass. Typical flow velocities

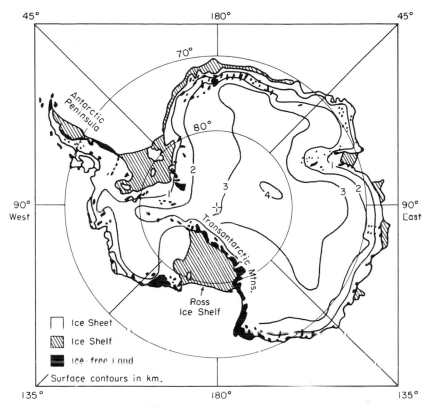

Fig. 1.   Map of Antarctica.

range from 1 m/yr or less in the central part of the ice sheet to about 200 m/yr near the coast [Budd *et al.* (1971a, Map 2/2)]. Higher velocities (up to 1400 m/yr) have, however, been measured in the large valley glaciers that drain into the Ross Ice Shelf and in the *ice streams* described in Section V.B. Whether the Ice Sheet is growing or shrinking at present is an important question. Several attempts have been made to answer it by calculating the total annual accumulation of snow and comparing it with estimates of the amount of ice lost as icebergs and by melting. At least seven figures for the mean annual thickness change, ranging from $+80$ to -30 mm in water equivalent, have been published in recent years [Dolgushin *et al.* (1962)]. Most estimates give a positive mass balance, but the results are inconclusive because none exceeds its standard error.

Further information about the Antarctic Ice Sheet can be found in summaries by Lliboutry (1965, pp. 505–514) and Shumskiy (1970) and in

Hatherton (1965). Budd *et al.* (1971a) have made a valuable compilation of data, with maps of measured quantities such as surface and bedrock topography, accumulation rates and surface temperatures plus deduced quantities such as ice velocities, ages, travel times, and basal temperatures.

The Greenland Ice Sheet (Fig. 2) has an area of about 1.7×10^6 km^2 and a volume of about 2.6×10^6 km^3 [Bader (1961)]. It is about 2400 km long from north to south and has a maximum width of about 1100 km. The surface has the form of two elongated domes; the larger and more northerly one reaches an elevation of about 3300 m. Both domes are displaced to the east of the center line of the ice sheet. The greatest ice thickness is about 3200 m. Bedrock in the interior of Greenland is close to sea level but the coastal areas are mountainous [Weidick (n.d.)] so the ice does not reach the sea over a broad front and there are no large ice shelves. Most of the ice flow is channeled into some 20 large outlet glaciers that penetrate the coastal mountains and discharge icebergs into the fjords. The glaciers in West Greenland are the fastest on earth, with velocities of up to 7 km/yr [Lliboutry (1965, p. 484)]. The mean annual accumulation is 340 mm water equivalent [Benson (1962)]. About half the mass is lost as icebergs and the remainder by ablation. Opinions differ as to whether the ice sheet is gaining or losing mass at present [Bader (1961); Bauer (1966)]; available data are not precise enough to give the answer. Summaries by Bader (1961) and Lliboutry (1965, pp. 481–487), and the book by Fristrup (1966) provide additional information about the Ice Sheet.

Ice caps are found in the Canadian and Soviet Arctic islands, in Iceland, Svalbard, and some sub-Antarctic islands, and in most major mountain ranges. The largest ice caps outside Greenland and Antarctica are on Ellesmere Island, where three ice caps have areas of more than 20,000 km^2. Further information about the distribution and characteristics of glaciers and ice caps throughout the world is given by Lliboutry (1965, Chapter 13), and, in more detail but only for the northern hemisphere, by Field (1975).

C. Ice-Age Ice Sheets

A remarkable feature of the past two and a half million years has been the repeated growth and decay of the world's ice cover. At the glacial maxima, ice covered about 30% of the earth's land surface, including almost the whole of Canada, the northern part of the United States, much of Europe, and parts of Siberia. Existing glaciers in other parts of the world were also greatly expanded. Moraines and other glacial deposits indicate

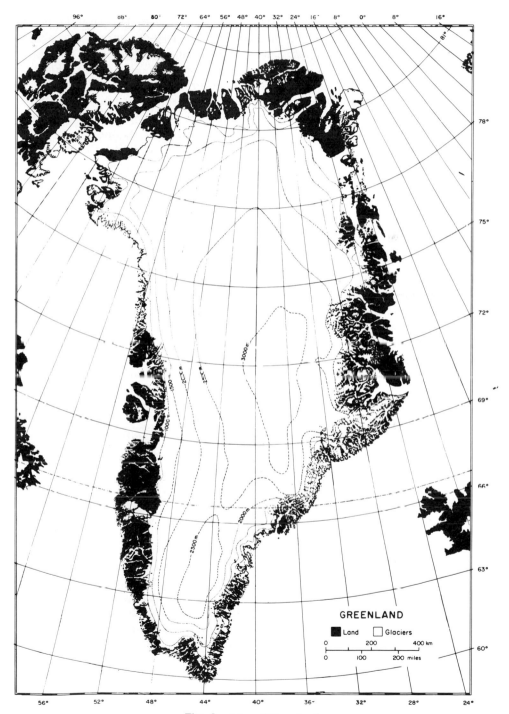

Fig. 2. Map of Greenland.

the extent of former ice sheets; the directions of glacial striae and the positions of erratic boulders provide some information about ice flow; and the amount of isostatic uplift can be related to ice thickness. The evidence can be interpreted in different ways, however, and the extent of even the most recent glaciation is still a matter of controversy. Compare, for example, the maps of Hughes *et al.* (1977) with that of Lliboutry (1965, Fig. 21.1). Hughes *et al.* show grounded ice sheets centered on the present Barents, Kara, and East Siberian Seas. They also postulate the existence of floating ice shelves covering most of the Arctic Ocean, the whole of Baffin Bay and the Greenland and Norwegian Seas, and extending into the North Atlantic as far south as an arc from the Grand Banks off Newfoundland to the continental shelf off Scotland. None of these features appears in Lliboutry's more conventional interpretation.

Because erosion and deposition by an ice sheet remove most signs of previous glaciations, much more is known about the last glaciation (the "late-Wisconsin" or "late-Wurm" Glaciation) than about earlier ones. At its maximum, the northern part of the North American mainland was covered by an ice sheet (the "Laurentide Ice Sheet") comparable in size with the present Antarctic Ice Sheet. An ice sheet also covered the Canadian Arctic Islands; although joined to the Laurentide Ice Sheet, it had separate centers of ice flow. Prest (1969, 1970) has mapped the boundaries of these ice sheets at their maximum and during their decay, Paterson (1972) has estimated their volumes during the same period, and Sugden (1977) has reconstructed the surface topography and estimated accumulation rates, ice velocities, and temperatures of the ice sheet at its maximum. The time of maximum extent varied from place to place. For most regions it is in the range 18,000–15,000 b.p. but in the North and Northwest it is about 14,000 b.p. Remnants of the ice sheet remained on the Canadian mainland until about 6000 b.p. [Prest (1969)].

Because calving of icebergs is the predominant form of ablation in Antarctica, the extent of that ice sheet is determined by the position of the *grounding line* where the ice starts to float and calving becomes possible. This in turn depends on world sea level. Growth of continental ice sheets in the northern hemisphere in the last glaciation caused sea level to fall by roughly 100 m [Shepard and Curray (1967)]. Thus an advance of the ice edge in Antarctica, irrespective of any climatic change, would be expected [Hollin (1962)]. A fall in sea level would also cause grounding of parts of present ice shelves, and in those areas, the ice could then thicken appreciably. Denton *et al.* (1975) have found evidence that this has happened repeatedly in the Ross Sea; the last time was during the Wisconsin Glaciation.

II. THEORETICAL BACKGROUND

A. Introduction

During the past 25 years, application of the methods of mathematical physics has greatly improved our understanding of glacier behavior. Many of these studies have involved application of continuum mechanics to the problem of glacier flow. Continuum mechanics deals with the motion of deformable bodies whose properties, although in fact determined by molecular structure, can be expressed in terms of a continuous theory. The fundamental problem is to set up a complete set of equations representing the laws of mechanics and the deformational properties of the material. In some cases the laws of thermodynamics are also required. Boundary conditions and, for non-steady-state problems, initial conditions, must also be set. The equations are then solved to give the distribution of stress, velocity, and, in some cases, temperature within the body. For an analytical solution to the problem, the glacier must be represented by a model, a body of simple shape such as a parallel-sided slab resting on an inclined rough plane. No single model serves all purposes; the aspect of glacier flow being investigated determines which model is appropriate.

Mathematical analyses of glacier flow have been made since at least the beginning of this century. However, all the early analyses assumed that ice deformed like a Newtonian viscous body. The turning point came in 1948, at a joint meeting of metallurgists and glaciologists in England, when it was realized that ice, a polycrystalline solid, would be expected to deform like other polycrystalline solids such as metals and rocks, at temperatures close to their melting points. Analyses by Orowan (1949) and Nye (1951), based on the assumption that ice behaves as a perfectly plastic material, clarified several aspects of glacier flow. Thereafter Glen (1955) established, by carefully controlled laboratory experiments, that the relationship between strain rate and stress in ice was nonlinear, at least for the range of stresses important to glacier motion. Thus the viscosity is not constant, but depends on the stress. Most subsequent analyses of glacier flow have been based on Glen's law. The 1950s also saw the revival, among physicists, of interest in the problem of glacier flow. This topic had provoked lively controversy among some of the most eminent physicists of the late 19th century. After that period, however, the majority of workers in the field were geographers or geologists and interest focused on describing and classifying glaciers rather than on understanding their mechanism.

The fundamental problems in glacier dynamics are

(1) given the rates of accumulation and ablation and the surface temperature, to calculate the distribution of ice thickness and velocity in the glacier that will maintain a steady state;

(2) to calculate how the system will react to a change in accumulation, ablation, or surface temperature.

The basic equations are set out in this section. Some of the mathematical analyses and their conclusions are discussed in subsequent sections.

B. Equation of Continuity

Throughout this chapter, except where stated otherwise, we take the origin on the bed at the center of the ice sheet; the x axis horizontal, the y axis vertical, positive upwards, and the z axis so as to make the system right handed. The velocity components in the x, y, z directions are u, v, w; t denotes time. The ice thickness in the y direction is h. The flux $q = \int_0^h u \, dy$ is the volume of ice flowing in unit time through a vertical cross section of unit length in the z direction. The thickness of ice added to the surface by accumulation (positive) or removed by ablation (negative) in unit time is b.

Ice is assumed to be incompressible. If we consider a column of ice of length δx in the x direction and unity in the z direction, for a unit of time, the change in thickness of the column is equal to the ice added at the surface less the difference between the fluxes entering the upstream side of the column and leaving the downstream side. Thus

$$\partial h/\partial t = b - \partial q/\partial x, \tag{1}$$

where the ice thickness h, the mass balance b, and the ice flux q are functions of x and t. This is the continuity equation for one dimension. For two dimensions, where the quantities are also functions of z, the equation is

$$\partial h/\partial t = b - \partial q_x/\partial x - \partial q_z/\partial z. \tag{2}$$

C. Stress Equilibrium Equations

The flow of a glacier is sufficiently slow that the acceleration term in Newton's second law can be neglected. The equations of motion thus reduce to equations of static equilibrium expressing the balance between the forces applied to the surface of the body and the forces such as gravity that act on all its parts. The equations can be derived by considering the equilibrium of a small cube whose sides are parallel to the coordinate axes [see, for example, Jaeger (1962, p. 115)]. In the coordinate system of the previous section, the equations are

$$\partial\sigma_x/\partial x + \partial\tau_{xy}/\partial y + \partial\tau_{xz}/\partial z = 0, \tag{3}$$

$$\partial\tau_{xy}/\partial x + \partial\sigma_y/\partial y + \partial\tau_{yz}/\partial z = \rho g, \tag{4}$$

$$\partial\tau_{xz}/\partial x + \partial\tau_{yz}/\partial y + \partial\sigma_z/\partial z = 0. \tag{5}$$

Here σ_x, σ_y, σ_z are the normal stresses, τ_{xy}, τ_{yz}, τ_{xz} the shear stresses, ρ the density (usually assumed to be a constant in calculations), and g the gravitational acceleration.

D. Flow Law of Ice

Numerous workers have deformed single crystals and polycrystalline aggregates of ice in the laboratory. The usual method is to apply constant uniaxial compression and measure the deformation as a function of time. Uniaxial tension or simple shear can also be used. If the load is suddenly applied, the ice instantaneously deforms elastically to a certain strain; permanent deformation ("creep") then begins and continues as long as the loading is maintained. For randomly oriented polycrystals, the deformation rate initially decreases with time (*transient creep*), but usually settles down to a steady value (*secondary creep*) for a while before starting to increase with time (*tertiary creep*). Because a single crystal deforms most easily if the applied shear stress is parallel to its basal plane, the initial decrease of deformation rate in polycrystals can be explained by interference between crystals with different orientations. Production, by recrystallization, of crystals more favorably oriented for deformation in the direction of the applied stress causes the final acceleration in strain rate. For secondary creep, over the rate of stresses usually found in glaciers (50–200 kPa), the relation between shear strain rate $\dot{\varepsilon}_{xy}$ and stress is found to have the form

$$\dot{\varepsilon}_{xy} = A\tau_{xy}^n. \tag{6}$$

This is known as Glen's law for ice after its discoverer [Glen (1955)].

Different experimenters have obtained different values of the constants A and n; measured strain rates for a given stress and temperature differ by almost an order of magnitude [Weertman (1973a, Fig. 4)]. Values of n vary from about 1.5 to 4.2 [Weertman (1973a, Table 2)] with a mean of about 3, the value usually adopted in glacier studies. The flow of ice thus differs markedly from that of a viscous fluid for which $n = 1$ and A^{-1} is the viscosity. Several workers have claimed that, at stresses below about 50 kPa, the value of n is reduced to near unity. Weertman (1969, 1973a) has argued, however, that these experiments were not carried on long enough to reach a steady creep rate. For randomly oriented polycrystalline ice at $-10°C$ and $n = 3$, a value $A = 3 \times 10^{-8}$ yr^{-1} kPa^{-3} gives a value of strain rate, for a stress of 100 kPa, in the middle of the range of values quoted by Weertman (1973a, Fig. 4).

The value of n does not depend on temperature; however, the term A follows the Arrhenius relation

$$A = A_0 \exp(-Q/RT). \tag{7}$$

Here A_0 is a "constant", T temperature, R the gas constant, and Q the activation energy for creep. Weertman (1973a, Table 2) has listed values of Q measured in laboratory experiments with polycrystalline ice. Values are 60 kJ/mole for temperatures below $-10°C$ (mean of 11 values) and 139 kJ/mole (mean of 4 values) for higher temperatures. Barnes *et al.* (1971) observed this change of value, which effectively means that the Arrhenius relation breaks down near the melting point. They suggested that grain-boundary sliding and a liquid phase at the grain boundaries contributed to creep at temperatures above about $-10°C$. Jones and Brunet (1978) confirmed this by showing that, for single crystals, Q maintains the lower value at all temperatures. The value of Q is such that at a temperature of $-25°C$ the strain rate produced by a given stress is about one-fifth of its value at $-10°C$. The value of A_0 depends on the grain size and impurity content of the ice; moreover, if the crystals have a preferred orientation in the direction of the applied stress, the value of A_0 is greater than for random orientation. However, no quantitative information on these effects is available. Further information on the creep of ice is contained in reviews by Weertman (1973a), Hobbs (1974, Chapter 4), and Glen (1975).

Replacing the true flow law by the assumption that ice is a perfectly plastic material is an approximation that is sometimes useful in analyses of glacier flow. In such a substance, there is no deformation for stresses less than some critical value (the *yield stress*). When the yield stress is reached the strain rate becomes very large. This corresponds to letting n tend to infinity, as may be seen by writing $A = A_1 \tau_0^{-n}$ in Eq. (6). Thus

$$\dot{\varepsilon}_{xy} = A_1(\tau_{xy}/\tau_0)^n. \tag{8}$$

If $n \to \infty$,

$$\dot{\varepsilon}_{xy} = 0, \qquad \tau_{xy} < \tau_0,$$

$$\dot{\varepsilon}_{xy} \to \infty, \qquad \tau_{xy} > \tau_0.$$

The three different flow laws are compared in Fig. 3.

In glaciers several stresses act simultaneously, so the flow law for un-

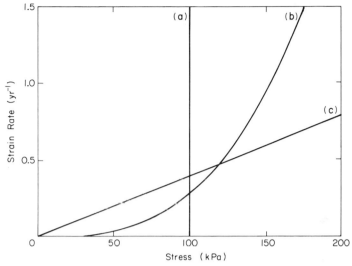

Fig. 3.   Different types of flow law: (a) Perfect plasticity with yield stress 100 kPa. (b) Glen's law with $n = 3$, $A = 0.28 \times 10^{-15}$ yr^{-1} Pa^{-3}. (c) Newtonian viscous flow with viscosity 8×10^{12} kg m^{-1} sec^{-1}.

iaxial stress has to be generalized. Nye (1953, 1957), following Odqvist (1934), has shown how to do this. Experiments by Rigsby (1958) showed that, to a good approximation, hydrostatic pressure does not affect the flow law, provided that temperature is measured relative to the freezing point. A flow law relating strain rates to stress deviators instead of stresses has this property. Stress deviators are obtained by subtracting the hydrostatic pressure from the normal stresses. Thus, for example,

$$\sigma_x' = \sigma_x - \tfrac{1}{3}(\sigma_x + \sigma_y + \sigma_z), \tag{9}$$

while shear stress components are unchanged.

A flow law must relate quantities that describe the overall state of stress and strain rate and, as it is a physical property of the material, it must not be affected by the choice of coordinate axes. The ice is assumed to be isotropic, and so it is reasonable to assume that the strain rates at any point are parallel to and proportional to the corresponding stress deviators. Thus

$$\dot{\varepsilon}_x = \mu \sigma_x', \qquad \dot{\varepsilon}_{xy} = \mu \tau_{xy}. \tag{10}$$

Similar relations hold for the other strain rate components, with μ a function of position. By definition, the six strain rate components can be expressed as derivatives of the ice velocity components. For example,

$$\dot{\varepsilon}_x = \partial u / \partial x, \tag{11}$$

$$\dot{\varepsilon}_{xy} = \tfrac{1}{2}(\partial u / \partial y + \partial v / \partial x). \tag{12}$$

Nye proposed relating the *effective strain rate* $\dot{\varepsilon}$ to the *effective shear stress* τ, defined by

$$2\dot{\varepsilon}^2 = \dot{\varepsilon}_x^2 + \dot{\varepsilon}_y^2 + \dot{\varepsilon}_z^2 + 2(\dot{\varepsilon}_{xy}^2 + \dot{\varepsilon}_{yz}^2 + \dot{\varepsilon}_{xz}^2), \tag{13}$$

$$2\tau^2 = \sigma_x'^2 + \sigma_y'^2 + \sigma_z'^2 + 2(\tau_{xy}^2 + \tau_{yz}^2 + \tau_{xz}^2), \tag{14}$$

with $\dot{\varepsilon}$ and τ taken to be positive quantities. These quantities are in fact the second invariants of the strain rate and stress deviator tensors; their values are unaffected by any rotation of the axes. The flow law is assumed to have the form

$$\dot{\varepsilon} = A\tau^n. \tag{15}$$

It follows from Eqs. (10), (13), and (14) that

$$\dot{\varepsilon} = \mu\tau, \tag{16}$$

and from Eq. (15)

$$\mu = A\tau^{n-1}. \tag{17}$$

From Eq. (10), the relations between the separate components are

$$\dot{\varepsilon}_x = A\tau^{n-1}\sigma_x', \qquad \dot{\varepsilon}_{xy} = A\tau^{n-1}\tau_{xy}. \tag{18}$$

The last relation shows that each strain rate component is proportional both to the corresponding stress deviator and, if $n = 3$, to the square of the effective shear stress. Thus an individual stress component acting on its own will produce a smaller strain rate than it would in the presence of other stresses. A tunnel dug in a glacier at the foot of an icefall, where it is subjected to a large longitudinal compressive stress in addition to the pressure of the ice above it, will close more rapidly than it would at the same depth elsewhere. Again, if τ_{xy} is the only nonzero stress deviator, $\tau = \tau_{xy}$ and $\dot{\varepsilon}_{xy}$ is proportional to τ_{xy}^3. If, on the other hand, there is also a longitudinal stress deviator σ_x', large compared with τ_{xy}, τ will be approximately equal to σ_x', and the relation between $\dot{\varepsilon}_{xy}$ and τ_{xy} will appear to be linear. These examples illustrate the complicated effects of a nonlinear flow law.

In the general case, Eqs. (3)–(5) and the six equations of the form of Eq. (18) are sufficient to determine the six unknown stresses and three unknown velocities [Eqs. (11) and (12) relate strain rates to velocities].

E. The Sliding Boundary Condition

The differential equations given in the previous sections must be supplemented by boundary conditions. At the surface, σ_y is equal to atmospheric pressure, and, because the surface is free, τ_{xy} and τ_{yz} must be zero. Moreover, for a "steady state" (ice thickness constant with time), v must be equal to the annual accumulation or ablation expressed as a thickness of ice. If the ice at the bed is below melting point, it is frozen to the bed and all velocity components should be zero, as has been confirmed by observations in tunnels [Goldthwait (1960); Holdsworth and Bull (1970)]. The stresses can have any value. If the basal ice is at its melting point, the condition becomes more complicated because the ice can slide over the bed. If the small amount of ice melted by geothermal and frictional heat is neglected, the velocity component perpendicular to the bed is zero. However, the velocity parallel to the bed may depend on the basal shear stress, the roughness of the bed, the pressure of the overlying ice, the interstitial pressure of water between ice and rock, and other quantities. The sliding mechanism is still a matter of controversy, largely as a result of the difficulty of observing it.

The problem is to explain how ice, assumed to be at its melting point, can move past bumps in the glacier bed. Weertman (1957a) was the first to analyze the process. He considered two mechanisms to be necessary for sliding: pressure melting and enhanced plastic flow. The resistance to glacier movement must occur at the upstream side of each bump. Relatively high pressure on the upstream side of bumps and low pressure on the downstream side will produce a net force tangential to the mean bed; that is, a nonzero average basal shear stress. The relatively high pressure on the upstream side of each bump lowers the freezing point there so ice can melt. The meltwater flows around the bump and refreezes on the downstream side where the pressure is reduced. The process is maintained by conduction of latent heat from the downstream to the upstream side of the bump. This process will not work for obstacles longer than about 1 m, because the heat conducted is negligible. All the ice deforms plastically but because the longitudinal stress is enhanced near a bump the strain rate there will be above average. The larger the bump the greater the distance over which the stress is enhanced, and therefore the greater the velocity because velocity is proportional to the product of strain rate and distance. Thus plastic deformation is more effective for large bumps than for small ones. Both processes are necessary for sliding; it follows that bumps of some intermediate size, the *controlling obstacle size*, must provide the main resistance to ice movement.

Weertman (1957a) modeled the glacier bed as a set of cubes on a flat plane. He calculated the velocity due to each mechanism separately, determined the controlling obstacle size as the size for which the two velocity components are equal, and then obtained the sliding velocity on the assumption that only obstacles of the controlling size affect the velocity. The equation is

$$u_b = C\tau_b^{(n+1)/2} S^{n+1} \tag{19}$$

Here u_b is sliding velocity, τ_b is basal shear stress, n is the usual flow law parameter, C depends on the flow law parameter A and the thermal properties of ice, and S is *bed smoothness* defined as the ratio of the distance between each cubical bump to its height.

As discussed in Chapter 2, sliding theory has been greatly developed since Weertman's original paper. However, Eq. (19) remains the most commonly used basal boundary condition.

F. Equation of Heat Transfer

For a given stress, the deformation rate of ice depends on temperature. Moreover, ice flow is a significant means of heat transfer within a glacier. Thus any theoretical analysis of glacier flow should treat ice flow and heat flow simultaneously. This makes the mathematics extremely complicated and few analyses of this type have yet been attempted [Shumskiy (1967); Grigoryan *et al.* (1976)]. The normal procedure in ice dynamics problems is to assume that the glacier is isothermal; the only real glaciers in this condition are the so-called *temperate glaciers* in which the ice is at its melting point everywhere. If this assumption is unrealistic, as in polar ice sheets, the problem is usually avoided by taking a value averaged over the ice mass for A the flow law parameter. Similarly, drastic simplifying assumptions regarding the ice dynamics are made in analyses of heat flow.

If temperature is introduced as a variable, an additional equation, the heat transfer equation, and appropriate boundary conditions are necessary. The heat transfer equation for a moving medium is derived by Carslaw and Jaeger (1959, p. 13). It can be written

$$DT/Dt - k\,\nabla^2 T = F/\rho c' \tag{20}$$

with

$$DT/Dt = \partial T/\partial t + u\,\partial T/\partial x + v\,\partial T/\partial y + w\,\partial T\partial z \tag{21}$$

and

$$JF = \dot{\varepsilon}_x \sigma_x + \dot{\varepsilon}_y \sigma_y + \dot{\varepsilon}_z \sigma_z + 2(\dot{\varepsilon}_{xy}\tau_{xy} + \dot{\varepsilon}_{yz}\tau_{yz} + \dot{\varepsilon}_{xz}\tau_{xz}). \tag{22}$$

In these equations, F is the internal heat generated per unit volume and time, J the mechanical equivalent of heat, k the thermal diffusivity, and c' the specific heat. Density and thermal properties are assumed constant. This is not true in the top few tens of meters of polar ice sheets, where the snow is not fully transformed into ice. Additional terms can be added to the equation to account for this case [Eq. (101)].

The surface boundary condition for the steady-state problem is that the temperature is fixed. For the non-steady-state problem, the surface temperature is a prescribed function of time. For temperate glaciers, and glaciers at the melting point in a basal layer of finite thickness, the basal boundary condition is fixed (melting point) temperature. If the basal temperature is below melting point, the boundary condition is that the basal temperature gradient is $-\phi/K$. Here ϕ is normally the geothermal heat flux. However, latent heat is added if water at the bed is being frozen to the basal ice [Weertman (1966)]. The condition is more complicated for a cold glacier that reaches melting point at its bed because only part of the geothermal heat may enter the ice; the rest is used to melt basal ice.

III.   FLOW OF ICE SHEETS

A.   Steady-State Profiles of Ice Sheets

As a result of the difference in scale, observations on polar ice sheets are usually much less detailed than those made on valley glaciers. For example, many measurements on how velocity varies with depth have been made in valley glaciers, but only one such set of data for a polar ice sheet has been published [Garfield and Ueda (1976)]. This distinction is maintained here. In this chapter we shall deal with questions relating to the surface profile of a polar ice sheet and how it is affected by bedrock topography or by changes in accumulation rate. Theoretical derivations and observations of how velocity varies with depth or over a cross section are discussed in Chapter 2.

We first derive the steady-state profile of an ice sheet with a horizontal bed, with given accumulation and ablation rates. Figure 4 shows a section of the ice sheet that is regarded as a long ridge perpendicular to the plane of the diagram. The higher parts of the ice sheet constitute an *accumulation zone* in which a net thickness of ice c is added to the surface each year and the remainder is an *ablation zone* where a net thickness a is lost annually. The boundary between the two is the equilibrium line at $x = L_1$. For simplicity accumulation and ablation rates are taken as constants. The surface slope α, if small, is $-dh/dx$.

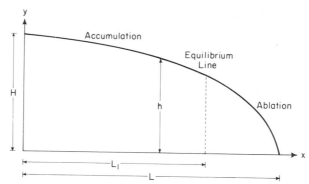

Fig. 4. Coordinate system for ice sheet.

The ice is assumed to deform in simple shear under the stress produced by its own weight; in this case $\dot{\varepsilon}_{xy} = \frac{1}{2}\, \partial u/\partial y$ is the only nonzero strain rate. Nye (1952b, 1969) and Collins (1968) have shown that the shear stress τ_{xy} is determined mainly by the surface slope and is

$$\tau_{xy} = \rho g(h - y)\alpha. \tag{23}$$

This formula is a good approximation where the top and bottom surfaces are almost parallel; this is true except near the edge. For simple shear, the flow law, Eq. (15), reduces to

$$\partial u/\partial y = 2A[\rho g(h - y)\alpha]^n. \tag{24}$$

We assume that the ice is not sliding over bedrock; the flux is then

$$q = 2\bar{A}(\rho g\alpha)^n(n + 2)^{-1}h^{n+2}. \tag{25}$$

Since temperature varies with x and y, the flow parameter is an average value $\bar{A}$ weighted in favor of the bottom layers, which are the warmest and in which most of the deformation occurs [Robin (1955); Nye (1959)].

For a steady state in the accumulation zone, we must have $q = cx$. Thus, from Eq. (25) with $-dh/dx$ written for α, we obtain

$$[c(n + 2)/2\bar{A}]^{1/n}x^{1/n}\, dx = -\rho g h^{1+2/n}\, dh. \tag{26}$$

Integration gives the equation of the surface profile

$$h^{2+2/n} = H^{2+2/n} - K_0 c^{1/n}x^{1+1/n}, \qquad 0 \le x \le L_1, \tag{27}$$

with

$$K_0 = 2^{1-1/n}(\rho g)^{-1}[(n + 2)/\bar{A}]^{1/n}. \tag{28}$$

In the ablation zone, the steady-state condition is $q = a(L - x)$ and the differential equation has the solution

$$h^{2+2/n} = K_0 a^{1/n}(L - x)^{1+1/n}, \qquad L_1 \leq x \leq L. \tag{29}$$

At $x = L_1$, both Eqs. (27) and (29) hold and the steady-state condition is $cL_1 = a(L - L_1)$. It follows that

$$H^2/L = K_0^{n/(n+1)}(ca)^{1/(n+1)}(c + a)^{-1/(n+1)}. \tag{30}$$

Thus for $0 \leq x \leq L_1$

$$(h/H)^{2+2/n} + (1 + c/a)^{1/n}(x/L)^{1+1/n} = 1 \tag{31}$$

and for $L_1 \leq x \leq L$

$$(h/H)^2 = (1 + a/c)^{1/(n+1)}(1 - x/L), \tag{32}$$

which is a parabola.

Equations (30) and (31) have been derived by Paterson (1972). Weertman (1961b) had previously deduced similar equations for an ice sheet in which all the motion is by sliding on the bed. In this case the velocity is given by the sliding relation, Eq. (19), instead of the flow law, Eq. (24), and thus the indices in the profile equations are slightly different.

As a special case, consider an ice sheet, such as that in Antarctica, in which all the mass is lost by iceberg calving at the edge. Equation (27) then applies everywhere and, because $h = 0$ at $x = L$,

$$H^2/L = K_0^{n/(n+1)}c^{1/(n+1)}, \tag{33}$$

and Eq. (27) can then be written [Vialov (1958)]

$$(h/H)^{2+2/n} + (x/L)^{1+1/n} = 1. \tag{34}$$

Because n, the flow law constant, has a value of about 3, Eq. (33) shows that the thickness of ice at the center of an ice sheet of given width is proportional to the eighth root of the accumulation rate c. In a steady-state ice sheet, the spreading of ice outwards under its own weight just balances the snowfall at the surface. The ice thickness determines the basal shear stress that causes the spreading. Because the flow law is non-linear, only small changes in equilibrium thickness are needed to accommodate a wide range of accumulation rates. Equations (30) and (32) show that the thickness and shape of an ice sheet with an ablation area is also insensitive to the ablation rate. From Eqs. (28) and (33), H is inversely proportional to the eighth root of $\overline{A}$, the value of which depends mainly on temperature. Thus for a given accumulation rate, the colder an ice sheet the thicker it will be; however, the dependence is not a sensitive one.

Observations support the conclusions of the preceding analysis. Figure 5 shows that Eq. (34) is a good fit to the profile of the Antarctic Ice Sheet for about 850 km inland from Mirny (lat 66.5° S, long 93° E) as measured by a Soviet Antarctic Expedition [Vialov (1958)]. Figure 6 shows profiles

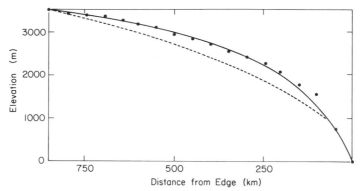

Fig. 5. Profile of Antarctic Ice Sheet from Mirny to Vostok (dots) compared with the profile given by Eq. (34) (solid line) and a parabola (broken line). [Data from Vialov (1958).]

along various ice sheets which cover a wide range of sizes, ice temperatures, and accumulation and ablation rates. The similarity between them supports the prediction that their shape is mainly determined by the plastic properties of ice.

B. Perfectly Plastic Ice Sheets

The profile equation can be simplified by assuming that, to a first approximation, ice behaves as a perfectly plastic material. The equation can then be derived by a simple balance-of-forces argument [Orowan (1949)]. In the central vertical plane of the ice sheet, there is no shear stress; the normal pressure at depth $H - y$ is the hydrostatic pressure $\rho g(H - y)$. Thus the total horizontal force, per unit length perpendicular to the plane of the diagram, is

$$\int_0^H \rho g(H - y)\, dy = \tfrac{1}{2}\rho g H^2.$$

The only other horizontal force on the central plane is the force due to the shear stress on the base. For a perfectly plastic ice sheet, the basal shear stress equals the yield stress τ_0 everywhere. Thus, for equilibrium,

$$\tfrac{1}{2}\rho g H^2 = \tau_0 L,$$

$$H^2/L = 2\tau_0/\rho g = \text{const.} \tag{35}$$

The equilibrium of a section of ice sheet of height h and length δx requires that

$$\frac{d}{dx}\left(\tfrac{1}{2}\rho g h^2\right)\delta x = \tau_0\,\delta x,$$

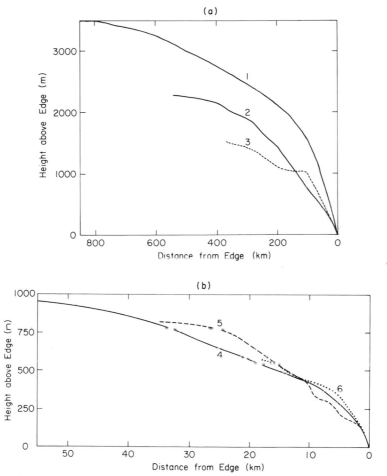

Fig. 6.   Surface profiles of ice sheets and ice caps: (a) 1, Antarctic Ice Sheet, Mirny to Vostok; 2, Greenland Ice Sheet, west side between latitudes 69° N and 71° N; 3, Greenland Ice Sheet, east side between latitudes 71° N and 72° N. (b) 4, Vatnajokull, Iceland; 5, Devon Island ice cap, Canada; 6, Barnes Ice Cap, Baffin Island, Canada.

which integrates to give the equation of the surface profile

$$h^2 = (2\tau_0/\rho g)(L - x),\qquad (36)$$

which is a parabola. It can also be written

$$(h/H)^2 = 1 - x/L \qquad (37)$$

and in this form follows from Eq. (34) by letting $n \to \infty$. This equation represents the extreme case in which the profile is determined solely by the plastic properties of ice (the yield stress) and is independent of accumulation and ablation rates. The ice thickness adjusts itself so that, at every point, the basal shear stress is equal to the yield stress.

A value of 100 kPa is often taken as the yield stress for ice; in this case $\tau_0/\rho g = 11$ m. With this value, Eq. (35) predicts the correct ice thickness in central Greenland (3200 m). Basal shear stresses in alpine valley glaciers are usually in the range 50–150 kPa [Nye (1952a)]. On the other hand, a mean of 100 kPa seems too high for most ice sheets [Paterson (1972)]. For Antarctica, Budd *et al.* (1971a) quote values ranging from near zero at the center to 100 kPa at the coast. A mean of 50 kPa gives $\tau_0/\rho g = 5.5$ m and a profile

$$h = 3.4(L - x)^{1/2}, \tag{38}$$

where all quantities are in meters.

Theoretical profiles are often used in reconstructions of past ice sheets based on the evidence of their extent provided by moraines [Paterson (1972); Sugden (1977)]. Such reconstructions are made, for example, to estimate the amount of isostatic depression or the change in world sea level produced by the former ice sheets in Europe and North America, or as boundary conditions in general circulation models of the earth's atmosphere at certain times in the past [CLIMAP (1976)]. If past accumulation and ablation rates and ice temperatures can be estimated, Eqs. (28) and (30) can be used to determine H from the known value of L and the profile can then be calculated from Eqs. (31) and (32). However, in the absence of such estimates, the simpler Eq. (38) should provide a reasonable approximation to the profile.

C. Other Steady-State Models

1. Conditions near Center of Ice Sheet

The steady-state profile equations we have derived are not valid near the center of an ice sheet. In that area, the surface slope and basal shear stress tend to zero, so the flow law cannot be expressed in the form of Eq. (24); the longitudinal stress becomes the dominant term. The parabolic profile [Eqs. (37) and (38)] has a nonzero slope at $x = 0$ and so is unrealistic there. Weertman (1961a) showed that taking longitudinal stress into account made the profile in the central region somewhat flatter than those previously derived. For ice sheets less than about 30 km wide, the correction term must be applied over an appreciable part of the profile but for larger ice sheets it can be neglected.

2. Conditions near Ice Edge

The equation for the basal shear stress [Eq. (23)] is a good approximation where surface and bed are almost parallel. This is not true near the edge of an ice sheet. Nye (1967) has given an exact solution for the distribution of stress and velocity in this region, on the assumption of perfect plasticity. This analysis predicts ice thicknesses about 15 m less than those of the standard profiles.

3. Effect of Depression of Bedrock

The theory described above applies to an ice sheet on a flat bed. In fact, the weight of ice will depress the bedrock and, if equilibrium is established, a thickness h of ice will depress the bedrock by an amount $h(\rho_i/\rho_r)$. Weertman (1961a) has shown that the same profile equations hold for the upper surface except that the ice thickness is replaced by the height above the original horizontal bed. The lower surface is the mirror image of the upper surface except that the depth of ice below the horizontal plane is scaled down by $\rho_i/(\rho_r - \rho_i)$. As $\rho_r \approx 3\rho_i$, the factor is about one-half (see Fig. 7). For the case of perfect plasticity, Eq. (35) still holds for the maximum ice thickness H. The equation for H_1, the maximum height above the original horizontal surface, as derived by Weertman (1964), is

$$H_1^2/L = 4\tau_0/3\rho g. \qquad (39)$$

4. Effect of Water at Bed

Weertman (1966) calculated the effect of a basal water layer on the thickness of an ice sheet. It is assumed that the water is produced solely by geothermal heat and frictional heat produced by sliding; no water penetrates from the surface. Weertman estimated that, if the water layer

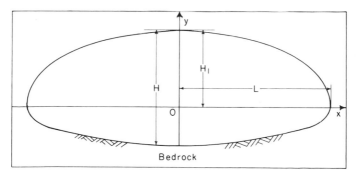

Fig. 7. Definition of symbols for analysis of ice cap on isostatically depressed bedrock.

is thick enough to submerge those bedrock bumps that provide the main hindrance to sliding (see Section II.E), the basal shear stress will be reduced by a factor of 2–4. Ice thickness is proportional to the square root of basal shear stress [Eq. (35)]. Thus, if the bed of an ice sheet, previously frozen, is warmed to the melting point and such a water layer forms, the thickness at the center could be halved. In an ice sheet bounded by the sea, like that in Antarctica, this would make the ice sheet spread and cause increased production of icebergs and a rise in world sea level. In an ice sheet with an extensive ablation area, as in the southern parts of the North American and Scandinavian ice-age ice sheets, a reduction in thickness would result in a reduction in lateral extent because more of the ice sheet would be in the ablation area.

5. Ice Sheet Fringed by Mountains

Weertman (1963) studied the case of an ice sheet, like that in Greenland, that is bordered by mountains and drained by valley glaciers cutting through them. He calculated theoretical profiles of an outlet glacier, the drainage area at its head, and the central part of the ice sheet. He concluded that the mountains had little effect on the profile of the central part of the ice sheet, provided the width of the mountain range and the spacing between outlet glaciers were small compared with the radius of the ice sheet.

6. Position of Ice Divides

In the previous analyses, accumulation rates were assumed to be the same on each side of the ice sheet. Weertman (1973b) has considered the effect of differing accumulation rates. His analysis shows that the ice divide is displaced towards the side with the higher accumulation. This can be seen from Eq. (33); H must be the same for both sides and L is then inversely proportional to $c^{1/(n+1)}$. Weertman also found that the position of an ice divide was insensitive to changes in accumulation rate; a change by a factor of 3 in the ratio of accumulation rates on the two sides would move the ice divide by about 10% of the dimensions of the ice sheet. Ice divides should thus be relatively stable features. Hence it is reasonable to assume that the ice in a core drilled at an ice divide originated near there. This simplifies calculations of the age of the ice.

D. Relation between Surface and Bedrock Topography

Although in many cases the theoretical profiles derived in the preceding sections successfully represent the large-scale features of an ice sheet, detailed comparisons over restricted areas reveal discrepancies [Robinson (1966)]. Comparison of ice thickness and surface slope provides an alternative to examining profiles. From Eq. (23), the basal shear stress for small slopes is

$$\tau_b = \rho g h \alpha. \tag{40}$$

For a perfectly plastic material, which is a reasonable approximation for ice, τ_b should be a constant, the yield stress; thus ice thickness should be inversely proportional to surface slope. Data from West Antarctica show this is not true [Bentley (1964)]. As an explanation, Robin (1967) suggested that the bedrock topography produced significant variations in longitudinal stress in the ice. Thus $\partial \sigma_x / \partial x$ cannot be ignored in comparison with $\partial \tau_{xy} / \partial y$, which contradicts the assumption made in deriving Eqs. (23) and (40). Correction terms have to be added. Orowan (1949) was the first to point out the importance of longitudinal stresses, and Lliboutry (1958) used such a correction term without deriving it rigorously. Since then Shumskiy (1961), Robin (1967), Budd (1968, 1970a, 1971), Collins (1968), and Nye (1969) have derived correction terms with varying degrees of rigor.

Consider the situation at an arbitrary point on the surface, taken as origin. The x axis is the surface tangent there, positive in the flow direction. The y axis is positive downwards. The x axis makes an angle α_0 with the horizontal. At $x = 0$, the slope of the bed is β_0 and the ice thickness is h_0. The stress equilibrium equations for plane strain are

$$\partial \sigma_x / \partial x + \partial \tau_{xy} / \partial y + \rho g \sin \alpha_0 = 0, \tag{41}$$

$$\partial \tau_{xy} / \partial x + \partial \sigma_y / \partial y + \rho g \cos \alpha_0 = 0. \tag{42}$$

Since the top surface is free, $\tau_{xy} = 0$ at $y = 0$. If α_0 and β_0 are small, the basal shear stress τ_b is approximately equal to $-\tau_{xy}$ at $x = 0$, $y = h_0$; the signs are opposite because the shear stresses are in opposite directions. Thus integration of Eq. (41) from surface to bed gives

$$\int_0^{h_0} \frac{\partial \sigma_x}{\partial x} \, dy = \tau_b - \rho g h_0 \alpha_0. \tag{43}$$

Since the surface is free, $\partial \sigma_y / \partial x = 0$ at $x = 0$. Differentiation of Eq. (42) with respect to x, followed by two integrations with respect to y, then gives

$$\int_0^{h_0} \frac{\partial \sigma_y}{\partial x} \, dy = - \int_0^{h_0} \int_0^y \frac{\partial^2 \tau_{xy}}{\partial x^2} \, dy \, dy'. \tag{44}$$

Because $dh_0/dx = 0$, the order of integration and differentiation may be interchanged on the left-hand sides of Eqs. (43) and (44). Subtraction then gives

$$\tau_b = \rho g h_0 \alpha_0 + \frac{\partial}{\partial x}\left[\int_0^{h_0} (\sigma_x - \sigma_y)\,dy\right] - \int_0^{h_0}\int_0^y \frac{\partial^2 \tau_{xy}}{\partial x^2}\,dy\,dy' \quad (45)$$

at $x = 0$.

In applications of the theory, it is convenient to use a coordinate system with the origin below the bed, the x axis horizontal and the y axis vertical and positive upwards. The surface is $y = h_s(x)$, the bed is $y = h_b(x)$; the slopes, if small, are $\alpha = -dh_s/dx$, $\beta = -dh_b/dx$. Equation (45) becomes

$$\tau_b \approx \rho g h \alpha + 2G - T', \quad (46)$$

where $h = h_s - h_b$, the ice thickness,

$$T' = \int_{h_s}^{h_b}\int_{h_s}^y \frac{\partial^2 \tau_{xy}}{\partial x^2}\,dy\,dy' \quad (47)$$

and

$$G = (\partial/\partial x)\int_{h_s}^{h_b} \sigma_x'\,dy = \frac{\partial}{\partial x}(h\tilde{\sigma}_x), \quad (48)$$

where $\sigma_x' = \frac{1}{2}(\sigma_x - \sigma_y)$ and the tilde denotes the average over the ice thickness.

The term G can be estimated from the flow law and measurements of surface strain rate $(\dot{\varepsilon}_x)_s$ if it is assumed that (1) the shear stress is small compared with the longitudinal stress deviator when these quantities are averaged over the ice thickness, and (2) the longitudinal strain rate does not vary with depth. In this case

$$(\dot{\varepsilon}_x)_s = A(\sigma_x')^n$$

and

$$G = A^{-1/n}(\partial/\partial x)[h(\dot{\varepsilon}_x)_s^{1/n}]. \quad (49)$$

The second assumption is inconsistent with the few available data from ice caps [Hooke (1973); Paterson (1976)] and it also fails in valley glaciers [Savage and Paterson (1963); Shreve and Sharp (1970); Raymond (1971); Harrison (1975)]. Moreover, determination of A, the flow law parameter averaged over the ice thickness, requires information on ice temperatures. Thus calculations of G are at best rough estimates. Observations on Wilkes Ice Cap [Budd (1968)] suggest that G can be neglected if it is averaged over horizontal distances of about 20 times the ice thickness.

There is no observational means of estimating the value of T', but a theoretical analysis by Budd (1968, 1970a) suggests that its value, averaged over distances greater than 3–4 times the ice thickness, is negligible.

Thus averages on three different scales have to be considered:

(1) *Large scale:* h and α are averaged over horizontal distances of the order of $20h$ or more. Both G and T' can then be neglected and

$$\bar{\tau}_b = \rho g \overline{h\alpha} \tag{50}$$

holds. At this scale, if the mean basal shear stress is constant, mean ice thickness can be predicted from mean surface slope and theoretical profiles would be expected to fit observations.

(2) *Intermediate scale:* Averages are taken over distances of about $4h$. Here T but not G is negligible and

$$\langle \tau_b \rangle = \rho g \langle h\alpha \rangle + 2\langle G \rangle \tag{51}$$

At this scale the surface slope is affected by changes in the longitudinal stress resulting from bedrock undulations. Equation (50) implies static equilibrium in which the downslope component of the weight of the ice exactly balances the friction of the bedrock. This is not always true however; there are regions where the component of the weight exceeds the friction; such a region will pull the ice from upstream and push the downstream ice. Similarly, regions where the friction exceeds the downslope component of the weight will be pushed by the upstream ice and pulled by the ice below. This is the origin of the longitudinal stress term G.

(3) *Small scale:* Slopes are measured over distances of about h, and each term in Eq. (46) must be taken into account.

Budd (1968), on the basis of observations on Wilkes Ice Cap, assumes that the true basal shear stress does not vary and is equal to the large-scale average. The observation that velocity on a temperate glacier is more highly correlated with large-scale average slope than with local slope [Bindschadler *et al.* (1977)] supports this hypothesis. In this case

$$\tau_b = \langle \tau_b \rangle = \bar{\tau}_b = \rho g \overline{h\alpha} \tag{52}$$

and Eq. (51) becomes

$$2G = \rho g (\overline{h\alpha} - \langle h\alpha \rangle). \tag{53}$$

On this hypothesis, variations in surface slope on the intermediate and small scales are entirely due to the longitudinal stress gradients.

Robin (1967) has analyzed the relation between surface and bedrock topography along a flow line for 46 km south of Camp Century in north-west Greenland. He assumed that Eq. (52) is true. He also considered that $\bar{\sigma}_x'(dh/dx) \ll h(d\bar{\sigma}_x'/dx)$ and so Eq. (49) becomes

$$G = A^{-1/n}h(\partial/\partial x)[(\dot{\varepsilon}_x)_s^{1/n}]. \tag{54}$$

Robin also took $\bar{h} = \langle h \rangle$ so Eq. (53) becomes

$$\langle \alpha \rangle - \bar{\alpha} = -2(\rho g)^{-1}A^{-1/n}(\partial/\partial x)[\langle(\dot{\varepsilon}_x)_s^{1/n}\rangle]. \tag{55}$$

This equation enables the fluctuations in slope about the mean slope to be predicted from the longitudinal strain rate. Robin did not measure $(\dot{\varepsilon}_x)_s$ but calculated it from the known mass balance on the assumption of a steady state. This implies that $bx = Uh$, where x is distance from the ice divide. It follows that $\dot{\varepsilon}_x = \partial U/\partial x = b/h$. Robin further assumed that $(\dot{\varepsilon}_x)_s = \dot{\varepsilon}_x$. Slope fluctuations predicted in this way were larger than those observed. To obtain satisfactory agreement the strain rates had to be halved. Robin suggested that this was plausible because the ice sheet might not be in a steady state or the flow lines might diverge, in which case the formula for the strain rate would be incorrect. In effect, there-fore, Robin only established that the two sides of Eq. (55) were pro-portional to each other. Figure 8 shows the results for part of Robin's traverse. It should be noted that the slope fluctuations are an order of magni-

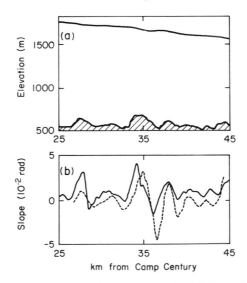

Fig. 8. (a) Surface and bedrock profiles along a flowline south of Camp Century, N.W. Greenland. (b) Observed surface slopes (solid line) and calculated surface slope deviations assuming half the steady-state velocity (broken line). [From Robin (1967), by courtesy of the publishers of *Nature*.]

tude greater than the mean slope. Beitzel (1970) analyzed some Antarctic data in the same way and obtained satisfactory agreement between observed and predicted slope fluctuations.

A complex theoretical analysis by Budd (1970b) predicts that bedrock undulations of wavelengths 3–4 times the ice thickness have more effect on the surface profile than undulations of other wavelengths. Although some of the underlying assumptions can be questioned [Raymond (1978)], field observations appear to confirm the predictions [Beitzel (1970); Budd and Carter (1971)]. Budd and Carter measured surface and bedrock elevations along a 100-km line on Wilkes Ice Cap, Antarctica. A parabolic profile was fitted to the surface and deviations of the actual slope from the predicted slope were calculated. For bedrock, slope deviations from the horizontal were used. Figure 9 shows the results of a power spectrum analysis of these slope deviations. Although the bedrock values have large amplitudes at short wavelengths (<2 km), the surface slope deviations have peaks at about 2.5 and 4 km or about 2.8 and 4.4 times the mean ice thickness of 900 m.

E. Time Scales for Ice Cores

As explained in Section 1, cores from polar ice sheets contain a detailed record of past climatic variations. Such records are of limited value, however, unless the age of the ice at each depth can be determined. The most

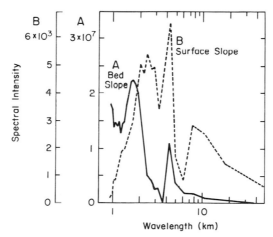

Fig. 9. Power spectra of (A) deviations of bedrock slope from a horizontal plane and (B) deviations of surface slope from a parabola, for a 105-km profile of Wilkes Ice Cap, Antarctica, from summit to edge. [From Budd and Carter (1971). Reproduced from the *Journal of Glaciology* by permission of the International Glaciological Society.]

commonly used method is calculation based on simplified models of flow in the ice sheet. As each year's snowfall is buried by that of subsequent years, it moves downward relative to the surface and also undergoes plastic deformation so that it becomes progressively thinner. (Because ice is incompressible, there is a corresponding extension in the horizontal plane.) This plastic thinning is in addition to the thinning resulting from the increase in density that occurs as the snow is slowly compacted into ice. We ignore compaction in this section by expressing each annual layer as its ice-equivalent thickness; in practice the thinning due to compaction is determined by measuring how density varies with depth in the core. At the center of an ice sheet in a steady state, the vertical velocity component, expressed as the distance an ice particle moves downwards in one year, must be equal to the thickness of one annual ice layer. Thus the age t of ice at ice-equivalent depth y is

$$t = \int_0^y v^{-1} \, dy. \tag{56}$$

Age is determined on the basis of some assumed relationship between v and y.

Consider the situation at the ice divide of a steady-state ice sheet on a horizontal bed. Take the origin on the surface with the y axis vertical, positive downwards. The horizontal velocity components are zero. The simplest assumption is that v is a linear function of y; in other words, the vertical strain rate $\partial v/\partial y$ is constant. For a steady state, the value of v at the surface must be equal to b, the ice-equivalent annual accumulation rate. If the ice is frozen to bedrock the value of v at the bed is zero. Thus

$$v = b(1 - y/h) \tag{57}$$

and from Eq. (56), since $t = 0$ at $y = 0$,

$$t = -(h/b) \ln(1 - y/h). \tag{58}$$

This is referred to as the Nye time scale because Nye makes the assumption of uniform vertical strain rate in several of his analyses of glacier flow. This relation will hold at places other than an ice divide if the values of b and h, and thus of v, do not vary much between the divide and the borehole.

The assumption of uniform vertical strain rate is not realistic if the ice sheet is frozen to its bed. If the bed is horizontal, the horizontal velocity components at all points on it are zero. Thus the horizontal strain rates are zero at the bed and, because ice is incompressible, the vertical strain rate must also be zero at the bed. To overcome this difficulty, Dansgaard and Johnsen (1969a) proposed a flow model in which the vertical strain rate is constant down to some depth h' and then decreases linearly to zero at the bed. Thus

$$h \geq y \geq h': \qquad dv/dy = K'(h - y)$$

$$v = -\tfrac{1}{2}K'(h - y)^2 \qquad (59)$$

$$y = h': \qquad v' = -\tfrac{1}{2}K'(h - h')^2 \qquad (60)$$

$$h' \geq y \geq 0: \qquad dv/dy = K'(h - h')$$

$$v - v' = K'(h - h')(y - h').$$

By Eq. (60)

$$v = \tfrac{1}{2}K'(h - h')(2y - h - h'). \qquad (61)$$

But $v = b$ when $y = 0$; thus

$$K' = -2b/(h - h')(h + h').$$

By Eq. (56)

$$h' \geq y \geq 0: \qquad t = \tfrac{1}{2}b^{-1}(h + h') \ln[(h + h')/(h + h' - 2y)] \quad (62)$$

$$y = h': \qquad t' = \tfrac{1}{2}b^{-1}(h + h') \ln[(h + h')/(h - h')] \qquad (63)$$

$$h \geq y \geq h': \qquad t - t' = (h + h')(y - h')/b(h - y). \qquad (64)$$

Equations (62)–(64) give the age, and Eqs. (59) and (61) the thickness of an annual layer, at depth y.

Dansgaard and Johnsen determined h' from the variation of u with y. They assumed that, at distance x from the ice divide,

$$u(y) = rxf(y),$$

where r is a constant. For two-dimensional flow ($\dot{\varepsilon}_z = 0$), the incompressibility condition is

$$\partial v/\partial y = -\partial u/\partial x = -rf(y),$$

which expresses the vertical strain rate as a function of y. Dansgaard and Johnsen applied their flow model to the borehole at Camp Century, Greenland, for which Weertman (1968) had computed $f(y)$ by integrating the flow law for simple shear [Eq. (6)]

$$\tfrac{1}{2}\partial u/\partial y = A\tau_{xy}^n,$$

with the shear stress given by Eq. (23). They found that Weertman's value of $f(y)$ could be approximated by the value unity down to 400 m above the bed and by the linear relation y/h below this. They therefore took $h - h' = 400$ m.

Philberth and Federer (1971) have also proposed a method of calcu-
lating time scales. As in the Dansgaard–Johnsen model, $\partial v/\partial y$ and u are
assumed to vary with y in the same way and u is obtained by integrating
Glen's law. However, Philberth and Federer use the function $u(y)$ instead
of approximating it by two straight line segments.

Figure 10 shows the three time scales for the 1367-m core through the
Greenland Ice Sheet at Camp Century. The Nye model gives younger
ages than the other models, the discrepancy increasing with age. The
Dansgaard–Johnsen model gives an age of 10,000 yr at 247 m above the
bed. A large change in oxygen isotope ratio, corresponding to the end of
the last glaciation, occurs at about this depth in the core. Because the last
glaciation is known from other evidence to have ended at about this time,
the Dansgaard–Johnsen time scale appears to be accurate down to this
depth.

These time scales are derived on the assumption that the ice sheet has
been in a steady state for the time period covered by the core (roughly
120,000 yr). It is also assumed that the vertical strain rate at the borehole
is typical of the flow path from the ice divide; this implies that the accumulation rate and ice thickness do not change upstream. This will only be
true for boreholes near ice divides. For other locations, computer calculations based on two-dimensional flow models can be used (e.g., Fig. 20).
However, the steady-state assumption is still made.

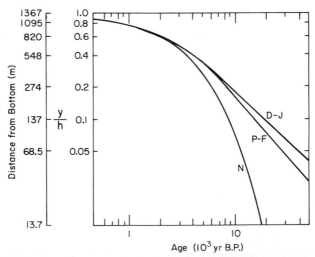

Fig. 10. Comparison of Dansgaard–Johnsen, Philberth–Federer, and Nye time scales
for the ice core from Camp Century, Greenland. [From Philberth and Federer (1971). Reproduced from the *Journal of Glaciology* by permission of the International Glaciological
Society.]

Weertman (1976) has pointed out that, if the temperature of the basal ice changes from the melting point to a value below it or vice versa, the distribution of vertical strain rate with depth also changes. Thus time scales calculated from flow models will be incorrect if the ice undergoes such a temperature change at any point upstream from the borehole.

Measurements of the vertical strain rate as a function of depth in the borehole provides an alternative to calculation. Paterson *et al.* (1977) have used this method on the Devon Island ice cap in the Canadian Arctic. As the borehole was only 900 m from the ice divide the measurements should be typical of the present flow path of the ice, but one still has to assume they are typical of past time.

The steady-state assumption may be a valid approximation for the last few thousand years. However, it is most unlikely that accumulation rates, ice thicknesses, and flow patterns during the last glaciation were the same as they are now. Thus theoretical time scales extending back beyond 10,000 B.P. should be viewed with suspicion. Other methods of dating ice based on seasonal variations in oxygen isotope ratio, concentration of microparticles and trace elements [Hammer *et al.* (1978)], or C^{14} dating of the CO_2 contained in air bubbles in the ice [Oeschger *et al.* (1976)] are being developed. So far none of these methods has been applied to ice older than about 15,000 yr; the problem of dating older ice remains unsolved.

F.   Measurement of Ice Thickness Changes

In view of the prevalence of the steady-state assumption in theoretical analyses, it is important to determine how close any given ice sheet is to this condition. The obvious way is to measure the change in surface elevation with time at a point fixed relative to bedrock. However, Nye (1975) pointed out that, while accumulation and ablation rates vary widely from year to year, ice flow does not adjust to such short-period changes. Thus an increase in elevation measured over a few years might merely result from one or two years of exceptionally heavy snowfall. For this method to give results relevant to the long-period state of balance of the ice sheet, the measurement interval would have to be very long. On the other hand, an elevation change measured at or below the depth at which snow has been transformed to ice (60–100 m in polar ice sheets) should reflect changes in ice flow, uninfluenced by short-period fluctuations in accumulation. Again measurements should be made at several places; a thickness increase at a single site might merely result from a bulge of increased thickness (a *kinematic wave*) passing through that site. Several of the methods described suffer from these limitations.

The straightforward way to measure elevation changes is by repeated leveling. This is simple on a valley glacier, but on an ice sheet it requires a long time and high precision. Nevertheless, a leveling line was carried across the ice sheet in central Greenland in 1959 and repeated in 1968. Elevation changes varied from place to place in a wavelike fashion; the amplitude varied between about 0.2 and 1.1 m and the wavelength was roughly 25 km [Seckel and Stober (1968)]. During this work the elevation change of a point in a shaft 40 m below the surface at Jarl-Joset Station was determined. Although this point was in firn rather than ice, measurements at that depth should give a better indication of the long-term behavior of the ice sheet than surface measurements. Moreover, a long accumulation record (97 yr) was obtained from stratigraphic studies. Comparison of the vertical velocity component, obtained by correcting the measured downward displacement for the movement of the ice downslope, with the mean accumulation rate indicated that the ice sheet at Jarl-Joset Station has a mass deficit of 0.08 m/yr water equivalent [Federer and von Sury (1976)].

Paterson (1976) measured the vertical velocity component at a point on the surface of the Devon Island ice cap by measuring the shortening, in one year, of a borehole that reached bedrock. Because the borehole was near the ice divide, the shear in the ice was very small and had a negligible effect on the length of the borehole. Comparison of the vertical velocity with the accumulation rate (11-yr mean) suggested that the ice cap was thickening slightly at present.

Measuring gravity is another way to determine ice thickness changes. Bentley (1971) observed an increase in gravity at the South Pole between 1957 and 1967. However, he concluded that most of the implied decrease in elevation resulted from sinking produced by the weight of snowdrifts that accumulated round the station; there was no evidence that the ice sheet was not in a steady state.

Nye *et al.* (1972) have proposed a method of measuring ice thickness changes by using a radio-echo sounder to establish a precise horizontal and vertical coordinate system, fixed relative to bedrock and independent of visible landmarks. When an echo sounder is moved over the surface of an ice sheet, the strength of the returned signal varies appreciably in a distance of the order of the radar wavelength. By mapping the fading pattern over a small area at two different times, the horizontal position of a point fixed relative to bedrock can be reestablished. The phase change between the transmitted and reflected signals depends on distance above bedrock; by measuring this phase change the vertical position can be established. The displacement of the snow surface, or of markers fixed in the firn, over a given time interval can be measured relative to this fixed coordinate system with a precision of a few centimeters. Although results have not yet been obtained, a phase-sensitive echo sounder has been built and one set of measurements made [Walford et al. (1977)]. This method has great potential.

Shumskiy (1965) proposed a method of determining thickness changes by comparing the accumulation rate with horizontal strain rates rather than with vertical velocity. The method is based on the equation of continuity. Take the x axis horizontal, the y axis vertical, positive upwards, and the origin below the bed of the ice sheet. Uniform density is assumed. Let the surface be

$$F_s(x, y, z, t) = y - h_s(x, z, t) = 0. \tag{65}$$

The change in surface elevation with time, relative to a point moving with the ice, must balance the accumulation rate c_s. Thus

$$DF_s/Dt = -u_s(\partial h_s/\partial x) + v_s - w_s(\partial h_s/\partial z) - \partial h_s/\partial t = -c_s, \tag{66}$$

where subscript s denotes a value at the surface. But, because ice is incompressible, the strain rates obey the relation

$$\dot{\varepsilon}_x + \dot{\varepsilon}_y + \dot{\varepsilon}_z = 0.$$

Integration of this relation with respect to y between surface and bed gives

$$v_s - v_b = -h(\langle \dot{\varepsilon}_x \rangle + \langle \dot{\varepsilon}_z \rangle),$$

where the angle brackets denote values averaged over the ice thickness and subscript b denotes the value at the bed. If it is assumed that the ice is frozen to its bed $v_b = 0$ and since $\partial h_s/\partial t = \partial h/\partial t$, Eq. (66) becomes

$$\partial h/\partial t = c_s - h(\langle \dot{\varepsilon}_x \rangle + \langle \dot{\varepsilon}_z \rangle) - u_s(\partial h_s/\partial x) - w_s(\partial h_s/\partial z). \tag{67}$$

The last term can be made zero by taking the x axis along the line of flow. The quantities $\langle \dot{\varepsilon}_x \rangle$ and $\langle \dot{\varepsilon}_z \rangle$ are assumed to be equal to the values measured at the surface or to a constant fraction of them; this is the main uncertainty in the calculation. Another uncertainty is the distance over which the surface slope $\partial h_s/\partial x$ should be measured. Shumskiy (1965) and Mellor (1968) have both applied this method to Jarl-Joset Station but use different data. Shumskiy concluded that the ice was thinning at a rate of more than 1 m/yr, while according to Mellor the amount is 0.37 m/yr. Comparison of these values with the previously quoted figure of 0.08 m/yr measured by Federer and von Sury (1976) suggests that this method is not reliable. Budd (1970c) has calculated thickness changes of Law Dome, Antarctica, by this method with the assumption that mean strain rates were 90% of the measured surface values. Results from different parts of the Dome varied from thickening of 0.6 m/yr to thinning of 0.8 m/yr [Budd (1970c, Fig. 10)], whereas measured changes in gravity indicated thinning at all points at rates from 0.5 to 2.2 m/yr [Budd and Radok (1971, Fig. 17d)].

For a floating ice shelf, equations similar to Eqs. (65) and (66) apply with subscripts s replaced by b, except that the sign of the right-hand side of Eq. (66) has to be changed because ice accretion (positive c_b) increases the value of F_b. Subtraction of this new equation from Eq. (66) gives an equation corresponding to Eq. (67). Because, as shown in Section V, velocities and strain rates are uniform throughout the thickness of an ice shelf, this equation reduces to

$$\partial h/\partial t = c_s + c_b - h(\dot{\varepsilon}_x + \dot{\varepsilon}_z) - u(\partial h/\partial x) - w(\partial h/\partial z). \tag{68}$$

Two methods of estimating long-term ice thickness changes from measured temperature distributions will be discussed in Section VI.

G. Non-Steady-State Ice Sheets

All the preceding analyses referred to an ice sheet in a steady state; that is, to one whose dimensions do not change with time. No real ice sheet is ever in a steady state; it is continuously adjusting to variations in accumulation and ablation rates resulting from climatic fluctuations. In most cases, however, the flow terms in the continuity equation [Eq. 2)] are approximately equal to the mass balance so the rate of change of thickness is small. In these circumstances, predictions of steady-state models should be good approximations.

To study the details of how an ice sheet responds to a change in climate requires a non-steady-state analysis. Modern work on this topic began with Weertman (1958) and Nye (1958). Subsequently Nye greatly developed the theory, the main features of which have been summarized by Paterson (1969, Chapter 11). The theory deals with the effect of a small perturbation in mass balance on a glacier initially in a stable equilibrium state. It is based on the continuity equation [Eq. (1)] and the basic assumption that the flux at any point is determined solely by the ice thickness and surface slope at that point. This theory has provided valuable insights into glacier behavior and explained many observed features. It is sufficiently well developed that, if enough data are available, the changes in ice thickness and terminus position resulting from a given change in mass balance can be predicted. Conversely, past mass balances can be calculated from a detailed history of terminus position as revealed by moraines. However, lack of data has so far prevented application of the theory to any ice sheet. Moreover, the ice is assumed to be isothermal, an assumption valid for temperate glaciers but not for ice sheets. For these reasons this theory is discussed in Chapter 2 rather than here.

Ice sheet response has been studied by numerical modeling using computers. This method has the advantage that it is not restricted to small perturbations or isothermal ice. This topic is discussed in Section IV.

The time required for a small nonequilibrium ice sheet to grow into a large ice-age ice sheet has been calculated by Weertman (1964). Consider an ice sheet on a horizontal bed with the coordinate system of Fig. 4. Ice is assumed to be perfectly plastic so Eq. (36) gives the profile

$$h = (2\tau_0/\rho g)^{1/2}(L - x)^{1/2}.$$

It is assumed that accumulation at rate c occurs over the whole ice sheet but not on the surrounding ground. Because there is no ablation area, the ice sheet must grow. The volume of half the ice sheet per unit length in the z direction is

$$\int_0^L h \, dx = \frac{2}{3} (2\tau_0/\rho g)^{1/2} L^{3/2}$$

and the rate of change of volume is $(2\tau_0 L /\rho g)^{1/2}(dL/dt)$. This must be equal to cL, giving a differential equation that integrates to

$$t = 2c^{-1}(2\tau_0/\rho g)^{1/2}(L^{1/2} - L_0^{1/2}),$$

where t is time required for an ice sheet of initial width L_0 to grow to width L.

Now suppose that, because of climatic change, accumulation stops and ablation at rate a begins all over the ice sheet. Ablation decreases the thickness and, if the slope initially remains unchanged, the basal shear stress will drop below the yield stress, deformation will cease, and the ice sheet will become stagnant. Thus the rate of change of thickness equals the ablation rate. The profile at time t is thus

$$h = (2\tau_0/\rho g)^{1/2}(L_0 - x)^{1/2} - at,$$

where L_0 is the initial half-width and negative values of h are disregarded. The half-width at time t, obtained by putting $h = 0$, is

$$L = L_0 - a^2 \rho g t^2 / 2\tau_0$$

and

$$t = a^{-1}[(2\tau_0/\rho g)(L_0 - L)]^{1/2}.$$

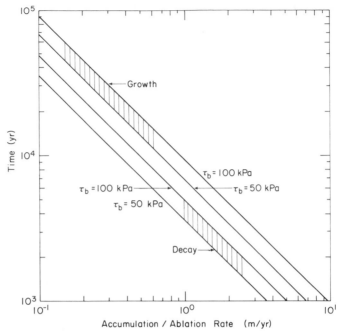

Fig. 11. Time for a small ice sheet to grow to a half-width of 1000 km, as a function of accumulation rate and basal shear stress, and time for an ice sheet of half-width 1000 km to disappear, as a function of ablation rate and basal shear stress. The shaded areas represent plausible values of accumulation and ablation. [From Weertman (1964). Reproduced from the *Journal of Glaciology* by permission of the International Glaciological Society.]

Comparison of these two expressions for t shows that, if $c = a$, the growth time is twice the decay time. In fact, ablation rates are normally 3 or 4 times greater than accumulation rates so the ice sheet is expected to decay much more rapidly than it grows. Figure 11 shows calculated growth and decay times of an ice sheet of half-width 1000 km. (It is assumed that for growth the initial width is small compared with the final width and for decay the final width is small compared with the initial width.) With plausible values of accumulation and ablation rates, growth times are in the range 15,000–50,000 yr while decay times range from 2000 to 4000 yr. Although the model is extremely simple, these times are in broad agreement with those of continental ice sheets during the last glaciation. Use of a more complex model of an ice sheet with accumulation and ablation areas made no appreciable difference to the conclusions [Weertman (1964)].

IV. NUMERICAL MODELING OF ICE-SHEET CHANGES

A.   Introduction

In this section we shall discuss the major problem in ice-sheet dynamics: how to determine the variation of ice-sheet dimensions with time in response to a given time-dependent distribution of accumulation and ablation. The initial dimensions of the ice sheet, the temperature distribution within it, and the flow properties of ice have to be known. Given these data, future glacier behavior can be predicted. The reverse problem, that of drawing inferences about past climate from evidence of the former extent of ice sheets provided by moraines and other features, can also be tackled. In Nye's analysis of these problems, described in Chapter 2, the perturbations from the steady-state dimensions of the glacier are assumed to be small. To remove this restriction and also to take account of factors that Nye neglected, such as longitudinal stress gradients, numerical methods have to be used. As in analytical treatments, simplifying assumptions are necessary in numerical modeling. Some of these may be questionable, and, in addition, the input data may be inadequate. However, numerical models provide quantitative predictions of ice-sheet behavior that can be compared with observations. Discrepancies suggest needed improvements in the model or the input data.

Several problems arise in numerical modeling. One is the basal boundary condition. If the ice is frozen to its bed, the basal velocity is zero. On the other hand, if the basal ice is at its melting point, the sliding velocity has to be prescribed at each point or a relation between sliding velocity and basal shear stress has to be assumed. Equation (19) is usually used, although the conditions under which it may apply are uncertain and the bed smoothness, being unknown, has to be assumed uniform. Another problem is that the values of the flow law parameters [Eqs. (7) and (15)] are not known with precision. Laboratory tests on ice can be carried on only to small total strains, and so the results may not be applicable to glaciers. Moreover, manufactured ice, which has different properties from glacier ice, has been used in most laboratory tests. The flow law parameters can be determined from field measurements of the tilting rates of boreholes [Shreve and Sharp (1970)], the closure rates of boreholes [Paterson (1977)] and tunnels [Nye (1953)], and the spreading rates of floating ice shelves [Thomas (1973b)]. However, there are various difficulties in interpretation, and the results show a wide scatter. In addition, the quantity A_0 in Eq. (7) depends on the shape, size, and orientation of the ice crystals and on the amount of soluble and insoluble impurities in the ice. As these are not usually known, A_0 has to be assumed constant. Again, the deformation rate of ice depends on its temperature and ice flow is a means of heat transfer within the ice sheet. Because simultaneous treatment of these two problems is very difficult, in most non-steady-state models the whole ice mass is assumed to be at some mean temperature.

By numerical modeling one could, in theory, determine the distribution of velocity throughout the ice mass. This approach is hardly practical and is also unnecessarily detailed. Normal procedure is to eliminate vertical variations by working with ice flux, or with velocity averaged over the ice thickness, as a function of position along flow lines.

B. Mahaffy Model

This model [Mahaffy (1976)] is essentially a generalization, to three dimensions and a nonsteady state, of the method used to calculate ice sheet profiles in Section III.A. The simplifying assumptions are similar to those made in the analysis of "laminar flow" in a valley glacier (see Chapter 2). Figure 12 shows the coordinate system. The ice sheet is assumed to deform by shear in the xz plane. Thus $\dot{\varepsilon}_{xy}$, $\dot{\varepsilon}_{yz}$ are the only nonzero strain rates, and so, by the flow law τ_{xy}, τ_{yz} are the only nonzero stress deviators. It then follows from the definition of the stress deviators that $\sigma_x = \sigma_y = \sigma_z$. With the further assumption that $\partial\sigma_y/\partial y \gg \partial\tau_{xy}/\partial x$, $\partial\tau_{yz}/\partial z$

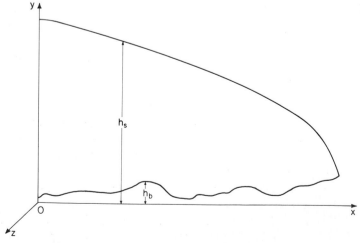

Fig. 12. Coordinate system for Mahaffy flow model.

the stress equilibrium equations [Eqs. (3)–(5)] reduce to

$$\partial\sigma_x/\partial x + \partial\tau_{xy}/\partial y = 0,$$

$$\partial\sigma_y/\partial y = \rho g,$$

$$\partial\tau_{yz}/\partial y + \partial\sigma_z/\partial z = 0.$$

The boundary conditions are that, at the surface $y = h_s$, σ_y is equal to atmospheric pressure (taken to be zero), and τ_{xy} and τ_{yz} are zero. The solutions are

$$\sigma_y = \sigma_x = \sigma_z = -\rho g(h_s - y), \tag{69}$$

$$\tau_{xy} = -\rho g(h_s - y)(\partial h_s/\partial x), \tag{70}$$

$$\tau_{yz} = -\rho g(h_s - y)(\partial h_s/\partial z). \tag{71}$$

These solutions hold only for small slopes $\partial h_s/\partial x$, $\partial h_s/\partial z$ in which case τ_{xy}, τ_{yz} are approximately equal to the shear components parallel to the surface. With the additional assumption that $\partial u/\partial y \gg \partial v/\partial x$ the flow law, Eq. (18), becomes

$$\partial u/\partial y = 2A(\tau_{xy}^2 + \tau_{yz}^2)^{(n-1)/2}\tau_{xy}. \tag{72}$$

Substitution from Eqs. (70) and (71) and two integrations with respect to y give

$$q_x(x, z, t) = -2\overline{A}(n + 2)^{-1}(\rho g)^n \alpha^{n-1}(\partial h_s/\partial x)h^{n+2} + hu_b(x, z, t). \tag{73}$$

Similarly, if $\partial w/\partial y \gg \partial v/\partial z$,

$$q_z(x, z, t) = -2\overline{A}(n + 2)^{-1}(\rho g)^n \alpha^{n-1}(\partial h_s/\partial z)h^{n+2} + hw_b(x, z, t). \tag{74}$$

Here $\overline{A}$ is the value of A averaged over the ice thickness, weighted in favor of the lower layers where most of the shear occurs, and assumed to be the same for all x and z, $\alpha = [(\partial h_s/\partial x)^2 + (\partial h_s/\partial z)^2]^{1/2}$ is the surface slope in the direction of flow, and $h = h_s - h_b$ is the ice thickness.

The continuity equation

$$\partial h/\partial t = b(x, z, t) - \partial q_x/\partial x - \partial q_z/\partial z \tag{2}$$

provides a third relation between q_x, q_z, and h. Equations (2), (73), and (74) can be solved numerically. The input data are the ice thickness at all points at time zero, the mass balance and sliding velocity at all points and times, and the flow law parameters.

The assumption that the normal strains are negligible compared with the shear strains does not hold at the center or the edge of the ice sheet. However, this does not introduce any serious error. The assumption may also break down if the bed slope is large or if a large part of the motion is by basal sliding. This restricts application of the model to certain kinds of ice sheets. Some sliding must be assumed at the edge, otherwise the flux there is zero [Eqs. (73) and (74)] and the ice sheet could not advance.

Mahaffy (1976) applied this model to predict the future behavior of the Barnes Ice Cap in Baffin Island using the present dimensions of the ice cap and the present relation between mass balance and elevation as input data. The sliding velocity was taken as zero, except at the edge. The model has also been used to study the growth of the Laurentide Ice Sheet over Canada during the last glaciation [Andrews and Mahaffy (1976)]. It appears to be a very useful model for studying the growth and decay of ice sheets on relatively smooth beds.

C. Budd and Jenssen Models

The model of Budd and Jenssen (1975) includes the effects of the longitudinal stress gradient G. The starting point is Eq. (53),

$$2\langle G \rangle = \rho g(\overline{h\alpha} - \langle h\alpha \rangle), \tag{53'}$$

where an overbar denotes averaging over a distance of 20 h, angle brackets denote averaging over a distance of $4h$, and G is given by

$$G = (\partial/\partial x)(h\sigma_x'), \tag{48}$$

with σ_x' now denoting the stress deviator averaged over the ice thickness. The x axis is horizontal along a flowline (the centerline in a valley glacier) and the y axis vertical. For Eq. (53) to hold in this coordinate system the surface and bed slopes must be small.

Budd and Jenssen assume that, when averaged over horizontal distances of $20h$ and over the ice thickness, the shear strain rate $\dot{\varepsilon}_{xy}$ is much greater than the longitudinal strain rate $\dot{\varepsilon}_x$ and that ice obeys the usual power flow law. Thus if U is the velocity averaged over the ice thickness and the sliding velocity is zero,

$$\overline{U} = 2A(n + 2)^{-1}(\rho g \overline{h\alpha})^n \overline{h} \tag{75}$$

and so

$$\rho g \overline{h\alpha} = [(n + 2)\overline{U}/2A\overline{h}]^{1/n}. \tag{76}$$

Budd and Jenssen also assume that, when averaged over distances of $4h$ and over the ice thickness, $\dot{\varepsilon}_x$ is much greater than $\dot{\varepsilon}_{xy}$ and that ice obeys a flow law of the form

$$\dot{\varepsilon}_x = B \sinh(\sigma_x'/\sigma_0), \tag{77}$$

where B and σ_0 are constants. This can be regarded as a generalization of the power law, and Budd and Jenssen state that this law covers a wider stress range. However, adoption of this law is also necessary to ensure computational stability [Budd and Radok (1971, p. 58)]. Equation (48) becomes

$$\langle G \rangle = \sigma_0 (\partial/\partial x)[\langle h \rangle \ \mathrm{arcsinh}(B^{-1}\langle \partial U/\partial x \rangle)]. \tag{78}$$

Substitution from Eqs. (76) and (78) in Eq. (53') then gives

$$[(n+2)\overline{U}/2A\overline{h}]^{1/n}$$
$$= \rho g \langle h\alpha \rangle + 2\sigma_0(\partial/\partial x)[\langle h \rangle \ \mathrm{arcsinh}(B^{-1}\langle \partial U/\partial x \rangle)]. \tag{79}$$

For valley glaciers, the first term on the right-hand side is multiplied by a "shape factor," less than unity, to allow for the drag of the valley walls.

Budd and Jenssen write the continuity equation [Eq. (1)] for a given glacier cross section in the form

$$(\partial/\partial x)(sZ\langle U \rangle h) = Z(b - \partial h/\partial t). \tag{80}$$

Here Z is the width of the cross section and s is a shape factor chosen so that sZU equals the value of ZU averaged over the cross section. For an ice sheet Z is interpreted as the distance between two flow lines.

The computational procedure is to solve Eq. (79) iteratively for U for a given value of h. Equation (80) is then solved to give $\partial h/\partial t$ and thus the value of h at the next time step, with b given. The new value of U can then be calculated. The procedure is applied to different cross sections. This model is stable; in other words, if b is kept constant in time, the ice sheet eventually reaches a steady-state profile.

This model has certain limitations. First, it is only two dimensional. The analysis is restricted to the centerline of a valley glacier or a single flow line in an ice sheet; transverse variations in ice thickness or surface elevation enter only through shape factors. The use of fixed values for the flow law constants implies uniform ice temperature. The use of different flow laws for longitudinal and shear stresses is difficult to justify. Moreover, the values chosen for the constants are such that the value of $\dot{\varepsilon}_x/\dot{\varepsilon}_{xy}$ varies from 50 for a stress of 50 kPa to 250 for a stress of 200 kPa; the ratio should be 0.22 [Paterson (1969, p. 86.)]. Thus the accuracy of any numerical predictions seems doubtful. Also, to ensure computational stability, the data have to be smoothed at certain time steps in the computations. This smoothing eliminates some features that may be real. Thus the additional precision obtained by including the longitudinal stress may be illusory and the model may be more complicated than is necessary to describe the large-scale behavior of ice masses.

Jenssen (1977) has developed a three-dimensional ice-sheet model with a temperature-dependent flow law that treats the ice flow and heat transfer problems simultaneously. The model incorporates melting and freezing at the base of the ice and mass loss due to iceberg calving. Jenssen used the model to study the buildup of the Greenland Ice Sheet. The results were unrealistic, however, because of the large grid size (200 × 100 km) used in the computations. Nevertheless, this type of model merits development.

V. FLOW OF ICE SHELVES

A. Theory

An ice shelf is a large ice sheet, from a few tens to hundreds of meters thick, floating on the sea but attached to land ice. Ice shelves surround much of the coast of Antarctica; in the northern hemisphere small ice shelves are found on the north coast of Ellesmere Island, in north Greenland, and in Franz Josef Land.

An ice shelf deforms plastically under its own weight; this process can be analyzed by methods similar to those used in the theory of flow in ice caps and glaciers. Figure 13 shows the coordinate system used; the z axis is perpendicular to the plane of the diagram. The main difference from previous analyses is in the basal boundary condition. Because the ice shelf is floating on the sea the shear stresses τ_{xy}, τ_{yz} are zero at the base as well as at the surface. Thus $\tau_{xy} = \tau_{yz} = 0$ everywhere, and by the flow law, $\dot{\varepsilon}_{xy} = \dot{\varepsilon}_{yz} = 0$. It follows that u and w are independent of y. The stress-equilibrium equations [Eqs. (3)–(5)] reduce to

$$\partial\sigma_x/\partial x + \partial\tau_{xz}/\partial z = 0, \tag{81}$$

$$\partial\sigma_y/\partial y = \rho_i g, \tag{82}$$

$$\partial\tau_{xz}/\partial x + \partial\sigma_z/\partial z = 0. \tag{83}$$

An additional boundary condition is $\sigma_y = 0$ at the surface if atmospheric pressure is ignored.

Weertman (1957b) was the first to analyze ice-shelf deformation. He considered a shelf of uniform thickness h, constant in time, and very small

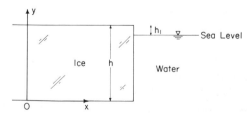

Fig. 13.   Coordinate system for analysis of ice shelf spreading.

compared with the horizontal dimensions. Far from the edges, it is reasonable to assume that all stress components are independent of x and z. It follows that $\tau_{xz} = 0$ and Eqs. (81)–(83) reduce to the single Eq. (82) which integrates to give

$$\sigma_y = -\rho_i g(h - y). \tag{84}$$

An expression for σ_x can be obtained from the fact that this stress, integrated over the ice thickness, must balance the pressure of sea water at the edge. Thus

$$\int_0^h \sigma_x \, dy = -\int_0^{h-h_1} \rho_w g(h - h_1 - y) \, dy = -\tfrac{1}{2}\rho_w g(h - h_1)^2, \tag{85}$$

where $h_1 = (1 - \rho_i/\rho_w)h$, the height of the ice surface above sea level. Most ice shelves are confined on each side by land or grounded ice and in this case $\dot{\varepsilon}_z = 0$. Thus $\sigma_z' = 0$ and $\sigma_x' = -\sigma_y' = \tfrac{1}{2}(\sigma_x - \sigma_y)$. Because all shear stresses are zero, the flow law [Eq. (15)] reduces to

$$\dot{\varepsilon}_x = (\tfrac{1}{2})^n A(\sigma_x - \sigma_y)^n$$

and $\dot{\varepsilon}_x$ is independent of y, so

$$2(\dot{\varepsilon}_x)^{1/n} \int_0^h A^{1/n} \, dy = \int_0^h (\sigma_x - \sigma_y) \, dy - \tfrac{1}{2}\rho_i g h h_1$$

by Eqs. (84) and (85). Thus

$$\dot{\varepsilon}_x = \overline{A}(\tfrac{1}{4}\rho_i g h_1)^n \tag{86}$$

with

$$\overline{A} = h^n \left(\int_0^h A^{-1/n} \, dy \right)^n. \tag{87}$$

Equation (86) gives the spreading rate of an ice shelf, far from the edge. Accumulation is necessary at the surface to maintain constant thickness; an ablating ice shelf will eventually disappear.

Weertman also considered the case of an ice shelf unconfined at the sides. In this case $\dot{\varepsilon}_z = \dot{\varepsilon}_x$ and any horizontal line in the shelf is lengthened by the same amount, irrespective of its orientation. For the case of $n = 3$, the strain rate is decreased by about 10% from the value given by Eq. (86).

In fact, almost all ice shelves are fed by glaciers and decrease in thickness from the *grounding line,* where the ice starts to float, to the outer edge. Thinning results from melting at the bottom as well as spreading. Because there is no shear stress at the base, the surface slope must be maintained by drag at the margins which may be caused by fjord walls, grounded ice, or thinner, more slowly moving parts of the ice shelf and by the bottom slope. Crary (1966) considered the case of an ice shelf, confined at the sides, with thickness varying in the direction of spreading. He obtained an integrated form of one of the stress-equilibrium equations by considering the balance of forces on a vertical section of ice shelf of length δx considered initially to be of unit width (perpendicular to the plane of the diagram in Fig. 14).

The force exerted on the element $ABCD$ by the difference between the longitudinal stresses on the two faces is balanced by the forces exerted by the shear on the sides of the element and the horizontal component of the water pressure on the bottom interface CD. If the ice is thick, the normal pressure on AC is given, to a first approximation, by the hydrostatic pressure, the mean value of which is $\frac{1}{2}\rho_i gh$. Thus the normal force is $\frac{1}{2}\rho_i gh^2$ and the difference in force between the two sides is

$$(d/dx)(\tfrac{1}{2}\rho_i gh^2)\,\delta x = \rho_i gh(dh/dx)\,\delta x.$$

The shear stress exerts a force $\tau_{xz}h\,\delta x$. The horizontal component of the water pressure on CD is

$$\rho_w g(h - h_1)[d(h - h_1)/dx]\,\delta x = (\rho_i^2/\rho_w)gh(dh/dx)\,\delta x.$$

Balance of forces then gives

$$\rho_i g(dh/dx) = \tau_{xz} + (\rho_i^2/\rho_w)g(dh/dx).$$

If the valley walls are vertical and parallel to each other, this equation can be extended across the total width Z of the ice shelf to give

$$\rho_i g(1 - \rho_i/\rho_w)Z(dh/dx) = \tau_s,$$

$$dh/dx = K_1\tau_s/Z,$$

(88)

where $1/K_1 = \rho_i g(1 - \rho_i/\rho_w)$. Crary thinks that values of τ_s should be comparable with basal shear stresses in glaciers. Thomas and Bentley (1978a) suggest a range of 40–100 kPa. Equation (88) shows that, with the approximations made in the analysis, the change in thickness of an ice shelf along the line of flow is independent of the thickness. In this respect ice shelves differ markedly from ice caps.

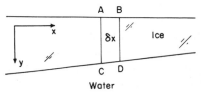

Fig. 14.   Section of floating ice shelf. See derivation of Eq. (88).

Thomas (1973a) obtained a solution of the stress-equilibrium equations for the case of an ice sheet flowing between parallel sides. He assumed:

(1) $\tau_{xy} = \tau_{yz} = 0$; this implies that all velocities and strain rates are independent of depth y and Eqs. (81)–(83) apply;

(2) $\partial\tau_{xz}/\partial z$ is independent of z;

(3) $\bar{\tau}_s$, the value of τ_{xz} at the sides averaged over the ice thickness, reaches some limiting value independent of x.

The aim of Thomas's work was to obtain values of the flow law parameters A and n [Eq. (15)] from measurements of strain rates at different places on ice shelves of known thickness. From the stress-equilibrium equations, the flow law, and the above assumptions, and with the ratios between the strain rates at a point known, Thomas showed that the value of n could be determined from the way in which $\dot{\varepsilon}_{xz}$ varied with distance from the side of the ice shelf. In an analogous way, the value of n can be determined from the variation in velocity across a valley glacier. With n known, the value of A can be found from the way in which the spreading rate $\dot{\varepsilon}_x$ varies for shelves of different thicknesses. Weertman's formula [Eq. (86)] can be used for the same purpose but the underlying assumptions are more restrictive.

Thomas (1973b) determined the flow law parameters over the stress range 40–100 kPa. These field data are extremely useful because, at such low stresses, it is almost impossible to carry on a laboratory experiment beyond the stage of transient creep. On the other hand, because the temperature difference between top and bottom of a typical ice shelf is between 20 and 30 deg, these measurements determine a value of A averaged over a large temperature range. However, Thomas's results, shown in Fig. 15 for the case in which A is assumed to have the same value for all ice shelves, are consistent with laboratory experiments in the stress range 10^2–10^3 kPa. In particular, they give a value $n = 3$. This is contrary to the results of some laboratory experiments that appear to show that n decreases to a value near unity for stresses below about 100 kPa [for references, see Weertman (1973a)]. The total strain in these experiments was very small however, so transient creep probably still prevailed at the end of the experiment.

Shumskiy and Krass (1976) have performed a complex analysis of flow and heat transfer in ice shelves. They consider both an unconfined ice shelf and one in which lateral spreading is prevented by fjord walls. The ice thickness is assumed to be uniform but can change with time as a result of accumulation or ablation at the upper and lower surfaces. The analysis is two dimensional, and ice flow and heat transfer are treated together. The mechanical boundary conditions are (1) shear stresses are

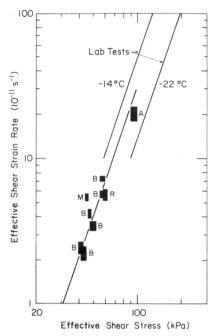

Fig. 15. Relation between effective shear strain rate and effective shear stress for ice shelves of differing thickness: A, Amery; B, Brunt; M, Maudheim; R, Ross. The error rectangles incorporate inaccuracies in strain rate, ice thickness, and surface elevation. Results of laboratory tests by Barnes *et al.* (1971) at two temperatures are also shown. The ice shelf mean temperatures lie between −11°C and −19°C. [From Thomas (1973b). Reproduced from the *Journal of Glaciology* by permission of the International Glaciological Society.]

zero at both boundaries and (2) hydrostatic equilibrium. The temperature is prescribed at both boundaries. Analytical expressions are obtained for the spreading rate and the temperature distribution.

B. Observations

Real ice shelves are much more complex than the simple theoretical models discussed above. As an example we shall describe the one studied in most detail, the Ross Ice Shelf. The total area of ice shelves in Antarctica is about 1.5×10^6 km²; the Ross Ice Shelf comprises about one-third of this [Robin (1975)].

Figure 16 shows thickness contours for this ice shelf. Ice thickness decreases from roughly 1000 m at the *grounding line* to about 250 m at the *ice front* where the ice breaks off to form icebergs. Ice velocity increases with distance from the grounding line, attaining values of up to 1 km/yr

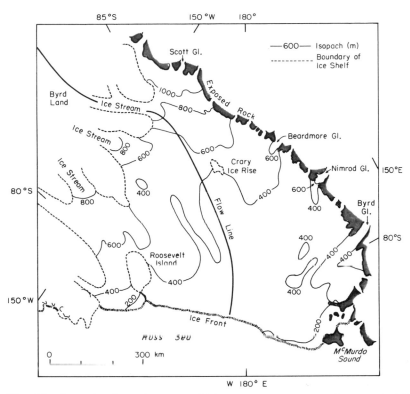

Fig. 16. Isopachs (lines of constant thickness) for the Ross Ice Shelf. Dashed lines indicate boundary of Ice Shelf. [From Robin (1975), by courtesy of the publishers of *Nature*.]

at the ice front [Robin, (1975)]. However, thickness and velocity do not vary uniformly over the ice shelf. As Fig. 16 shows, valley glaciers draining ice from the East Antarctic Ice Sheet through the Transantarctic Mountains persist as streams of thicker, faster-moving ice for some distance into the ice shelf. This is especially marked in the case of Byrd Glacier. Regions of rapidly moving ice within an ice sheet are called *ice streams;* their edges are clearly marked by intensely crevassed zones (Fig. 17). As ice streams flow into ice shelves they are retarded by the drag of the more slowly moving ice on either side, so their thickness decreases in the downstream direction. The slower ice, on the other hand, is dragged along by the ice stream and tends to thicken in the downstream direction. The shear is gradually reduced as the ice stream moves farther into the ice shelf and thickness and velocity become more nearly uniform. Most of the ice of the West Antarctic Ice Sheet in Byrd Land also drains through ice streams that form in the grounded ice and flow out into the ice shelf.

Fig. 17.   Ice stream in West Antarctica (83°50′ S, 150° W). [U.S. Navy photo.]

Figure 16 also shows *ice rises,* the largest of which is known as Roosevelt Island. These are regions where the ice is grounded. Thus the surface, instead of being almost horizontal as on the ice shelf, has the parabolic profile characteristic of ice caps. Shelf ice tends to flow around ice rises; their boundaries are often marked by crevasses.

The surface mass balance is positive everywhere on the ice shelf, except for a small albation area near McMurdo Sound [Swithinbank (1970)]. Accumulation rates vary from about 0.08 to 0.25 m/yr water equivalent. Mass change c_b at the lower surface is difficult to measure. However, it can be estimated from the continuity equation [Eq. (68)] if a steady state is assumed. In this case

$$c_b = h(\dot{\varepsilon}_x + \dot{\varepsilon}_z) + u(\partial h/\partial x) - c_s, \qquad (89)$$

where the x axis is taken along the flow line. The quantities c_s, h, $\dot{\varepsilon}_x$, and $\dot{\varepsilon}_z$ are easily measured, but measurement of u was difficult until the recent introduction of accurate positioning systems based on Doppler microwave measurements to satellites. Normal practice has been to estimate the total annual accumulation over the drainage basin, which is usually poorly defined, and take u to be the value necessary to carry this away with the known ice thickness. At Little America Station, near the front of the ice shelf, Crary and Chapman (1963) calculated a bottom melting rate of 0.6 m/yr from measurements of change of surface elevation with time. Estimates by Crary et al. (1962, Table 11), based on Eq. (89), suggest that the basal melting decreases with increasing distance from the ice front, reaching zero at about 150 km from it. Paige (1969) measured a bottom melting rate of 0.81 m/yr close to the ice front near Scott Base. Giovinetto and Zumberge (1968) estimate that there is accretion, at a mean rate of 0.15 m/yr, at the bottom of the southern part of the ice shelf. Robin (1968) however, while agreeing that accretion is likely there, thinks that this estimate is far too high because the ice shelf could not absorb all the latent heat released. Observations of sediment-laden ice near Crary Ice Rise (Fig. 16), where fracturing has probably overturned part of the ice shelf, suggest bottom accretion upstream [Thomas (1976)]. However, the few available measurements suggest that melting occurs under most Antarctic ice shelves and Thomas and Coslett (1970) think that significant accretion is only possible under thin ice shelves or where a body of seawater is trapped underneath the shelf. Wakahama and Budd (1976), on the other hand, estimate that freezing at a rate of about 0.5 m/yr occurs at the base of the Amery Ice Shelf.

Figure 18 shows ice thickness, velocity, and the sum of the two horizontal components of strain rate along what is believed to be a flow line from the crest of the West Antarctic Ice Sheet to the front of the Ross Ice Shelf (see Fig. 16). The central part of the flow line follows an ice stream. Downstream from the grounding line, ice thickness decreases rapidly from about 1000 m to the value of a few hundred meters typical of ice shelves. The thickness is halved in about the first 300 km. The thickness does not change much for the next 350 km, then decreases for the last 150 km. These observations do not agree with the predictions of Eq. (88) that

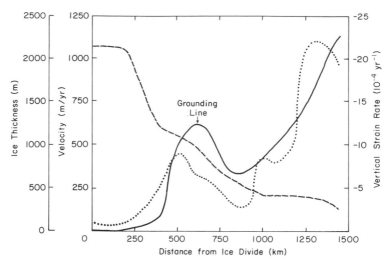

Fig. 18. Ice thickness (dashed line), velocity (solid line), and vertical strain rate (dotted line) along a flow line, shown on Fig. 16, from the ice divide of the West Antarctic Ice Sheet to the ice front of the Ross Ice Shelf. [From Thomas (1976), courtesy of the publishers of *Nature*.]

the rate of thickness change is independent of the thickness. In general, ice velocity increases along the flow line as expected in an ice shelf confined at the sides and with accumulation over its surface. An exception is the decrease in velocity in the first 200 km from the grounding line. This could result from damming of the flow by the Crary Ice Rise; an alternative explanation might be that the profile does not in fact follow a flow line in this region. The strain rate increases towards the ice front in spite of the decrease in thickness, contrary to the predictions of Eq. (86). These discrepancies between theory and observation result from breakdown of the assumptions underlying the theory and illustrate the differences between a real ice sheet and the simple theoretical models. The theory of the flow of ice streams has still to be developed fully, although Hughes (1977a) has made a start in this direction.

The past history and present state of the Ross Ice Shelf will be discussed in Section VII.B.

VI. TEMPERATURES IN ICE SHEETS AND ICE SHELVES

A. Introduction

The distribution of temperature in an ice sheet deserves study both on its own account and because of its relation to other processes. Past variations of surface temperature can be deduced from the present variation of temperature with depth. The rate of ice deformation is strongly temperature dependent. The basal temperature is important in considering erosion; bedrock is protected when the ice is frozen to it. If the temperature of an ice sheet, previously frozen to its bed, reaches the melting point, the ice can start to slide and so its velocity will increase; this could result in a major advance of the ice edge. Again, properties such as the absorption of radio waves and the velocity of seismic waves, on which depend techniques of ice thickness measurement, vary with temperature. Different drilling techniques are required according to the ice temperature.

Several factors interact to produce the temperature distribution. Short-wave solar radiation, long-wave radiation from water vapor and carbon dioxide in the atmosphere, and condensation of water vapor provide heat to the surface. The surface loses heat by outgoing long-wave radiation and sometimes by evaporation. Turbulent heat exchange with the atmosphere may warm or cool the surface according to the temperature gradient in the air immediately above. Ice deformation and, in some cases, refreezing of meltwater warm the interior of the ice sheet while geothermal heat and friction (if the ice is sliding) warm the base. Heat is transferred within the ice sheet by conduction, ice flow, and, in some cases, water flow. The relative importance of these factors varies from place to place.

B. Surface Temperatures

Seasonal variations of surface temperature are rapidly damped with increasing depth in the ice sheet so that at a depth of 10 or 15 m the temperature remains constant throughout the year. In regions where the maximum air temperature is less than 0°C, no snow melts even in summer and the temperature at a depth of 10 m is generally assumed to be equal to the mean annual air temperature at the surface. Loewe (1970) has compared 10-m temperatures with mean annual temperatures measured in meteorological screens at 16 stations on the Greenland and Antarctic Ice Sheets. Temperatures were within 2 deg of each other, except at one station, the coldest, where the difference was nearly 4 deg. The discrepancies can probably be explained by differences between screen temperature and surface temperature plus the fact that, at many of the stations, screen temperatures were measured over only 1 or 2 yr and so may not represent the long-term mean.

This simple relationship breaks down when there is summer melting. The meltwater percolates into the snow and, when it reaches a subfreezing layer, it refreezes. The latent heat released is a powerful means of warming the snow and may raise the 10-m temperature several degrees above mean anual air temperature. Müller (1976, Fig. 7) found differences

as high as 10 deg on an arctic valley glacier. Moreover, the 10-m temperature may change from year to year as a result of variations in the amount
of melt; Müller observed an increase of 7 deg in 5 yr at one station.
Although air temperature, and thus the amount of melting, normally increases with decrease of elevation, there is a certain critical elevation
(which varies from year to year) below which this warming effect is reduced. Below this elevation there is enough meltwater to form a continuous layer of ice in the snowpack; this prevents further meltwater penetration. In the ablation area, refreezing meltwater forms a layer of *superimposed ice* at the snow–ice interface early in the summer and some of the
latent heat released will warm the ice below. Later in the summer, however, ablation removes the superimposed ice and some of the ice underneath, so there is no lasting heating effect. Thus 10-m temperatures may
be higher in the accumulation area than in the ablation area, even though
air temperatures are lower [see Paterson (1969, p. 175)].

C. Steady-State Temperature Distributions

The general equation of heat transfer and the boundary conditions have
been discussed in Section II.F. To obtain analytical solutions, drastic simplifying assumptions are necessary. A major simplification is to set
$\partial T/\partial t = 0$ in Eq. (21) creating a thermal steady state. However, because
quantities such as ice thickness, ice velocity, and mass balance occur as
constants in the equation, the ice sheet is also assumed to be in a steady
state dynamically.

Robin (1955) derived a solution applicable to the central part of an ice
sheet. The base is assumed to be horizontal. Near the center, both u and
$\partial T/\partial x$ are small; thus the term $u \, \partial T/\partial x$ in Eq. (21) can be neglected. It is
also assumed that $\partial T/\partial z = 0$ and that the heat produced by internal deformation can be treated as a flux, additional to the geothermal flux, at the
base. Thus $F = 0$ in Eq. (20). This is plausible because most of the
shearing takes place near the bed where the shear stress is greatest and
the ice is warmest. This extra flux is equal to a normal geothermal flux if u
is about 20 m/yr. The vertical velocity v has to be expressed as a function
of y. The simplest relation is a linear one (constant vertical strain rate) as
used in the derivation of the Nye time scale in Section III.E. The relation
$v = -by/h$ satisfies the boundary conditions $v = 0$ at $y = 0$ and $v = -b$
at $y = h$. Equations (20) and (21) then become

$$k(\partial^2 T/\partial y^2) + (by/h)(\partial T/\partial y) = 0. \tag{90}$$

The boundary conditions are

$$y = h, \qquad T_s = \text{const}$$

$$y = 0, \qquad (\partial T/\partial y)_b = \text{const} = -\phi/K.$$

The solution is

$$T - T_b = (\partial T/\partial y)_b \int_0^y \exp(-y^2/l^2) \, dy \tag{91}$$

with

$$l^2 = 2kh/b.$$

When b is positive the solution can also be written

$$T - T_s = \tfrac{1}{2}\pi^{1/2} l (\partial T/\partial y)_b [\text{erf}(y/l) - \text{erf}(h/l)] \tag{92}$$

where

$$\text{erf } z = 2\pi^{-1/2} \int_0^z \exp(-Y^2) \, dY.$$

Zotikov (1963) has extended Robin's solution to cover the case of basal melting.

Figure 19 shows this solution for an ice sheet 2500 m thick with various values of b. The values $k = 1.33 \times 10^{-6}$ m^2 sec^{-1}, $K = 2.51$ W m^{-1} deg^{-1}, and $\phi = 4.2 \times 10^{-2}$ W m^{-2} (the average flux for continental shields) were used. The curves apply to other heat fluxes if the temperature scale is adjusted so that $(T_s - T_b)/h = -\phi/K$. For accumulation rates of about 0.3 m/yr or more, the upper half of the ice sheet is approximately isothermal.

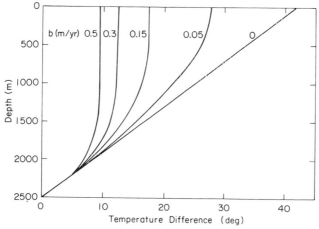

Fig. 19.   Steady-state temperature distribution in an ice sheet 2500 m thick for various ice-equivalent accumulation rates b.

For low accumulation rates, the temperature difference between surface and bed is very sensitive to the accumulation rate.

Robin's steady-state solution applies only near the center of an ice sheet, because only there can the horizontal advection term $u(\partial T/\partial x)$ be neglected. Except at an ice divide, the pattern of flow is such that, with increasing depth in a borehole, the ice has orginated as surface snow at progressively higher elevations and therefore at progressively lower temperatures. Thus a decrease of temperature with increase of depth is expected. This argument applies only in the upper layers where the geothermal flux has little or no effect on temperatures. To treat this case, Robin (1955) made the drastic simplifying assumption that heat conduction could be neglected. The heat transfer equation then becomes

$$u(\partial T/\partial x) + v(\partial T/\partial y) = 0.$$

At the surface, $v_s = -b$ and $(\partial T/\partial x)_s = \lambda\alpha$. The surface slope is taken to be positive. Typical values of λ (the increase in air temperature per unit decrease in surface elevation) lie between 0.6 and 1.4 deg per 100 m. It follows that

$$(\partial T/\partial y)_s = u_s\alpha\lambda/b. \tag{93}$$

Note that our coordinate system has the origin on the bed and the y axis positive upwards. Thus Eq. (93) represents an increase in temperature with distance above the bed. This is usually called a *negative* temperature gradient because temperature decreases with increasing depth. Because both u_s and α increase with distance from the ice divide, the negative near-surface temperature gradient should increase also. French observations in Greenland, quoted by Robin (1955), and also data from Antarctica [Budd *et al.* (1971a, Map 1/7)] show this trend. A negative temperature gradient could also result from recent climatic warming because each annual layer would be warmer than the one below it. This may explain some observed discrepancies between measured gradients and those calculated from Eq. (93).

In Robin's analysis, the term $u(\partial T/\partial x)$ does not occur in the heat transfer equation, because it is assumed to be negligible, which should be true near an ice divide. In a steady state, on this basis, the temperature profile along a given vertical line, fixed in space, does not change as the ice flows past. An alternative point of view is to assume that u does not vary with depth so that an initially vertical column in the ice remains vertical. Then $\partial T/\partial t$ can be regarded as a local derivative moving with the column and implicitly incorporating the term $u(\partial T/\partial x)$. However, the assumption that the column remains vertical is unrealistic at large distances from the ice divide. Budd has used this *moving-column model* extensively [e.g., Budd and Radok (1971); Budd *et al.* (1971a)]. Radok *et al.* (1970) have generalized Robin's steady-state solution by solving the equation

$$k(\partial^2 T/\partial y^2) + (by/h)(\partial T/\partial y) = W, \tag{94}$$

where $W = \partial T/\partial t$, the warming or cooling rate, is taken as constant. It represents a combination of warming as the ice moves down slope, temperature change due to change of ice thickness with time, and climatic change. It can be written

$$W = \lambda(u_s \alpha - \partial h/\partial t) + \dot{T}_c, \tag{95}$$

where $\dot{T}_c$ is the climatic effect. The separate terms are not distinguished in the analysis however. The solution is

$$\partial T/\partial y = (\partial T/\partial y)_b \exp(-y^2/l^2) + (Wl/k)I(y/l), \tag{96}$$

where

$$I(z) = \exp(-z^2) \int_0^z \exp(Y^2)\, dY.$$

This is Dawson's integral, a tabulated function [Abramowitz and Stegun (1965)]. The first term on the right-hand side of Eq. (96) is Robin's solution. Radok *et al.* (1970) give temperature profiles for various values of accumulation rate, ice thickness, surface warming rate, and basal heat flux. However, the underlying assumption that any change in surface temperature produces the same change at all depths is not realistic.

Budd *et al.* (1971a), following the example of Bogoslovski (1958), have used a computer to calculate temperature distributions along flow lines in the Antarctic Ice Sheet. Budd *et al.* use Eq. (94) with $W = \lambda U \alpha$. The values of b, h, α, T_s, and basal heat flux must be known along the flow line; U is obtained from the condition for a dynamic steady state, namely, that the ice flux through a vertical section between two flow lines must balance the upstream accumulation. Particle paths and ages of the ice are computed from the velocities. Figure 20 is a typical result. The negative temperature gradient in the upper layers increases in magnitude and penetrates to greater depths with increasing distance from the ice divide.

Diagrams such as Fig. 20 are valuable for showing the general features of the temperature distribution. However, detailed predictions may be inaccurate as a result of defects in the model or of inaccuracies in the input data. The underlying assumption that an initially vertical column of ice remains vertical becomes increasingly unrealistic with increasing distance from the ice divide. As Fig. 19 shows, the temperature difference between surface and bed is very sensitive to the accumulation rate when the rate is as low as in the interior of Antarctica. Again, the geothermal

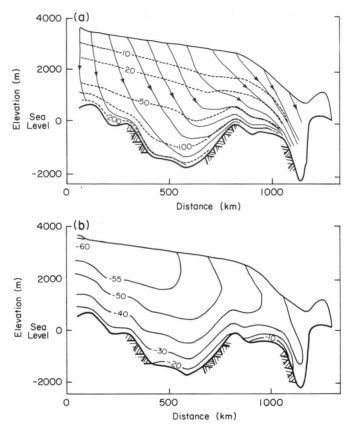

Fig. 20. Surface and bedrock profiles of the flowline from Vostok to Wilkes, East Antarctica, showing (a) particle paths (solid lines) and isochrones (10³ yr, broken lines) and (b) isotherms (°C). All calculations assume steady state. [From Budd *et al.* (1971a).]

heat flux has not been measured in Antarctica. Thus Oswald and Robin (1973), on the basis of the nature of radar reflections, have identified "lakes" beneath the ice sheet in places where the original calculations of Budd *et al.* (1971a) predicted basal temperatures of −20 to −30°C. The discrepancy arose largely because the original estimates of the accumulation rate were too high, although the value taken for the geothermal heat flux may also have been wrong. Budd *et al.* (1976) have also applied the same steady-state moving-column model to Law Dome, Antarctica. Temperatures measured in boreholes conformed closely to the predictions, except for a slight warming observed in the upper layers. Climatic warming in the last century, or a slight lowering of the ice surface, or some combination of the two, could have caused this.

D.   The Camp Century Temperature Profile

The first temperature measurements to bedrock in a polar ice sheet were made by Hansen and Langway [Weertman (1968)] in the 1387-m borehole at Camp Century, Greenland. Several authors have fitted steady-state temperature distributions to these data. We shall outline the different models to show that there is no unique solution to the problem. Calculations based on different assumptions can provide satisfactory fits to the observations. Figure 21 shows measured temperatures and the differences between observation and prediction for the different models.

The Robin steady-state model with present values of accumulation rate, surface temperature, and basal temperature gradient fails to predict the negative temperature gradient near the surface and gives a basal temperature that is 2.7 deg too low.

Weertman (1968) derived another steady-state solution. He took account of horizontal advection and also included a term for heating due to shear within the ice instead of regarding it as an additional heat flux at the bed. The equation is thus

$$k(\partial^2 T/\partial y^2) + (by/h)(\partial T/\partial y) - u(\partial T/\partial x) + \varepsilon_{xy}\tau_{xy}/J\rho c' = 0, \quad (97)$$

where the symbols are as in Eqs. (20) and (22). The shear stress is calculated from Eq. (23) and, because temperature is known as a function of

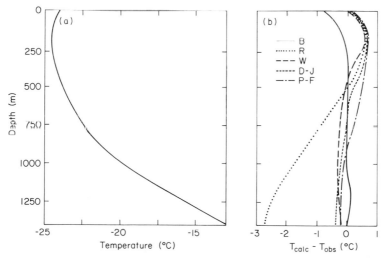

Fig. 21: (a) Measured temperature–depth profile through the Greenland Ice Sheet at Camp Century. (b) Deviations from observed temperatures of those calculated on the steady-state flow models of Robin (1955), Weertman (1968), Dansgaard and Johnsen (1969b), and Philberth and Federer (1971), and by the method of Budd *et al.* (1971b).

depth, the strain rate $\dot{\varepsilon}_{xy} = \frac{1}{2}(\partial u/\partial y)$ can be calculated from the flow law, Eq. (6). The velocity is then obtained by numerical integration. An approximate relation for $\partial T/\partial x$ in terms of known quantities was derived by differentiating Robin's solution. Weertman found that the two new terms could be approximated by a constant in the upper 1000 m of the borehole and by a linear function of y in the remainder. Equation (97) can then be solved analytically. This solution has a negative temperature gradient near the surface, but its value is too small. Moreover, predicted and observed basal temperatures differ by 2.0 deg. Weertman then suggested that the present value of accumulation rate may not have been typical of the past. He found that with $b = 0.224$ m/yr instead of the present value of 0.35 m/yr, the calculated temperature–depth curve fitted the observations to 0.5 deg. However, subsequent measurements of thicknesses of annual layers in the core from the borehole have shown that the present accumulation rate is indeed typical of the past 10,000 yr [Hammer et al. (1978)]. Weertman also postulated a climatic warming of 0.5 deg in the past 1000 yr to explain the observed value of the negative temperature gradient.

Dansgaard and Johnsen (1969b) used the simple equation

$$k(\partial^2 T/\partial y^2) - v(\partial T/\partial y) = 0, \tag{98}$$

but, instead of taking a linear relation between v and y at all depths (uniform vertical strain rate), they assumed the vertical strain rate was uniform from the surface to 400 m above the bed and then decreased linearly with depth to a value of zero at the bed. This is the same flow model they used to derive a time scale for the Camp Century core (see Section III.E). The resulting temperature profile does not have a negative gradient near the surface but, below a depth of 400 m, it is within 0.4 deg of the measured profile.

Philberth and Federer (1971) also used Eq. (98) but derived $v(y)$ as follows. From the flow law they obtained a relation of the form $u = Uf(y)$. If x is the distance from the ice divide, $U = bx/h$ for a steady state. The incompressibility condition for two dimensions is $\partial v/\partial y = -\partial u/\partial x$. Thus v is given by

$$v = h^{-1}[U(\partial h/\partial x) - b]\int_0^y f(y)\,dy.$$

This model gives a good fit to observations in the lower half of the borehole.

Budd et al. (1971b) used a computer to solve Eq. (94) for various values of b and W and chose the values that gave the minimum root-mean-square deviation between observed and computed temperatures. They neglected the top 200 m as being disturbed by recent climatic changes, but obtained an excellent fit to the remainder (standard deviation 0.04 deg) with $b = 0.36$ m/yr and $W = -0.37$ deg per 1000 yr. The value of b is very close to that observed (0.35 m/yr) and the surface cooling is in reasonable agreement with that for the past 2500 yr as deduced from the oxygen isotope profile [Robin (1976, Fig. 8)]. The surface cooling could be purely climatic or due to an increase in ice thickness of about 0.05 m/yr or to some combination of the two effects. This technique provides a much better fit to the data than any other method. It has also been used to fit data from Byrd Station and Vostok, Antarctica [Budd *et al.* (1973, 1975)].

E.   Ice Thickness Changes Deduced from Temperature Gradients

Budd (1969, p. 96) and Robin (1970) have suggested that the rate of change of ice thickness with time can be estimated from the deviation of the measured near-surface temperature gradient from that predicted by steady-state theory. Although this is not a precise method, it does determine long-term trends whereas several of the methods described in Section III.F apply only to short periods.

If b is the accumulation rate and W the rate of surface warming, a layer initially on the surface at temperature T_s is at depth b after one year, at which time the surface temperature is $T_s + W$. Thus, if y is depth below the surface, $(\partial T/\partial y)_s = -W/b$ and Eq. (95) becomes

$$\partial h/\partial t = (b/\lambda)(\partial T/\partial y)_s + \dot{T}_c/\lambda + u_s\alpha. \tag{99}$$

If the climatic warming $\dot{T}_c$ is known, or assumed to be zero, $\partial h/\partial t$ can be determined.

For studying long-period changes, Robin took $(\partial T/\partial y)_s$ to be the temperature gradient measured over the depth interval 150 to 300 or 400 m and noted that this was approximately constant at Site 2 in Greenland and Byrd Station in Antarctica. Robin assumed that past fluctuations of surface temperature had the form $T(0, t) = T_0 \cos(\omega t - \delta)$ and used the solution of the equation

$$k(\partial^2 T/\partial y^2) - v(\partial T/\partial y) = \partial T/\partial t$$

with this expression as surface boundary condition [Carslaw and Jaeger (1959, p. 389)]. The medium is assumed to be semi-infinite with v constant and equal to the accumulation rate. From this solution, Robin calculated the temperature gradient $(\partial T/\partial y)_c$ over the depth range 150–300 m for various values of ω. He found that the temperature gradient was approximately constant for $2\pi/\omega = 1000$ yr and then chose $\delta = 3\pi/2$ so that a surface temperature minimum occurred 250 yr ago as climatic data show.

If a value is also chosen for the amplitude T_0 of the temperature fluctuation, $\partial h/\partial t$ can be found from Eq. (99) with $\dot{T}_c = -b(\partial T/\partial y)_c$. For Byrd Station, $\partial h/\partial t = +6, -7$, or -32 mm/yr for $T_0 = 0.25, 0.5$, or 1 deg, respectively. The conclusions thus depend strongly on the value chosen for T_0. Moreover, there is the implicit assumption that the surface temperature has varied according to the cosine law for several cycles.

F. Non-Steady-State Temperature Distributions

Because no real ice sheet is ever in a steady state, models based on this assumption have obvious limitations. The main reason for their extensive use has been lack of knowledge about the past history of the ice sheet. However, the recently developed oxygen isotope method [Dansgaard *et al.* (1973)] now provides detailed information about past surface temperatures. On the other hand, a dynamic steady state usually has to be assumed for lack of information about the past dimensions and flow pattern of the ice sheet. Few analytical solutions of the heat transfer equation for the nonsteady state have been obtained; computers have to be used for all but the simplest cases.

Benfield (1951) studied the effect of accumulation on temperatures in a snowfield. He used the equation

$$\partial T/\partial t = k(\partial^2 T/\partial y^2) - v(\partial T/\partial y) \tag{100}$$

applied to a semi-infinite medium with downward velocity v uniform with depth y and equal to the accumulation rate. Note that the origin is now taken on the surface and y measured positive downwards. The boundary condition was that the surface temperature followed a sine wave with a period of one year. The initial temperature was assumed to be uniform. Benfield showed that snowfalls of 4 m/yr or more considerably enhanced the depth of penetration of the winter "cold wave."

Carslaw and Jaeger (1959, p. 388) give solutions for T of Eq. (100) with v constant, for the conditions

$$T(y, 0) = T_1 + C_1 y, \qquad T(0, t) = T_2 + C_2 t,$$

with T_1, T_2, C_1, and C_2 constants. Radok (1959) and Mellor (1960) have given expressions for $\partial T/\partial y$. These authors used this solution to investigate the validity of Robin's equation for the negative temperature gradient near the surface [Eq. (93)], originally derived under the unrealistic assumption that heat conduction could be neglected. Use of Eq. (100) is based on the "moving-column model" described previously. Mellor concluded that Eq. (93) could justifiably be used to calculate near-surface temperature gradients in thick ice sheets.

The only published analysis to use the temperature record obtained from oxygen-isotope data as surface boundary condition is that of Johnsen (1977). The data were temperatures measured in a 50-m borehole at Crête near the highest point of the Greenland Ice Sheet. Because the borehole was entirely in firn, the change in density, and thus in thermal properties, with depth was significant. Heating due to firn compaction was also taken into account. Horizontal velocity could be ignored because the borehole was near the ice divide. Johnsen's equation is effectively

$$\partial T/\partial t = k(\partial^2 T/\partial y^2) + [(\rho c')^{-1}(dK/d\rho)(d\rho/dy) - v](\partial T/\partial y) + F/(J\rho c'),$$
(101)

which is the basic heat transfer equation [Eqs. (20)–(22)] with a term added to take into account the varying thermal properties of the firn. Here y is depth below the surface. The vertical velocity v was calculated from the accumulation rate and the depth–density curve; F was also determined from the depth–density curve. The oxygen-isotope record, determined from a 404-m core, extended to about 1400 B.P. The regression line of 10-m temperature on oxygen-isotope ratio for the Greenland Ice Sheet was used to convert oxygen-isotope ratio to temperature. This is probably satisfactory for Greenland samples, where the 12 data points show very little scatter about the regression line, but might be less satisfactory in other areas. The age of each oxygen-isotope sample was determined by counting annual layers. Equation (101) was solved numerically with this boundary condition. The initial condition was isothermal. Observed and predicted temperatures differed by up to 0.1 deg between 10 and 20 m but, below this, the discrepancy did not exceed 0.025 deg.

G. Temperatures in Ice Shelves

Temperature distributions in ice shelves have received much less attention than those in ice sheets. The two cases differ in the basal boundary condition; at the base of an ice shelf the temperature is fixed at the freezing point of seawater and there may be melting or accretion. Wexler (1960) calculated the non-steady-state temperature distribution in an ice mass of uniform thickness, with uniform temperature at time $t = 0$. Thereafter the surface was maintained at this temperature and the basal temperature at the freezing point of seawater. There was no accumulation or ablation at top or bottom. This would approximate conditions in an initially land-based ice mass from the time it reaches the sea and starts to float. Wexler also derived the non-steady-state temperature distribution in a very thick ice shelf melting at its base. Jenssen and Radok (1961) used

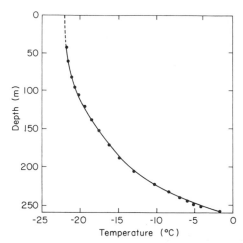

Fig. 22. Measured temperature–depth profile through the Ross Ice Shelf at Little America V. [From Bender and Gow (1961), courtesy of the International Association of Hydrological Sciences.]

a computer to obtain solutions of the non-steady-state Eq. (100). However their treatment of the lower boundary condition for an ice shelf is not satisfactory. Figure 22 shows the temperature distribution in the Ross Ice Shelf at Little America V [Bender and Gow (1961)].

VII. INSTABILITIES IN ICE SHEETS

A. Instability against Change in Mass Balance

Glaciers and ice sheets are normally regarded as relatively stable features of the landscape, making slow advances and retreats in response to variations in climate. However, catastrophic advances or *surges*, described in Chapter 2, show that some glaciers have inherent mechanical instabilities. A surge in a large ice sheet could have a major effect on world sea level and climate. This is only one of several possible types of instability which will now be discussed.

Bodvarsson (1955) pointed out that if the equilibrium line (the boundary between the areas of net accumulation and net ablation) is always at a fixed elevation, the steady-state profile of an ice sheet may be unstable. Figure 23 illustrates the idea. For the steady-state profile S–S, the mass of snow that accumulated in one year over the area above the equilibrium line equals the amount of ice lost in one year from the area below it. If the ice thickness starts to increase, the accumulation area and hence the total

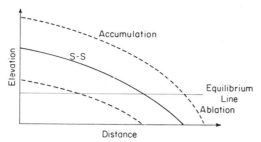

Fig. 23.   Ice sheet with equilibrium line at fixed elevation. The solid line represents the steady-state profile. The other profiles may be unstable.

accumulation will increase so that it may now be greater than the total ablation. If so, the ice sheet will continue to grow rather than return to the steady-state profile. In the same way, if the ice sheet surface drops below the profile $S-S$, the total ablation may exceed the total accumulation and the ice thickness will continue to decrease.

Weertman (1961b) has analyzed the problem mathematically and concluded that if a small ice cap exceeds a certain critical size, a moderate increase in accumulation rate may cause it to grow unstably. Similarly, a moderate increase in ablation rate, or decrease in accumulation rate, may cause a large ice sheet to become unstable and eventually disappear. In his analysis, Weertman assumed that the ice sheet lies on land extending to infinity. Thus the analysis is more applicable to an ice-age ice sheet than to the Antarctic Ice Sheet, which is bounded by the sea. Instability of this kind might help to explain the repeated growth and decay of the large ice-age ice sheets.

B.   Instability of Junction between Ice Sheet and Ice Shelf

In the ice sheet profiles derived in Section III the ice thickness at the edge is zero. This is true, however, only for those ice sheets that are grounded on bedrock that is above sea level. There is another type of ice sheet, sometimes called a *marine ice sheet,* in which the underlying bedrock is below sea level and would remain so even after the completion of isostatic uplift following removal of the ice sheet. Such ice sheets are fringed by floating ice shelves. The West Antarctic Ice Sheet is of this type; most of it drains through the Ross and Filchner–Ronne Ice Shelves.

The extent of an ice shelf seems to depend largely on the existence of land or grounded ice to protect its sides, or on shoals on which the outer margin may be grounded. Because ice shelves hold back the flow of ice streams draining the ice sheet, they may be essential to prevent the complete disintegration of a marine ice sheet. All Antarctic ice shelves have positive mass balances at present but if the climate became warmer and ablation set in, the ice shelves, and possibly the whole West Antarctic Ice Sheet, would disintegrate. Mercer (1968) suggested that this happened about 120,000 yr ago when world sea level was about 6 m above present.

The *grounding line,* where the ice starts to float, is the boundary between ice sheet and ice shelf. The ice thickness at the grounding line is greater than the equilibrium thickness of the ice shelf, however, and so the ice shelf thins toward its outer margin. The position of the grounding line is largely determined by sea level. Because the characteristic profile of an ice sheet is very steep at the edge, an advance of the grounding line can result in a large change in ice thickness. Denton and Borns (1974) and Denton *et al.* (1975) have obtained geological evidence of repeated expansions of the West Antarctic Ice Sheet when the grounding line was in roughly the present position of the outer margin of the Ross Ice Shelf (see Fig. 16). Denton *et al.* (1975) believe that this happened during the Wisconsin Glaciation, that thinning began before 10,000 B.P., and that the ice in McMurdo Sound became ungrounded about 5000 B.P.

Hughes (1973) has suggested that the position of the junction of the West Antarctic Ice Sheet and the Ross Ice Shelf is unstable. He made an indirect calculation that suggested that the grounding line is retreating and speculated that this would continue until the whole ice sheet disintegrated.

Weertman (1974) has made a theoretical analysis of the problem for an ice shelf unconfined at the sides and not grounded on any shoals. Figure 24 shows the coordinate system and symbols used. The accumulation rate is $c(x)$. Equation (86) gives the spreading rate of an ice shelf. The power flow law used in that derivation can be transformed to the approximation of perfect plasticity by replacing A by A_1/τ_0^n and letting $n \to \infty$. Equation

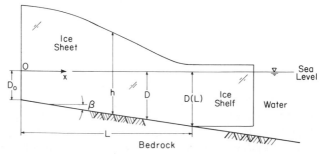

Fig. 24. Definition of symbols for analysis of stability of ice-sheet–ice-shelf junction.

(86) can then be written

$$h = \rho_w h_1/(\rho_w - \rho_i) = 4\tau_0 \rho_w/\rho_i(\rho_w - \rho_i)g. \tag{102}$$

This gives the shelf thickness as a function of yield stress and the densities of ice and water. The ice-sheet–ice-shelf junction is at $x = L$, where the depth from sea level to bedrock is

$$D(L) = D_0 + \beta L. \tag{103}$$

But the distance from sea level to the bottom of the ice shelf is $\rho_i h/\rho_w$. Thus

$$L = \beta^{-1}[4\tau_0 g^{-1}(\rho_w - \rho_i)^{-1} - D_0]. \tag{104}$$

If $D_0 > \rho_i h/\rho_w$, no ice sheet can exist. For a flat bed $\beta = 0$ and $L \to \infty$ if $D_0 < \rho_i h/\rho_w$; in other words, the ice sheet will extend to the edge of the continental shelf.

Weertman also calculated L using the more realistic power flow law. In this case the shelf thickness is not unique, its value at the junction is

$$h(L) = K_2(cL)^{2/9}, \tag{105}$$

and so

$$L = \beta^{-1}[K_2(cL)^{2/9} - D_0] \tag{106}$$

According to the values of β and D_0 there may be one, two, or no real values of L satisfying Eq. (106). If $\beta = 0$, there is only one solution—an unstable one. That is, if $L > c^{-1}(D_0/K_2)^{9/2}$ the ice sheet will continue to grow, whereas if L is less than this value the ice sheet will shrink to zero. When there are two solutions, the larger one is stable and the smaller one unstable. Estimated values for the West Antarctic Ice Sheet, $\beta = 4 \times 10^{-4}$, $D_0 = 200$ m, give $L = 700$ km, which is about the observed value. On the other hand, if $\beta = 6 \times 10^{-4}$, $D_0 = 250$ m, values well within the precision of observation, no stable ice sheet can exist.

Thomas and Bentley (1978a) extended Weertman's analysis to the more typical case of an ice shelf confined in a bay and grounded on shoals near its outer margin. They found that the grounding line was more stable than for an unconfined ice shelf. They also considered the effect of ice streams and estimated the rate of retreat of the grounding line under different conditions.

Observational tests of Hughes's suggestion have also been attempted. Direct observation of movement of the grounding line is extremely difficult; because the sea floor is rough there is no precise line but rather an area where the ice is grounded on the bumps but not on the hollows. Instead, efforts have been largely concentrated on assessing the state of balance of the West Antarctic Ice Sheet and Ross Ice Shelf. Whillans (1976)

examined the stability of the ice sheet along the 160-km flowline from the
ice divide to Byrd Station, some 600 km upstream from the grounding
line, by studying the pattern of internal radio echoes. The data suggest
that there has been no major change in flow pattern in the last 30,000 yr.
Whillans (1977) also compared the mass flux at Byrd Station, calculated
from the velocity–depth profile derived from borehole measurements,
with the accumulation rate integrated from the ice divide. He concluded
that the ice sheet was thinning at the rate of about 0.03 m/yr, which confirms that it is not far from a steady state at present.

Thomas and Bentley (1978b) used the continuity equation [Eq. (68)],
integrated over the ice thickness, to estimate thickness changes in the
Ross Ice Shelf. They concluded that the ice shelf is thickening by at least
0.3 m/yr upstream of the Crary Ice Rise (Fig. 16) but that it is probably
close to a steady state elsewhere. Bentley and Shabtaie (1978), from a
study of temperature profiles and electrical resistivity measurements, also
concluded that the ice shelf is close to a steady state in the area where
they made their measurements. Greischar and Bentley (1978) estimated
that, if the grounding line retreated to its present position between 5000
and 10,000 yr ago, as geological evidence suggests, the sea bed there will
rise by roughly another 100 m before isostatic equilibrium is restored.
This uplift should be causing an advance of the grounding line of between
10 and 100 m/yr. Thus existing field data do not support the idea that the
West Antarctic Ice Sheet is disintegrating.

The major ice-age ice sheets such as the Laurentide Ice Sheet centered
over the present Hudson Bay, the Innuitian Ice Sheet over the Canadian
Arctic Islands north of latitude 74° N, and Fennoscandian Ice Sheet centered over the present Baltic Sea were marine ice sheets. Hughes *et al.*
(1977) have postulated that, to maintain their stability, these ice sheets
must have been surrounded by floating ice shelves covering much of the
Arctic Ocean, the whole of Baffin Bay and the Greenland and Norwegian
Seas, and extending into the North Atlantic. There is little geological evidence for or against the existence of such extensive ice shelves, and
whether they were essential to the stability of the ice sheets is unclear.
However, the decay of parts of these ice sheets was so rapid that some
form of instability must have been involved.

For example, at 9000 B.P. the terminus of the outlet glacier from the
Laurentide Ice Sheet in Hudson Strait was only about 250 km back from
its maximum position; yet 800 yr later the sea entered Hudson Bay [Prest
(1969)]. This represents an ice retreat of 1200 km and reduction in ice
thickness from an estimated 2300 m to zero [Paterson (1972)]. The explanation probably lies in an inherent instability in tidewater glaciers in
fjords. Post and LaChapelle (1971, p. 77) think that headlands, constric-

tions in the channel, and shoals anchor the terminus. A slight retreat from such a position results in a catastrophic breakup of the ice and the terminus retreats to the next anchor point. This happens even if the water is not deep enough to float the ice. Paterson (1972) and Hughes *et al.* (1977) think that the ice retreated from Hudson Strait in this way and that once the ice front was inside Hudson Bay there were no further anchor points to prevent complete disintegration. This idea is similar to the *calving bay* hypothesis, put forward by Hoppe (1948) to explain the rapid decay of the Fennoscandian Ice Sheet.

C. Surges

A surge is the most impressive type of instability that occurs in valley glaciers. A glacier, after having shown little sign of unusual activity for some years, starts to move at speeds perhaps a hundred times normal for a period ranging from a few months to 2 or 3 yr. The rapid movement then suddenly stops. Surges in any one glacier occur at regular intervals, typically between 10 and 100 yr. Surging glaciers are restricted to certain regions and glaciers in the one region do not all surge at the same time. These facts suggest that surges result from some mechanical instability in the particular glacier. This appears to permit very rapid sliding by decoupling the glacier from its bed. However, the nature of the instability is unclear.

Surges occur in certain ice caps in Iceland and Svalbard [Thorarinsson (1969); Liestøl (1969); Schytt (1969)]; whether they can occur in Antarctica is not known. A surge of a major portion of the Antarctic Ice Sheet would raise world sea level by a few tens of meters and submerge all the world's ports. Wilson (1964) suggested that ice ages are initiated by Antarctic surges; an outflux of ice would greatly increase the ice cover of the Southern Ocean, significantly increase the earth's albedo, and thus start the growth of continental ice sheets in the northern hemisphere. The buildup of the Antarctic Ice Sheet after a surge would be slow because of the large area and low precipitation rate. Thus surges, if any, are expected to occur only at intervals of many thousands of years. Budd (1975) has developed a numerical model to illustrate the typical behavior of surge-type glaciers: a long period of buildup, a rapid ice discharge during the surge, followed by another slow buildup. By choosing suitable values of an ice viscosity parameter and a "basal friction lubrication factor" the behavior of various glaciers in which surges have been observed can be reproduced. Budd also finds that, by suitable choice of parameters, he can model surges of the East Antarctic Ice Sheet. This of course leaves open the question of whether such surges in fact occur.

D. Creep Instability

This type of instability depends on the fact that the deformation rate of ice, for a given stress, increases with temperature. Thus a small increase in deformation rate will increase the deformational heating and hence the temperature; this will cause a further increase in deformation rate. This is a system with positive feedback, and, in certain circumstances, there may be a runaway increase in temperature. Robin (1955) introduced this idea into glaciology by suggesting that it was a possible mechanism for surges; creep instability could warm the basal ice, previously frozen to its bed, to the melting point so that it could start to slide. However, he did not investigate the conditions under which this could happen.

Clarke *et al.* (1977) have studied the question in detail; their model is purely thermal, however, and does not take the ice dynamics into account. They concluded that conditions for creep instability are likely to occur in certain glaciers and ice sheets and that ablation increases the chance of instability while accumulation decreases it. However the time for instability to develop is of the order of $10^2 - 10^3$ yr for glaciers and $10^3 - 10^4$ yr for ice caps. Thus creep instability is eliminated as a mechanism for most surges. Moreover, as these times are longer than the residence time of ice in ablation zones, instability can only occur in accumulation areas. Low accumulation rates and great ice thicknesses favor instability. These conditions prevail in Antarctica and also prevailed in the central parts of the large ice-age ice sheets. Thus Clarke *et al.* think that creep instability might cause surges in these ice sheets.

E. Thermal Convection

Hughes (1970, 1976) suggested that there may be thermal convection in parts of the Antarctic Ice Sheet. This idea was based on the observation of a density inversion at Byrd Station by Gow (1970); density had a maximum value of 920.6 kg/m^3 at 1000-m depth and decreased to 917 kg/m^3 at the bottom (2164 m). This inversion is probably a general feature; the depth at which it begins will depend on the temperature distribution. Hughes originally envisaged convection in the form of typical convection cells but later modified this to a dyke-sill type of convection. Hughes believes that convection begins in unstable ripples in the isotherms near the bed of the ice sheet. Some ripples develop into upward bulges of basal ice that shrink rapidly in their horizontal dimensions and grow vertically to become ascending dykes of warm basal ice. Sills of this ice are injected as thin horizontal layers within the cold ice that is gradually sinking between the dykes. Convection starts under the domes of thick ice near the center of the ice sheet but flow carries this ice toward the margins so the dykes converge toward the heads of the major ice streams. To satisfy the criterion for convection, Hughes had to assume that the initial upward movement of the ice results from rapid transient creep and that the ice crystals then develop a preferred orientation in a direction that maintains the convective flow. For further details, the reader is referred to the original papers, listed in Hughes (1976).

It is very difficult to test this idea observationally. Three observations that Hughes quotes as indirect support, namely, (1) anisotropic ice fabrics in West Antarctica, indicated by seismic data [Hughes (1972)], (2) high velocities in ice streams [Hughes (1976)], and (3) the occurrence of short sharp "cold spikes" in deep oxygen-isotope profiles [Hughes (1977b)], can all be explained in other ways.

Thermal convection, and most of the other instabilities, remain speculative ideas. In view of the number of possible instabilities, however, it seems unwise to assume that the great ice sheets are necessarily stable. Further investigation of this question is one of the major problems of glaciology.

ACKNOWLEDGMENTS

This chapter was written while the author was enjoying the hospitality of the Department of Geophysics and Astronomy at the University of British Columbia (R. D. Russell, Head). C. F. Raymond, R. H. Thomas, and J. Weertman made valuable comments on the draft manuscript.

REFERENCES

The following special abbreviations are used:
CRREL   U.S. Army Cold Regions Research and Engineering Laboratory, Hanover, New Hampshire.
IAHS 104   International Association of Hydrological Sciences Publication 104. (Similarly for other numbers.)
IASH 86   International Association of Scientific Hydrology Publication 86. (Similarly for other numbers.) (IASH is the former name of IAHS.)
SIPRE   U.S. Army Snow, Ice and Permafrost Research Establishment, Wilmette, Illinois. (The former name for CRREL.)

Abramowitz, M., and Stegun, I. A. (1965). "Handbook of Mathematical Functions." Dover, New York.
Andrews, J. T., and Mahaffy, M. A. W. (1976). *Quat. Res.* (*N.Y.*) **6**, 167–183.
Bader, H. (1961). *In* "Cold Regions Science and Engineering" (F. J. Sanger, ed.), Part 1, Sect. B2, pp. 1–18. CRREL, Hanover, New Hampshire.
Bardin, V. I., and Suyetova, I. A. (1967). *In* "Proceedings of the Symposium on Pacific-Antarctic Sciences" (T. Nagata, ed.), pp. 92–100. Dep. Polar Res., Nat. Sci. Mus., Tokyo.

Barnes, P., Tabor, D., and Walker, J. C. F. (1971). *Proc. R. Soc. London Ser. A* **324**, 127–155.

Bauer, A. (1966). *IASH Bull.* **11**, 8–12.

Beitzel, J. E. (1970). *IASH* **86**, 191–203.

Bender, J. A., and Gow, A. J. (1961). *IASH* **55**, 132–141.

Benfield, A. E. (1951). *Mon. Not. R. Astron. Soc. Geophys. Suppl.* **6**, 139–147.

Benson, C. S. (1962). *SIPRE Res. Rep.* **70**.

Bentley, C. R. (1964). *In* "Research in Geophysics" (H. Odishaw, ed.), Vol. 2, pp. 335–389. MIT Press, Cambridge, Massachusetts.

Bentley, C. R. (1971). *Antarct. Res. Ser.* **16**, 191–198.

Bentley, C. R., and Shabtaie, S. (1978). *EOS, Trans. Am. Geophys. Union* **59**, 309 (abstr. only).

Bindschadler, R., Harrison, W. D., Raymond, C. F., and Crosson, R. (1977). *J. Glaciol.* **18**, 181–194.

Björnsson, H. (1974). *Jökull* **24**, 1–26.

Bodvarsson, G. (1955). *Jökull* **5**, 1–8.

Bogoslovski, V. N. (1958). *IASH* **47**, 287–305.

Budd, W. F. (1968). *IASH* **79**, 58–77.

Budd, W. F. (1969). *Aust. Nat. Antarct. Res. Exped. Sci. Rep.* **108**.

Budd, W. F. (1970a). *J. Glaciol.* **9**, 19–27.

Budd, W. F. (1970b). *J. Glaciol.* **9**, 29–48.

Budd, W. F. (1970c). *IASH* **86**, 414–446.

Budd, W. F. (1971). *J. Glaciol.* **10**, 177–195.

Budd, W. F. (1975). *J. Glaciol.* **14**, 3–21.

Budd, W. F., and Carter, D. B. (1971). *J. Glaciol.* **10**, 197–209.

Budd, W. F., and Jenssen, D. (1975). *IAHS* **104**, 257–291.

Budd, W. F., and Radok, U. (1971). *Rep. Prog. Phys.* **34**, 1–70.

Budd, W. F., Jenssen, D., and Radok, U. (1971a). *Univ. Melbourne, Meteorol. Dept., Pub.* **18**.

Budd, W. F., Jenssen, D., and Radok, U. (1971b). *Nature (London), Phys. Sci.* **232**, 84–85.

Budd, W. F., Jenssen, D., and Young, N. W. (1973). *Proc. Australas. Conf. Heat Mass Transfer, 1st, 1973,* Sect. I, pp. 17–24.

Budd, W. F., Jenssen, D., and Young, N. W. (1975). *Sov. Antarct. Exped. Inf. Bull.* **90**, 50–58.

Budd, W. F., Young, N. W., and Austin, C. R. (1976). *J. Glaciol.* **16**, 99–110.

Carslaw, H. S., and Jaeger, J. C. (1959). "Conduction of Heat in Solids," 2nd ed. Oxford Univ. Press (Clarendon), London and New York.

Clarke, G. K. C., Nitsan, U., and Paterson, W. S. B. (1977). *Rev. Geophys. Space Phys.* **15**, 235–247.

Clausen, H. B. (1973). *J. Glaciol.* **12**, 411–416.

CLIMAP (1976). *Science* **191**, 1131–1137.

Collins, I. F. (1968). *J. Glaciol.* **7**, 199–204.

Cragin, J. H., Herron, M. M., Langway, C. C., and Klouda, G. (1977). *In* "Polar Oceans" (M. J. Dunbar, ed.), pp. 617–631. Arct. Inst. North Am., Calgary, Alberta.

Crary, A. P. (1966). *Geol. Soc. Am. Bull.* **77**, 911–930.

Crary, A. P., and Chapman, W. H. (1963). *J. Geophys. Res.* **68**, 6064–6065.

Crary, A. P., Robinson, E. S., Bennett, H. F., and Boyd, W. W. (1962). *IGY Glaciol. Rep.* **6**.

Crozaz, G. (1969). *Earth Planet. Sci. Lett.* **6**, 6–8.

Dansgaard, W. (1961). *Medd. Groenl.* **165**, 1–120.

Dansgaard, W., and Johnsen, S. J. (1969a). *J. Glaciol.* **8**, 215–223.
Dansgaard, W., and Johnsen, S. J. (1969b). *J. Geophys. Res.* **74**, 1109–1110; correction **74**, 2795.
Dansgaard, W., Johnsen, S. J., Clausen, H. B., and Gundestrup, N. (1973). *Medd. Groenl.* **197**, 1–53.
Denton, G. H., and Borns, H. W. (1974). *Antarct. J. U.S.* **9**, 167.
Denton, G. H., Borns, H. W., Grosswald, M. G., Stuiver, M., and Nichols, R. L. (1975). *Antarct. J. U.S.* **10**, 160–164.
Dolgushin, L. D., Yevteyev, S. A., and Kotlyakov, V. M. (1962). *IASH* **58**, 286–294.
Federer, B., and von Sury, H. (1976). *J. Glaciol.* **17**, 531.
Field, W. O., ed. (1975). "Mountain Glaciers of the Northern Hemisphere," Vols. 1 and 2. CRREL, Hanover, New Hampshire.
Flint, R. F. (1971). "Glacial and Quaternary Geology." Wiley, New York.
Fristrup, B. (1966). "The Greenland Ice Cap." Rhodos, Copenhagen.
Garfield, D. E., and Ueda, H. T. (1976). *J. Glaciol.* **17**, 29–34.
Giovinetto, M. B., and Zumberge, J. H. (1968). *IASH* **79**, 255–266.
Glen, J. W. (1955). *Proc. R. Soc. London Ser. A* **228**, 519–538.
Glen, J. W. (1975). *In* "Cold Regions Science and Engineering" (T. C. Johnson, ed.), Part II, Sect. C2b. CRREL, Hanover, New Hampshire.
Goldthwait, R. P. (1960). *SIPRE Tech. Rep.* **39.**
Gow, A. J. (1970). *IASH* **86**, 78–90.
Gow, A. J., and Williamson, T. (1971). *Earth Planet. Sci. Lett.* **13**, 210–218.
Greischar, L. L., and Bentley, C. R. (1978). *EOS, Trans. Am. Geophys. Union* **59**, 309 (abstr. only).
Grigoryan, S. S., Krass, M. S., and Shumskiy, P. A. (1976). *J. Glaciol.* **17**, 401–417.
Hammer, C. U., Clausen, H. B., Dansgaard, W., Gundestrup, N., Johnsen, S. J., and Reeh, N. (1978). *J. Glaciol.* **20**, 3–26.
Harrison, W. D. (1975). *J. Glaciol.* **14**, 31–37.
Hatherton, T. ed. (1965). "Antarctica." Reed & Reed, Wellington, New Zealand.
Herron, M. M., Langway, C. C., Weiss, H. V., and Cragin, J. H. (1977). *Geochim. Cosmochim. Acta* **41**, 915–920.
Hobbs, P. W. (1974). "Ice Physics." Oxford Univ. Press (Clarendon), London and New York.
Holdsworth, G., and Bull, C. (1970). *IASH* **86**, 204–216.
Hollin, J. T. (1962). *J. Glaciol.* **4**, 173–195.
Hooke, R. L. (1973). *J. Glaciol.* **12**, 423–438.
Hoppe, G. (1948). *Geogr. Skr. Uppsala Univ. Geogr. Inst.* **20**, 1–112.
Hughes, T. J. (1970). *Science* **170**, 630–633.
Hughes, T. J. (1972). *Geophys. J. R. Astron. Soc.* **27**, 215–229.
Hughes, T. J. (1973). *J. Geophys. Res.* **78**, 7884–7910.
Hughes, T. J. (1976). *J. Glaciol.* **16**, 41–71.
Hughes, T. J. (1977a). *Rev. Geophys. Space Phys.* **15**, 1–46.
Hughes, T. J. (1977b). *IAHS* **118**, 336–340.
Hughes, T. J., Denton, G. H., and Grosswald, M. G. (1977). *Nature (London)* **266**, 596–602.
Jaeger, J. C. (1962). "Elasticity, Fracture and Flow," 2nd ed. Wiley, New York.
Jenssen, D. (1977). *J. Glaciol.* **18**, 373–389.
Jenssen, D., and Radok, U. (1961). *IASH* **55**, 112–122.
Johnsen, S. J. (1977). *IAHS* **118**, 388–392.
Johnsen, S. J., Dansgaard, W., Clausen, H. B., and Langway, C. C., Jr. (1972). *Nature (London)* **235**, 429–434; corrigendum **236**, 249.

Jones, S. J., and Brunet, J.-G. (1978). *J. Glaciol.* **21**, 445–455.
Liestøl, O. (1969). *Can. J. Earth Sci.* **6**, 895–897.
Lliboutry, L. (1958). *IASH* **47**, 125–138.
Lliboutry, L. (1965). "Traité de Glaciologie," Vol. 2. Masson, Paris.
Lliboutry, L. (1969). *J. Geophys. Res.* **74**, 6525–6540.
Lliboutry, L. (1971). *Recherche* **2**, 417–425.
Lliboutry, L. (1975). *IAHS* **104**, 353–363.
Lliboutry, L., Morales Arnao, B., Pautre, A., and Schneider, B. (1977). *J. Glaciol.* **18**, 239–254.
Loewe, F. (1970). *J. Glaciol.* **9**, 263–268.
Lorius, C. (1973). *Recherche* **4**, 457–472.
Mahaffy, M. W. (1976). *J. Geophys. Res.* **81**, 1059–1066.
Meier, M. F. (1965). *In* "The Quaternary of the United States" (H. E. Wright, Jr., and D. G. Frey, eds.), pp. 795–805. Princeton Univ. Press, Princeton, New Jersey.
Meier, M. F. (1974). *In* "Encyclopædia Britannica," 15th ed., pp. 175–186. Encyclopædia Britannica, Chicago, Illinois.
Mellor, M. (1960). *J. Glaciol.* **3**, 773–782.
Mellor, M. (1968). *IASH* **79**, 275–281.
Mercer, J. H. (1968). *IASH* **79**, 217–225.
Müller, F. (1976). *J. Glaciol.* **16**, 119–133.
Nye, J. F. (1951). *Proc. R. Soc. London Ser. A* **207**, 554–572.
Nye, J. F. (1952a). *J. Glaciol.* **2**, 82–93.
Nye, J. F. (1952b). *J. Glaciol.* **2**, 103–107.
Nye, J. F. (1953). *Proc. R. Soc. London Ser. A* **219**, 477–489.
Nye, J. F. (1957). *Proc. R. Soc. London Ser. A* **239**, 113–133.
Nye, J. F. (1958). *Nature (London)* **181**, 1450–1451.
Nye, J. F. (1959). *J. Glaciol.* **3**, 493–507.
Nye, J. F. (1967). *J. Glaciol.* **6**, 695–715.
Nye, J. F. (1969). *J. Glaciol.* **8**, 207–213.
Nye, J. F. (1975). *J. Glaciol.* **14**, 49–56.
Nye, J. F., Berry, M. V., and Walford, M. E. R. (1972). *Nature (London), Phys. Sci.* **240**, 7–9.
Odqvist, F. K. G. (1934). "Plasticitetsteori Med Tillämpningar." Ingenjörsvetenskap-sakademien, Stockholm.
Oeschger, H., Stauffer, B., Bucher, P., and Moell, M. (1976). *J. Glaciol.* **17**, 117–133.
Orowan, E. (1949). *J. Glaciol.* **1**, 231–236.
Oswald, G. K. A., and Robin, G. de Q. (1973). *Nature (London)* **245**, 251–254.
Paige, R. A. (1969). *J. Glaciol.* **8**, 170–171.
Paterson, W. S. B. (1969). "The Physics of Glaciers." Pergamon, Oxford.
Paterson, W. S. B. (1972). *Rev. Geophys. Space Phys.* **10**, 885–917.
Paterson, W. S. B. (1976). *J. Glaciol.* **17**, 3–12.
Paterson, W. S. B. (1977). *Rev. Geophys. Space Phys.* **15**, 47–55.
Paterson, W. S. B., Koerner, R. M., Fisher, D., Johnsen, S. J., Clausen, H. B., Dansgaard, W., Bucher, P., and Oeschger, H. (1977). *Nature (London)* **266**, 508–511.
Philberth, K., and Federer, B. (1971). *J. Glaciol.* **10**, 3–14.
Post, A., and LaChapelle, E. R. (1977). "Glacier Ice." Univ. of Toronto Press, Toronto.
Prest, V. K. (1969). *Geol. Surv. Can., Map* **1257A.**
Prest, V. K. (1970). *In* "Geology and Economic Minerals of Canada" (R. J. W. Douglas, ed.), pp. 676–764. Geol. Surv. Can., Ottawa.
Radok, U. (1959). *Nature (London)* **184**, 1056–1057.

Radok, U., Jenssen, D., and Budd, W. (1970). *IASH* **86,** 151–165.
Raymond, C. F. (1971). *J. Glaciol.* **10,** 39–53.
Raymond, C. F. (1978). *In* "Rockslides and Avalanches" (B. Voight, ed.), Vol. 1, pp. 793–833. Am. Elsevier, New York.
Raynaud, D., and Delmas, R. (1977). *IAHS* **118,** 377–381.
Rigsby, G. (1958). *J. Glaciol.* **3,** 273–278.
Robin, G. de Q. (1955). *J. Glaciol.* **2,** 523–532.
Robin, G. de Q. (1967). *Nature (London)* **215,** 1029–1032.
Robin, G. de Q. (1968). *IASH* **79,** 265–266.
Robin, G. de Q. (1970). *IASH* **86,** 141–151.
Robin, G. de Q. (1975). *Nature (London)* **253,** 168–172.
Robin, G. de Q. (1976). *J. Glaciol.* **16,** 9–22.
Robinson, E. S. (1966). *J. Glaciol.* **6,** 43–54.
Röthlisberger, H. (1974). *Neuen Zürch. Ztg.* **196,** 1–15.
Savage, J. C., and Paterson, W. S. B. (1963). *J. Geophys. Res.* **68,** 4521–4536.
Schytt, V. (1969). *Can. J. Earth Sci.* **6,** 867–873.
Seckel, H., and Stober, M. (1968). *Polarforschung* **6,** 215–221.
Shepard, F. P., and Curray, J. R. (1967). *Prog. Oceanogr.* **4,** 283–291.
Shreve, R. L., and Sharp, R. P. (1970). *J. Glaciol.* **9,** 65–86.
Shumskiy, P. A. (1961). *IASH* **55,** 142–149.
Shumskiy, P. A. (1965). *Dokl. Akad. Nauk SSSR* **162,** 320–322.
Shumskiy, P. A. (1967). *Phys. Snow Ice Conf. Proc., 1966,* Vol. 1, Part 1, pp. 371–384.
Shumskiy, P. A. (1970). *IASH* **86,** 327–347.
Shumskiy, P. A., and Krass, M. S. (1976). *J. Glaciol.* **17,** 419–432.
Sugden, D. (1977). *Arct. Alp. Res.* **9,** 21–47.
Suyetova, I. A. (1966). *Polar Rec.* **13,** 344–347.
Swithinbank, C. (1970). *IASH* **86,** 472–487.
Thomas, R. H. (1973a). *J. Glaciol.* **12,** 45–53.
Thomas, R. H. (1973b). *J. Glaciol.* **12,** 55–70.
Thomas, R. H. (1976). *Nature (London)* **259,** 180–183.
Thomas, R. H., and Bentley, C. R. (1978a). *Quat. Res. (N.Y.)* **10,** 150–170.
Thomas, R. H., and Bentley, C. R. (1978b). *J. Glaciol.* **20,** 509–518.
Thomas, R. H., and Coslett, P. H. (1970). *Nature (London)* **228,** 47–49.
Thompson, L. G. (1974). *Antarct. J. U.S.* **9,** 249–250.
Thorarinsson, S. (1969). *Can. J. Earth Sci.* **6,** 875–882.
Untersteiner, N., and Nye, J. F. (1968). *J. Glaciol.* **7,** 205–213.
Vialov, S. S. (1958). *IASH* **47,** 266–275.
Wakahama, G., and Budd, W. F. (1976). *J. Glaciol.* **16,** 245 (abstr. only).
Walford, M. E. R., Holdorf, P. C., and Oakberg, R. G. (1977). *J. Glaciol.* **18,** 217–229.
Weertman, J. (1957a). *J. Glaciol.* **3,** 33–38.
Weertman, J. (1957b). *J. Glaciol.* **3,** 38–42.
Weertman, J. (1958). *IASH* **47,** 162–168.
Weertman, J. (1961a). *J. Glaciol.* **3,** 953–964.
Weertman, J. (1961b). *J. Geophys. Res.* **66,** 3783–3792.
Weertman, J. (1962). *J. Geophys. Res.* **67,** 1133–1139.
Weertman, J. (1963). *IASH* **61,** 245–252.
Weertman, J. (1964). *J. Glaciol.* **5,** 145–158.
Weertman, J. (1966). *J. Glaciol.* **6,** 191–207.
Weertman, J. (1968). *J. Geophys. Res.* **73,** 2691–2700.
Weertman, J. (1969). *J. Glaciol.* **8,** 494–495.

Weertman, J. (1973a). *In* "Physics and Chemistry of Ice" (E. Whalley, S. J. Jones, and L. W. Gold, eds.), pp. 320–337. Royal Society of Canada, Ottawa.

Weertman, J. (1973b). *J. Glaciol.* **12**, 353–360.

Weertman, J. (1974). *J. Glaciol.* **13**, 3–11.

Weertman, J. (1976). *Quat. Res. (N.Y.)* **6**, 203–207.

Weidick, A. (n.d.). "Quaternary Map of Greenland." Geological Survey of Greenland, Copenhagen.

Wexler, H. (1960). *J. Glaciol.* **3**, 626–645.

Whillans, I. M. (1976). *Nature (London)* **264**, 152–155.

Whillans, I. M. (1977). *J. Glaciol.* **18**, 359–371.

Wilson, A. T. (1964). *Nature (London)* **201**, 147–149.

Zotikov, I. A. (1963). *IASH Bull.* **8**, 36–44.

2 TEMPERATE VALLEY GLACIERS

C. F. Raymond

Geophysics Program AK-50
University of Washington
Seattle, Washington

DYNAMICS OF SNOW AND ICE MASSES
Copyright © 1980 by Academic Press, Inc.
All rights of reproduction in any form reserved.
ISBN 0-12-179450-4

I. INTRODUCTION

Although most glaciers are remote from the principal centers of human population, there is a practical need to deal with them as an important dynamic part of the mountain environment. Glacier advances, sudden releases of water in large floods, and large ice avalanches represent important natural hazards affecting resource extraction, recreation, and local habitations [e.g., Untersteiner and Nye (1968); Post and Mayo (1971); Bjornsson (1974); Lliboutry et al. (1977, Chapter 7)]. Glaciers also provide a dependable supply of water for irrigation and hydropower [e.g., Meier (1972)]. Both the hazards and the benefits are becoming of ever increasing importance as a result of the increasing habitation of mountainous regions and development of adjacent land.

Scientific interest in glaciers focuses on a number of intriguing questions. Glacial erosion and deposition are powerful geomorphic processes [Embleton and King (1968)]. The structural changes occurring in deforming glacier ice are analogous to some high-temperature metamorphic processes of interest to geologists [Kamb (1964)]. Of major concern is the interaction of glaciers with the atmosphere and their relationship to climate [Hoinkes (1964)]. Glaciers have been a proving ground for fundamental concepts about the dynamic behavior of large natural ice masses and add to the understanding of large ice sheets, both present and past (see Chapter 1).

A distinguishing characteristic of a glacier is that it moves. Most of the principal concepts about glacier motion were developed by the early part of this century, mostly as a result of observations on glaciers in the Alps [e.g., Agassiz (1847); Forbes (1846); Tyndall (1896); Blümcke and Finsterwalder (1905); Deeley (1895); Deeley and Parr (1913, 1914)]. It was known that glaciers moved continuously in time and space with faster motion in the center than near the edges and that velocity decreased with depth below the surface. Although the motion was continuous, variations in rate were found to occur on daily, seasonal, and longer time scales. Based on these observations, it was accepted that glacial ice deformed as a thick viscous fluid, although there was much dispute between proponents of this view and an alternative hypothesis which proposed ice deformed by displacement on crack surfaces in a process of alternate fracture and re-healing. Based on the concept of viscous deformation, mathematical models were made that considered effects of ice thickness, slope, and channel geometry on deformation rate and velocity. Some of the motion was attributed to sliding made possible by freezing and melting around small bumps and viscous deformation around large bumps on the bed. Seasonal variation of velocity was explained by water penetrating to the bed and reducing frictional resistance to the sliding motion.

These same views are held today, but there has been a substantial sophistication of understanding. Since the early 1950s, there has been an infusion of modern concepts of mechanics and physics, which has yielded a quantitative body of experimental and observational knowledge interrelated by mathematical and physical theories. Modern information and concepts have been described in several important books [Lliboutry (1964, 1965); Paterson (1969); Shumskiy (1969); Vivian (1975)] and general articles [Kamb (1964); Lliboutry (1971a)]. The purpose of this chapter is to incorporate the most recent advances in knowledge about temperate valley glaciers into a general description of their dynamic behavior.

II. GENERAL DESCRIPTION OF PROCESSES

The existence of a perennial ice mass depends on the interplay of accumulation and ablation processes. Over a time scale of a year or more on a glacier, accumulation processes dominate in the upper reaches and ablation processes dominate in the lower reaches. Downslope transport by motion is essential to maintain this configuration. The geometry of a glacier and the adjustments it makes to varying climate depend on both the climatically controlled mass exchange with the external environment (mass balance) and internally controlled redistribution of mass by glacier motion according to simple concepts of mass conservation (continuity).

A. Mass Balance

Interaction with the atmosphere in a zone within several meters of the upper surface is the major factor in mass balance. The dominant contribution to accumulation is deposition of snow. The pattern of deposition is strongly influenced by orographic effects and horizontal transport by wind. Heat is exchanged with the atmosphere by short-wave solar radiation, long-wave thermal radiation, and sensible and latent heat transfers (see Chapter 6). This is the dominant factor in ablation whenever there is net heat income and the surface reaches the melting point. Heat loss and subsurface cooling can be an indirect factor in mass balance by delaying the onset of melting, retarding its rate, or freezing downward percolating water to form superimposed ice. In comparison to atmospheric heat sources, internal heat sources from mechanical dissipation of heat and geothermal heat are usually so small as to be negligible for mass balance. These accumulation and ablation processes are complicated by mountain topography and have complex relationships to standard meteorological parameters. The quantitative understanding of the relationship between mass exchange on a glacier and large and local scale climates is an important meteorological problem reviewed by Hoinkes (1964).

The nourishment of a glacier depends on the balance between accumulation and ablation. The balance b at a point can be defined as the net addition of ice mass on unit horizontal area of reference surface, assuming no horizontal straining or motion. It is a function of horizontal position and time. Figure 1 shows typical situations with net accumulation and net ablation. Actual measurements can be complicated by the presence of liquid

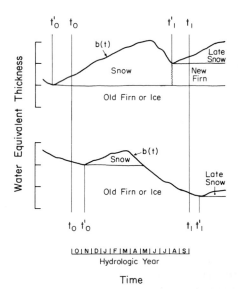

Fig. 1. Balance $b(t)$ is mass added per unit area since an arbitrary reference time. At any given time this may exist as snow, firn, and/or ice. Annual balance is defined as $b(t_1) - b(t_0)$. Net balance is defined as $b(t_i') - b(t_0')$. [Simplified from Mayo et al. (1972).]

water, complex density distributions, difficulty in recognition of strati-graphic horizons, settling of marker stakes, and freezing in the firn.

Balance tends generally to increase with altitude but also can show variations along elevation contours, especially on glaciers flowing east-west. Locally, balance can decrease with altitude in response to topo-graphic effect on wind drifting or exposure to solar radiation. Typically, each year there is a line, above which the net balance is positive and below which the net balance is negative, termed the equilibrium line, which separates the accumulation and ablation areas.

From the point of view of glacier dynamics it is useful to consider the total mass added above a given location X on the glacier per unit time averaged over one or more years. This is evaluated by

$$Q_b(X, t) = \int_{A(X)} \frac{\partial b}{\partial t} \, dA, \tag{1}$$

where $A(X)$ is the horizontal area of the glacier above position X. The quantity $Q_b(X, t)$ represents the ice mass flux which the glacier would have to transport by motion to keep the total ice mass above X constant. It has been called the balance flux by Budd *et al.* (1971). When evaluated at the terminus, it gives the rate of change of total mass of the glacier.

B. Motion

The smooth manner in which velocity varies in glaciers indicates that glaciers flow with a fluidlike behavior. In addition to deforming internally, glaciers slip at the boundaries. Such sliding motion is observed at the margins of glaciers and beneath the surface (as a result of access provided by natural cavities, excavated tunnels, and boreholes). Also, short-term variations in surface velocity which cannot be explained by any other mechanism are believed to be caused by slip.

Speeds measured at the surface of typical glaciers range from several to several hundred meters per year. The shear stress at the base of these glaciers is restricted to a narrow range near 1 bar (10^5 Pa). Although there is no clear relationship between surface speed and basal shear stress (Fig. 2a), a systematic relationship is found when account is taken of ice depth (Fig. 2b). Ice depth is typically in the range of one hundred to several hundred meters and tends to have an inverse relationship with slope. This behavior is the normal mode of flow characteristic of most temperature glaciers. Although sliding may enlarge the range of velocity shown in Fig. 2, the range of velocities serves to identify upper limits to rates which are possible by internal deformation for characteristic ice rheology and range of base stress.

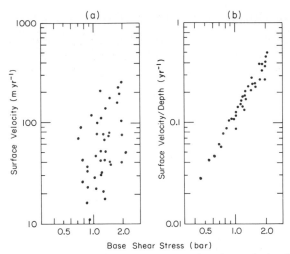

Fig. 2. Measured surface velocity (a) and measured surface velocity divided by depth (b) versus estimated base shear stress from a number of glaciers. [After Budd and Radok (1971).]

Knowledge about rates of sliding and the physical control of the process is quite limited but it is known that slip rates cover an extremely broad range. The principal evidence for very high sliding rates comes from certain unusual glaciers which move at speeds of kilometers per year. Although it has been suggested that unusual properties of the ice associated with special texture or loss of coherence by fracturing may be responsible, it is generally accepted that this rapid motion occurs by rapid sliding. Some outlet glaciers from Antarctica and Greenland maintain continuous fast motion [Carbonnell and Bauer (1968); Swithinbank (1963)]. In mountain glaciers, however, extremely fast motion is usually restricted to a short interval of months to a few years, called a surge. Glaciers which surge do so quasi-periodically with quiescent intervals of tens of years in which the normal mode of flow persists [Meier and Post (1969)].

In a valley glacier it is convenient to consider the total ice mass flux across a transverse section of the valley. This is computed as

$$Q_v(X) = \int_{S(X)} \rho U \, dA, \tag{2}$$

where X represents the horizontal location of a vertical section of area $S(X)$, U is velocity normal to the section, and ρ is ice mass per unit volume.

C. Continuity and Geometry Changes

Mass transfer across external surfaces by mass balance and across internal surfaces by motion act in combination to control the geometry of a glacier. The difference between the balance and the dynamic fluxes at a given position X along the glacier gives the rate of mass increase above X; thus

$$dM(X)/dt = Q_b(X) - Q_v(X), \tag{3}$$

where $M(X)$ is the mass above position X. Differentiation gives

$$\partial \bar{\rho} S / \partial t = \partial Q_b / \partial X - \partial Q_v / \partial X, \tag{4}$$

where S is the cross-sectional area and $\bar{\rho}$ the average ice mass density over S. In Eq. (4), $\partial Q_b / \partial X$ represents the rate of balance change integrated across the width of the glacier and $\partial Q_v / \partial X$ represents the net mass inflow to a slab of unit width across the glacier. These together determine the mass change and the corresponding volume change of the slab. Equation (4) represents the starting point for modeling the adjustment of a valley glacier to changing climate or internal conditions.

Normal glaciers flow at relatively steady rates and tend to adjust their geometry to achieve rates of motion such that Q_h and Q_v are about equal when averaged over several years. These glaciers achieve a nearly steady state in which geometry and motion evolve slowly with time. In contrast, a surge-type glacier apparently cannot maintain a steady value of Q_v equal to Q_b, and is forced to oscillate between low flux and surges of high flux to achieve a balance on a time scale of a surge period.

III. STRUCTURE OF A GLACIER

A. Thermal Structure

Temperate glaciers are loosely defined as those glaciers that are entirely at the melting point except possibly for a thin surface layer that is transiently cooled during the winter. Because of impurities and the grain structure, glacier ice does not have a distinct melting point [Lliboutry (1971b)]. The essential feature is that the ice contains some liquid phase; hence heat exchange results in a change in the liquid content of the ice but has little effect on the temperature [Harrison (1972)]. Actual water contents found for glacier ice are of the order of 0.1–1% fractional volume [Raymond and Harrison (1975); Vallon et al. (1967)]. The water exists primarily in a network of veins lying along three grain edges [Nye and Frank (1973)]. The veins are less than several tens of microns in diameter, and as a result, the permeability of glacier ice is low and only very small fractions of the surface meltwater can penetrate through the glacier by this means [Raymond and Harrison (1975)].

Because the ice is at the melting point, the temperature decreases with depth as a result of the depression of the melting point by pressure. This amounts to $-0.0074°C/bar$ of pressure which corresponds to $-0.00065°C/m$ of ice equivalent. The temperature is several hundredths of a degree colder because of the presence of impurities and their concentration into the small amount of liquid [Harrison (1972)].

The base of a temperate glacier is a heat sink because geothermal heat flowing to it from below cannot be conducted upward and typically results in basal melting of about 0.01 m yr^{-1}.

B. Grain Structure

Snow accumulates as a loose, porous aggregate of complexly shaped ice particles and undergoes various metamorphic changes that are greatly accelerated by the introduction of liquid water. Rapid rounding and grain growth causes a mean grain size of about 1 mm (see Chapter 6) and a density of about $0.5-0.6$ Mg m^{-3} by the end of the summer. In temperate conditions, the firn undergoes compaction from the weight of the overlying firn and snow to reach a density of $0.8-0.9$ Mg m^{-3} at depths up to several tens of meters and ages of several years [Anderson and Benson (1963)]. Densification occurs especially rapidly above impermeable boundaries where the snow or firn is saturated with liquid [Vallon *et al.* (1976)]. Because of the rapid densification, it is usually reasonable to assume a characteristic density of about 0.9 Mg m^{-3} for nearly all of a temperate glacier.

Because air permeability is lost at a density of about 0.8 Mg m^{-3} most glacier ice is bubbly. There is a substantial variation in size, spacing, and shape of bubbles. Bubble-free ice can be found in thin layers centimeters thick, or less, but larger pods are also found. Grain size of the ice slowly increases by growth processes controlled by grain boundary energy, but dynamic metamorphic processes produce a variety of types of texture [e.g., Allen *et al.* (1960)]. Most abundant is a coarse-grained bubbly ice composed of grains with interdigitating cross sections about $10-100$ mm across [Kamb (1959)] and a complex branching shape extending over large volumes in three dimensions [Rigsby (1968)]. A fine-grained ice composed of regularly shaped grains of millimeter size is also found. Intermediate grain size is uncommon, but rather mixtures of coarse and fine grains are found. The association of ice of different texture and bubble content is typically in a layerlike pattern referred to as foliation.

Preferred orientations of the crystallographic c axis is a particularly important structural characteristic of glacier ice. Firn and newly formed fine-grained ice in the accumulation area sometimes has a broad vertical c-axis orientation [Vallon *et al.* (1976)]. The fine-grained ice included in foliation also shows a single maximum, but it is perpendicular to the plane of foliation [Kamb (1959)]. In marked contrast, the coarse-grained interdigitating texture has multiple-maximum c-axis fabrics [Rigsby (1960); Kamb (1959)]. Three or four maxima were found to be characteristic of ice experiencing long continued shear stress. Preferred orientation of the a axes has also been discovered in association with multiple-maximum c-axis fabrics [Matsuda *et al.* (1976)].

C. Large-Scale Structural Patterns

Systematic features in the spatial variation of texture in glaciers have been identified by aerial photography, structural mapping of the upper surfaces, and a few observations in crevasse walls, tunnels, and core samples. Average grain size tends to increase with depth, except near the base where the grain size may decrease [Kamb and Shreve (1963); Vallon *et al.* (1976)]. It also increases down glacier [Allen *et al.* (1960)]. Foliation exists at a depth with the relative amount of fine ice decreasing with depth [Kamb and Shreve (1963)]. Foliation is typically subparallel to the bed at the base and margin but can be warped by folding and offset by crevasses or shear planes [Allen *et al.* (1960)]. In some cases, the pattern of foliation on the surface shows primarily a longitudinal trend with strongest development near the margin [e.g., Meier (1960)]. However, the most prominent foliation exists below ice falls, and in the center it is steeply dipping and transverse to the flow [e.g., Allen *et al.* (1960)]. Transverse foliation sometimes shows a periodic variation in strength that is revealed on the surface as a striking sequence of alternating dark and light bands called internal ogives, or sometimes Forbes bands [e.g., Atherton (1963)]. In trunk glaciers a variety of structural patterns determined by different tributaries can be found in juxtaposition (Fig. 3). The structural pattern on surge-type glaciers with tributaries is especially complicated due to folding induced by asynchronous oscillations in the flow rates of the tributaries [Post (1969)].

The surfaces of glaciers are cracked in many locations. These cracks, termed crevasses, form perpendicular to a principal axis of surface strain rate when it is extensile. Remarkable patterns of consistent orientation and sometimes regular spacing illustrate features of the surface strain rate field [Meier (1960)]. Normally, these crevasses extend only a limited

Fig. 3. View of Sherman Glacier, Alaska, showing structures in ice streams originating from different tributaries. At upper right and in the stream traversing the center left to right, a residual primary sedimentary layering is visible. In the lower part an ice stream shows arcuate foliation formed as transverse features at the base of an icefall out of sight to the left. The dark and light banding represent internal ogives associated with variations of foliation character. Also evident are crevasse patterns which reflect orientation of maximum extensile strain rate and medial moraines which approximately trace out stream lines on this glacier. [Photo by Austin Post, University of Washington.]

depth of several tens of meters below the surface [Nye (1952)], but filling with water may cause them to penetrate to greater depths [Weertman (1973a)].

D. The Bed

Core samples, photography at the base of boreholes, observations in tunnels and natural cavities, exposures at ice margins, and morphology of beds of past glaciers give some picture of the structure of the base of a glacier. Because of difficulty of access, observations come from only a few locations on existing glaciers; hence it is difficult to establish typical conditions. The available evidence indicates the existence of a structurally distinct layer of ice up to several meters thick extending over wide areas

[Vivian and Bocquet (1973)]. It is characterized by a dark color, lack of bubbles, layers of fine rock debris, dispersed pebbles, occasional larger rocks, and pods of water. The contact itself appears to be quite variable in character. A rather simple contact with relatively clean basal ice contacting solid rock across a distinct interface has been observed at the base of rapidly moving icefalls [Kamb and LaChapelle (1964)]. Exposed glaciated bedrock shows characteristic marks such as striations and crescentic marks, which indicate important interactions between rock particles in the ice and the bed [Flint (1971)]. The erosion products must remain at the contact, be incorporated into the basal ice, or be mobilized in a water film. These features suggest a possibly important role of rock debris at the interface. Observations at the bottom of boreholes at locations more typical than icefalls show the existence of a subsole drift of several tenths of a meter thickness [Engelhardt *et al.* (1978)].

One important property of the contact zone discovered under icefalls and near the ice margins is the existence of air-filled gaps between the ice and the rock. These are found on the downglacier side of bumps and can be meters long in the flow direction [Vivian and Bocquet (1973)]. Although not observed directly, it is inferred that cavities of pressurized water exist in some places under thick ice.

IV. NOTATION AND BASIC EQUATIONS

The fluidlike flow of ice is described by components of velocity u_i and deformation rate

$$d_{ij} = \tfrac{1}{2}(\partial u_i/\partial x_j + \partial u_j/\partial x_i), \tag{5}$$

where i and j each represent one of the three Cartesian coordinate directions. The state of stress is described by components of stress σ_{ij} or alternatively by components of deviatoric stress τ_{ij} and mean compressive stress p (pressure) where

$$p = -\sigma_{ii}/3 \tag{6}$$

and

$$\tau_{ij} = \sigma_{ij} + p\delta_{ij}, \tag{7}$$

where δ_{ij} is the Kroenicker delta and a repeated index implies summation. By definition, τ_{ii} is zero. Other important quantities are the effective shear strain rate

$$\dot{\varepsilon} = (\tfrac{1}{2}d_{ij}d_{ij})^{1/2} \tag{8}$$

and shear stress

$$\tau = (\tfrac{1}{2}\tau_{ij}\tau_{ij})^{1/2}. \tag{9}$$

Deformation rate and stress are constrained by basic conservation laws of mass, momentum, and energy. It is convenient to consider these in a right-handed Cartesian coordinate system defined with origin at the upper surface, the x axis downslope parallel to the mean upper suirface, the y axis downward, and z axis horizontal. The corresponding components of velocity will be denoted u, v, and w, respectively. Since a temperate glacier has nearly constant density, conservation of mass implies incompressible deformation and

$$\partial u/\partial x + \partial v/\partial y + \partial w/\partial z = d_{xx} + d_{yy} + d_{zz} = 0. \tag{10}$$

Conservation of momentum is expressed by

$$\partial \tau_{xx}/\partial x + \partial \tau_{xy}/\partial y + \partial \tau_{xz}/\partial z - \partial p/\partial x + \rho g \sin \alpha = 0, \tag{11a}$$

$$\partial \tau_{xy}/\partial x + \partial \tau_{yy}/\partial y + \partial \tau_{yz}/\partial z - \partial p/\partial y + \rho g \cos \alpha = 0, \tag{11b}$$

$$\partial \tau_{xz}/\partial x + \partial \tau_{yz}/\partial y + \partial \tau_{zz}/\partial z - \partial p/\partial z = 0, \tag{11c}$$

where α is the slope of the x axis. Since the ice temperature in temperate glaciers is the melting point, conservation of energy becomes primarily a statement about the rate of water production by mechanical dissipation of heat. However, locally near the base of glaciers, stress-controlled temperature gradients lead to important conductive heat flow with consequent freezing and melting in a regelationlike process.

V. EXPERIMENTAL CREEP BEHAVIOR OF ICE

A. Behavior of Single Crystals

Single crystals of ice shear easily on the basal planes perpendicular to the c axis of the hexagonal crystalline structure, but within these planes there is no particular direction in which shear occurs most easily [Glen and Perutz (1954)]. Shear can occur on other planes, but the strain rate is one or more orders of magnitude smaller than on the basal planes for equivalent shear stress [Higashi (1969)]. These results show ice crystals have a strong plastic anisotropy. This is the most importar.t property of single crystals to keep in mind; it alerts us to the possible mechanical significance of the c-axis preferred orientations commonly found in glacier ice.

B. Uniaxial Deformation of Polycrystalline Ice at the Melting Point

Numerous experimental studies have been done on the creep behavior of polycrystalline ice [Weertman (1973b)]. Because the temperature dependence of creep near the melting point is uncertain [Barnes *et al.* (1971)] and our focus is on temperate ice, we shall consider only those experiments done within several hundredths of a degree of the melting point. These have been uniaxial creep experiments. The first extensive experiments were done by Glen (1955), who deformed bubblefree, artificially prepared samples made by freezing water-saturated crushed ice. Experiments were done at various temperatures including one set at −0.02°C. In these experiments, stress ranged from 0.91 to 9.2 bars and total axial strains ranged from 10^{-2} to 10^{-1}. Colbeck and Evans (1973) deformed natural fine-grained, bubbly glacier ice in a tunnel cut into the glacier where the temperature could be controlled very precisely at 0°C. In these experiments, total axial strain reached a maximum of 0.015 and the

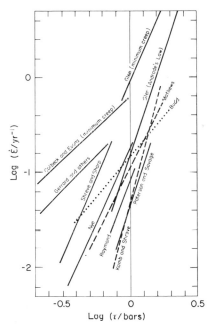

Fig. 4. Effective shear strain rate versus effective shear stress for laboratory experiments [Glen (1955); Colbeck and Evans (1973)] from analysis of borehole deformation [Mathews (1959); Paterson and Savage (1963); Shreve and Sharp (1970); Kamb and Shreve (1966); Kamb (1970), and Raymond (1973)], surface velocity [Budd and Jenssen (1975)], and tunnel closure [Nye (1953)] in glaciers. Lines show fit of data to power law form and the range of $\dot{\varepsilon}$ of the data and the corresponding range of τ (inferred in the case of glacier measurements).

axial stress used ranged from 0.07 to 1 bar. These and other results are plotted in terms of effective shear strain rate $\dot{\varepsilon}$ and effective shear stress τ on Fig. 4. A straight line on this plot corresponds to a power law

$$\dot{\varepsilon} = A\tau^n, \tag{12}$$

where the power n is given by the slope of the line.

Interpretation of the creep experiments at the low stresses appropriate to glacier flow is ambiguous. Two curves are shown on Fig. 4 for Glen's experimental data. For axial stress below about 4 bars (effective stress below 2.3 bars) an initial primary phase of decelerating creep rate continued to the termination of the experiments without the appearance of a secondary stage of time-independent creep rate. The higher curve was found by taking the minimum creep rate observed in each experiment. The results can be fit by a power law with n equal to 3.2. The lower curve was obtained by extrapolating the strain rate versus time for the low-stress experiments to infinite time using Andrade's law, which gives a strain rate significantly lower than the minimum value observed at the end of a low-stress experiment, as can be seen by the lower position of the quasi-viscous line shown in the Fig. 4. This gives n equal to 4.2. Colbeck and Evans (1973) interpreted their experiments in a way similar to the minimum creep analysis of Glen (1955) and extend his minimum creep curve to lower stress. When fitted to a power law over the stress range 0.1–1 bar, their data give n about 1.3.

C. Generalized Flow Law for Three-Dimensional Deformation

To use results from uniaxial experiments for general stress states, Nye (1953) proposed a generalized flow law relating the deformation rate to deviatoric stress

$$d_{ij} = \zeta(\tau)\tau_{ij}, \tag{13a}$$

where τ is effective shear stress defined in Eq. (9). For the particular case of a power law,

$$\zeta(\tau) = A\tau^{n-1}. \tag{13b}$$

By Eq. (13a) strain rate is incompressible [Eq. (10)] and independent of mean stress which for temperate ice is consistent with experimental information on ice [Rigsby (1958)] and other materials [Weertman (1970)]. Since ζ is a scalar function of the invariant τ, Eq. (13a) gives isotropic behavior.

Equations (13) can be expressed in an equivalent inverse form,

$$\tau_{ij} = 2\eta d_{ij}, \tag{14a}$$

where 2η is the inverse of ζ. For the case of a power law

$$\eta = 1/(2A\tau^{n-1}) = B\dot{\varepsilon}^{-(1-1/n)}, \tag{14b}$$

where

$$B = \tfrac{1}{2}A^{-1/n}. \tag{14c}$$

In this form, the flow law is analogous to that for a viscous fluid, for which the effective viscosity η depends on stress or on strain rate. More complicated three-dimensional behavior than this is theoretically possible for a nonlinear flow law but experiments show that this form of the flow law is adequate for the secondary stage of creep of initially isotropic samples [Byers *et al.* (1973)].

D. Applicability to Glaciers

Kamb (1972) observed structural changes in samples deformed under laboratory conditions similar to those used in the determination of experimental flow laws. He found that, during secondary creep, the texture and c-axis orientation fabrics were changing progressively and were different from those characteristic of glacier ice. This indicates that the secondary creep rate in experiments may not be indicative of the creep rate occurring in a glacier. The existence of preferred orientation of c axes and foliation in glaciers casts doubt on the isotropic generalized flow law of Eq. (13). An isotropic law would be reasonable if the structure of the ice always adjusted to the local stress. However, there is considerable evidence that much of the structure seen in glaciers is produced in zones of high deformation rate and is carried passively down glaciers (see, for example, Fig. 3). Thus, in much of a glacier, the ice is deforming under a structure imposed by stress patterns different than that which exists locally. The grain size may also be important [Baker (1978)]. Furthermore, the flow law may depend on the amount of liquid water present in the ice. Because the water is primarily at three grain intersections [Ketcham and Hobbs (1969)], there is apparently no dramatic loss of strength associated with small amounts of water, but there is some effect on creep rate [Duval (1977)]. Effects of bubbles and included rock particles [Hooke *et al.* (1972)] represent other potential complications. Although the complexities are potentially important, there is very little information about their quantitative effect.

VI. VARIATION OF VELOCITY AND STRESS WITH DEPTH

A. Observations of Velocity versus Depth

Depth profiles of velocity through glaciers have been observed by measurement of tilting rate in boreholes. Figure 5 summarizes profile shapes obtained at various locations. Normally there is a characteristic blunt profile in which there is only a small variation with depth for a significant range (down to about one-half of the full depth) and a zone of relatively strong shear near the bottom. Results from locations of exceptionally large surface strain rate or surface rotation rate associated with curvature of the surface show deviations from this typical pattern.

B. Planar Slab Model

The variation of velocity with depth in glaciers has been studied analytically using a model of a uniformly sloping parallel-sided ice slab [Nye (1952, 1957)]. Even though this model is extremely simplified, it is remarkably successful in predicting the observed shape of the depth profile of velocity. It is treated most easily using the coordinate system described in Section IV. On the upper surface ($y = 0$), stress is determined by atmospheric pressure ($\tau_{xy} = \tau_{zy} = 0$ and $\sigma_{yy} = -p_a$). On the bottom surface the normal velocity component v is zero assuming the ice moves parallel to the boundary. Also, it is assumed that there is no motion across the slope, so the transverse velocity component w is zero.

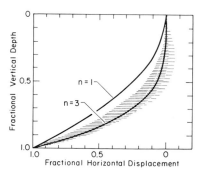

Fig. 5. Horizontal displacement versus vertical depth measured in boreholes. Shaded envelop includes data from 12 holes [Gerrard *et al.* (1952); Mathews (1959); Savage and Paterson (1963); Shreve and Sharp (1970); Raymond (1971).] Fractional value is defined as difference between surface value and value at depth divided by difference between surface and bottom values. Dark shaded envelop includes 11 of the 12 holes. Data from holes which were not measured to more than 80% of the full ice depth were not included. Heavy solid curves show predictions from "laminar flow" theory (Eq. 18).

Because of the uniform thickness of the slab, a solution is sought such that stress components are independent of x and z. Conservation of momentum [Eqs. (11)] reduces to

$$\partial \tau_{xy}/\partial y = -\rho g \sin \alpha, \qquad \partial \sigma_{yy}/\partial y = -\rho g \cos \alpha, \qquad \partial \tau_{yz}/\partial y = 0. \quad (15)$$

Integration down from the upper surface gives the statically determined stress components:

$$\tau_{xy} = \tau_{yx} = -\rho g y \sin \alpha, \tag{16a}$$

$$\sigma_{yy} = -\rho g y \cos \alpha - p_a, \tag{16b}$$

$$\tau_{zy} = \tau_{yz} = 0. \tag{16c}$$

The other components of the stress tensor are not determined by these assumptions alone.

C. "Laminar Flow"

The additional assumption that velocity is independent of x and z leads to the simple "laminar flow" theory of Nye (1952). With the assumption about velocity, the distribution of stress [Eqs. (16)], the flow law [Eq. (13)], and continuity [Eq. (10)], it follows that the only nonzero component of deformation rate is d_{xy} $(= d_{yx})$. The boundary conditions described above constrain the motion to be downslope and parallel to the surface ($v = w = 0$), and downslope velocity u depends only on depth as if the ice were composed of laminae moving rigidly over one another.

This deformation pattern and the flow law [Eqs. (13)] imply that the only nonzero component of deviatoric stress is τ_{xy} $(= \tau_{yx})$. It follows that the effective shear stress τ defined by Eq. (9) is equal to $|\tau_{xy}|$. Furthermore, Eqs. (7) and (16b) give

$$-p = \sigma_{xx} = \sigma_{zz} = \sigma_{yy} = -\rho g y \cos \alpha - p_a.$$

Velocity versus depth is found from the low law [Eq. (13)], which gives

$$\tfrac{1}{2} \partial u/\partial y = d_{xy} = A \tau^{n-1} \tau_{xy}. \tag{17}$$

When combined with Eq. (16a) and integrated down from the surface, this gives

$$u(y) = u_s - [2A/(n + 1)](\rho g \sin \alpha)^n y^{n+1}. \tag{18}$$

When n is 1, which corresponds to a Newtonian viscous fluid, a parabolic depth profile of velocity corresponding to the well-known Poiseuille flow is found. When n is 3, a value appropriate for ice, the deformation tends to be concentrated in a basal shear layer. This is a definite feature of observed profiles (Fig. 5). The extreme case of pluglike flow is approached as n approaches infinity, which corresponds to perfect plasticity analyzed by Nye (1951). The theoretical shape of the depth profile of velocity can be understood in terms of the depth variation of effective viscosity. For the power flow law, effective viscosity decreases with increasing deviatoric stress. Therefore, deep layers are softer than shallow layers, and the contrast in shear strain rate is enhanced in comparison to expectations for a linearly constant viscosity fluid.

If the full thickness of the slab is h, the basal shear stress τ_b and total velocity difference between surface and bed $u_s - u_b$ can be found from Eqs. (16a) and (18), respectively. When combined, these give

$$u_s - u_b = [2A/(n + 1)](\tau_b)^n h. \tag{19}$$

The ice volume flux per unit width (q) is found by integrating the velocity u over the entire ice thickness; hence

$$q = u_b h + [2A/(n + 2)](\rho g \sin \alpha)^n h^{n+2}. \tag{20}$$

From this the velocity averaged over depth can be found. The contribution from internal deformation is

$$\bar{u}_{\text{def}} = \frac{q - u_b h}{h} = \frac{2A}{n + 2}(\rho g \sin \alpha)^n h^{n+1} = \frac{n + 1}{n + 2}(u_s - u_b). \tag{21}$$

From Eq. (19) this may be expressed as

$$\bar{u}_{\text{def}} = [2A/(n + 2)](\tau_b)^n h. \tag{22}$$

The profile shapes predicted from the "laminar flow" theory for two values of n are plotted on Fig. 5. The parabolic profile appropriate to a linear fluid ($n = 1$) deviates significantly from observations. There is good agreement when n is about 3, which is consistent with expectations from experimental flow laws. Some of the deviations of observation from the precise "laminar flow" shape may arise from softening due to nonzero strain rate components other than d_{xy} (similar to that discussed below) or rotation caused by curvature of the surface in the direction of flow [e.g., Savage and Paterson (1963)]. The distinction between horizontal component of velocity versus vertical depth sensed by boreholes and surface parallel velocity versus surface normal depth of the theory is not significant as long as the surface slope is not large and the deformation is predominantly simple shear parallel to the surface.

D. "Extending" and "Compressing" Flow

Nye (1957) showed that another simple solution is possible for planar slab geometry, in which u may vary with longitudinal distance x. His assumptions constrain longitudinal strain rate $\partial u/\partial x$ to be a constant r. From incompressibility [Eq. (10)] and the bottom boundary condition, an immediate consequence is that

$$v = \int_h^y \frac{\partial v}{\partial y} \, dy = r(h - y). \tag{23}$$

When r is positive, which corresponds to extending flow, a layer of ice parallel to the slab surfaces is being extended in the longitudinal direction and thinned in the y direction in addition to the shear deformation present in the "laminar flow" model. Although the deformation is more complex than in the "laminar flow" model, the deformation rate components and the corresponding deviatoric stress components remain independent of x and z. Therefore, the values of τ_{xy}, σ_{yy}, and τ_{yz} are equal to those for "laminar" flow. However, the stress state has the new feature that τ_{xx} and $\tau_{yy} = -\tau_{xx}$ are not zero. This also means that the horizontal stress component σ_{xx} is not equal to the vertical component σ_{yy} [Eq. (7)]. A further consequence is that τ_{xx} contributes to τ. Therefore, the y profile of shear strain rate d_{xy} given by Eq. (17) is modified in comparison to laminar flow even though the driving shear stress τ_{xy} is the same. Because of the nonlinear flow law and the coupling between stress components it implies, the ice is softened by a nonzero value of τ_{xx} and the shear strain rate (d_{xy}) is larger than predicted for laminar flow. The velocity profile therefore shows correspondingly larger differences between any two depths. Nye integrated Eq. (17) to find the depth profile of velocity. It may be expressed as

$$u = u_s - [2A/(n + 1)]|\tau_{xy}|^n yF, \tag{24a}$$

where

$$F = (\tau/\tau_{xy})^{n+1}[1 + n(\tau_0/\tau)^{n+1} - (n + 1)(\tau_0/\tau)^{2n}] \tag{24b}$$

and where τ_0 is the effective stress at the surface equal to $(|r|/A)^{1/n}$ and both τ_{xy} and τ depend on y. The quantity F is a factor that corrects the laminar flow profile to account for the longitudinal strain rate. It can be thought of as a function of $\tau_0/|\tau_{xy}|$ or of τ_0/τ and is plotted in Fig. 6. As long as the effective stress at the surface is less than about one-half the basal shear stress, the velocity difference between surface and bed is not much different than that of the laminar flow model.

Although the longitudinal strain rate in glaciers depends on position, [e.g., Raymond (1971)], this simple model of extending and compressing flow captures the essence of important processes. First, there is a component of velocity normal to the glacier surface [Eq. (23)] which allows the slab model to achieve a steady state with a nonzero net balance. Net accumulation can be balanced by downward motion of extending flow and net

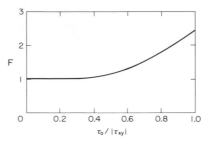

Fig. 6. Correction factor for velocity difference between surface and depth associated with softening of ice by a depth independent longitudinal strain rate. The factor F is defined in Eqs. (24) and is based on the analysis of Nye (1957).

ablation can be balanced by upward motion of compressive flow. Second, the variation of surface normal component of velocity v with depth y causes the thickness of surface parallel ice layers to change with time and is important in deducing the age of ice at various depths. Third, there is a softening effect on the ice.

E. Unbalanced Lateral and Longitudinal Forces

Even when the conditions of the "laminar flow" model do not hold, Eq. (19) can be used to estimate the velocity difference between the surface and bed, as long as a corrected value of the base shear stress is used. This is suggested by the good agreement between observed and theoretical shapes of profiles of velocity versus depth. This approach should be valid from a theoretical point of view as long as τ_{xy} is the dominant contribution to τ and is linear with depth. The results described above suggest that the first requirement is often reasonable; since shear stress magnitude is likely to almost always increase monotonically downward, the second condition is also reasonable.

The integrated force on the sides of a column passing through the complete thickness of the glacier can be used to get a correct value of basal stress. In the planar slab model, these forces add to zero, but when there are lateral and longitudinal variations in thickness and stress, Eq. (11a) is integrated over depth to give

$$\tau_{xy}|_{\text{bed}} - \tau_{xy}|_{\text{surf}} = -\bar{\rho}gh \sin \alpha - h \left\langle \frac{\partial \tau_{xx}}{\partial x} \right\rangle_h + h \left\langle \frac{\partial p}{\partial x} \right\rangle_h - h \left\langle \frac{\partial \tau_{xz}}{\partial z} \right\rangle_h.$$

$$(25)$$

The first term is the shear stress in the absence of both lateral and longitudinal gradients. The second and third terms express a correction arising from unbalanced longitudinal forces and the fourth a correction from unbalanced side forces. These corrections are considered in following sections.

VII. VARIATION OF VELOCITY AND STRESS IN A CROSS SECTION

A. General Features

Since mountain glaciers flow in valleys, the ice is supported from the sides as well as from the bottom. The effect of the sides is illustrated by the velocity variation across the surface of a valley glacier that is maximum in the central part and decreases toward the margins [e.g., Meier (1960)]. The lateral profile of velocity from the center toward either margin is much like the depth profile (Fig. 5) with a blunt rigid central portion and shear concentrated near the margin. Some lateral profiles across the surface indicate a maximum shear strain rate somewhat to the center of the margin [e.g., Raymond (1971)].

Measurements in a line of boreholes across the Athabasca Glacier, Canada, gave one example of the velocity distribution over the area of a valley cross section (Fig. 7). The shape of this valley is fairly typical. It is roughly parabolic with a ratio of half-width to depth of about two. In this particular case, there is a strong lateral variation of velocity across the base of the glacier from high values of about 80% of the maximum surface velocity in the center to negligible values at the edges. Evidence from other glaciers suggests a similar pattern of lateral variation of slip velocity, where beneath the center slip is inferred to be significant, but little or no slip is found at the margins. Typically marginal slip is small except where the margin is steep and formed by smooth solid rock [e.g., Glen and Lewis (1961); Meier et al. (1974)]. Although the lateral variation of slip on Athabasca Glacier may be more extreme than is typical, it illus-

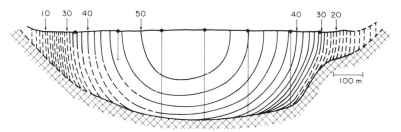

Fig. 7. Contours of equal downglacier component of velocity observed in Athabasca Glacier, Canada [Raymond (1978)]. Measurement locations are shown as vertical lines (boreholes) and circles (surface markers). Units of contours are meters per year.

trates that the pattern of flow in a cross section depends not only on the shape of the cross section but also on the basal boundary condition. Notably, the largest shear strain rates are near the margin rather than near the base in the center, indicating support from the sides is very important. This feature is probably associated with rapid central sliding that shifts more support to the edges than would exist with more uniform sliding.

B. Theory of Flow in Cross Sections

Features of flow in the cross section of a valley can be examined assuming rectilinear flow in a cylindrical channel in which longitudinal velocity u is parallel to the axis of the channel and depends only on position in the plane of the cross section. The only known analytical solution for nonlinear flow law corresponds to contours of constant velocity which are semicircular in shape [Nye (1965b)]. This is valid for a channel of semicircular cross section with no boundary slip and is reasonably correct near the central portion of any channel shape independent of slip boundary condition [Raymond (1974)]. Solutions for other channel shapes have been obtained using numerical methods [e.g., Nye (1965b); Reynaud (1973)].

Results for the symmetric parabolic channel cross section characteristic of glaciated valleys are most relevant. These are summarized by Nye (1965b) for various half-width to depth ratios assuming no slip and n equal to 3 in the power flow law (Fig. 8). At the center line the shape of the profile of velocity versus depth is close to that for laminar flow and shear stress deviates only slightly from being linear with depth for half-width to depth ratios greater than one. The shape of profiles of velocity and shear stress versus distance across the surface are much more sensitive to the specific shape of the channel cross section than are the depth profiles at the center. Shear stress and corresponding shear strain rate magnitudes are zero in the center and at the margins and reach maxima on the surface somewhat centerward from the margin. Absolute maxima of shear stress and strain rate magnitude are reached at the bottom in the center and support from the bottom is the dominant contribution in equilibration of the ice. Reynaud (1973) shows how major support from the sides and rapid sliding in the central part of the valley, which are more consistent with observations from Athabasca Glacier, can be achieved using a basal boundary condition where basal shear stress is proportional to effective normal stress.

Nye (1952, 1965b) defined several parameters that quantify the effect of valley sides on the flow in a cross section. One parameter of interest concerns the effect of side drag on the shear stress at the bottom, particularly

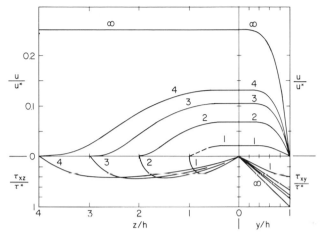

Fig. 8. Variations of velocity u and shear stress components τ_{xy} and τ_{xz} versus depth (y) at the centerline and width (z) at the surface in symmetric parabolic channels for various half-width to depth ratios and flow law exponent of 3. Unit of velocity is $u^* = 2A(\rho g h \sin \alpha)^3 h$ and of stress is $\tau^* = \rho g h \sin \alpha$, where α is channel slope and h is centerline ice thickness. [After Nye (1965b).]

in the center. If $h\langle \partial \tau_{xz}/\partial x \rangle$ in Eq. (25) is expressed as a fraction of $\rho g h \sin \alpha$, where h is the center-line depth and α is surface slope, the remaining fraction defined as f is supported by the shear stress at the bottom. This gives

$$\tau_{xy}|_{\text{bed}} = -f\rho g h \sin \alpha. \tag{26}$$

The term f is called the shape factor; it is dimensionless and depends on the proportions of the channel cross section and not its actual size. It will also depend on the pattern of basal slip or the corresponding pattern of basal shear stress across the width of the channel.

A simple estimate of shape factor has been made by Nye (1952) based on the hydraulic radius of the channel defined as S/P, where S is the cross-section area and P is the length of the bed across the width of the channel. Gross equilibrium requires that

$$\langle \tau_b \rangle P = \rho g S \sin \alpha, \tag{27}$$

where $\langle \tau_b \rangle$ is the base shear stress averaged over the channel width. By assuming the basal shear stress in the center of the channel is equal to the average, comparison of Eqs. (26) and (27) leads to

$$f = S/Ph. \tag{28}$$

Values of f for typical parabolic cross-section shapes lie between 0.5 and

0.6 but smaller values are characteristic of V-shaped profiles and larger values are found for flat-bottomed U-shaped valleys.

Another estimate of f is obtained assuming no slip at the bed (or equivalently uniform slip). Based on his numerical calculations, Nye (1965b) chooses f so that Eqs. (26) and (19) give the correct velocity difference between surface and bed. In the center of a channel, this is not much different from the value of f in Eq. (26) that gives the correct base shear stress. This estimate of f is up to 20% larger than that based on hydraulic radius for typical channels. If sliding is important and much larger in the center than the margins, then shape factor can be a similar percentage less than estimated from hydraulic radius (Raymond, 1971).

From the point of view of mass exchange by flow and modeling changes in mass distribution in valley glaciers, the total mass flux Q_v through a cross section is of primary importance. This can be expressed in terms of the velocity u averaged over the area S of the section as

$$Q_v = S\langle \rho u \rangle_S \approx \rho S \langle u \rangle_S, \tag{29}$$

assuming density ρ is constant in the section. For convenience of measurement, it is advantageous to relate $\langle u \rangle_S$ to quantities that are easily measured at the surface such as maximum surface velocity at the center-line u_s or velocity averaged over the surface width $\langle u \rangle_w$. If the dimensionless ratios between them were known theoretically, Q_v could be measured as

$$Q_v \approx \rho S \langle u \rangle_w [\langle u \rangle_S / \langle u \rangle_w] \tag{30}$$

or

$$Q_v = \rho S u_s [\langle u \rangle_S / u_s], \tag{31a}$$

or could be calculated as

$$Q_v = \rho S[u_b + [2A/(n + 1)](f\rho gh \sin \alpha)^n h][\langle u \rangle_S / u_s] \tag{31b}$$

when u_s is calculated from Eqs. (19) and (26). For a particular valley cross section, S is a definite function of h. A parabolic shape gives

$$W = \kappa h^{1/2} \quad \text{or} \quad S = \tfrac{2}{3}\kappa h^{3/2}, \tag{32}$$

where κ is constant. If there is also no sliding ($u_b = 0$), then

$$Q_v = \rho K (\sin \alpha)^n h^{n+5/2}, \tag{33a}$$

where

$$K = [4A/3(n + 1)]\kappa(\rho g)^n f \langle u \rangle_S / u_s. \tag{33b}$$

From the results of Nye (1965b), $f\langle u\rangle_{\rm S}/u_{\rm s}$ is insensitive to h, so K is approximately constant. For a parabolic channel of width ratio two, the ratio $\langle u\rangle_{\rm S}/\langle u\rangle_{\rm w}$ is close to one when there is no slip [Nye (1965b)], but it can be as much as 15% greater when there is fast slip [Raymond (1971)]. The corresponding comparison for $\langle u\rangle_{\rm S}/u_{\rm s}$ is about 0.63 for no slip and values up to 30% larger for fast central slip.

VIII. LONGITUDINAL VARIATION OF VELOCITY AND STRESS

A. General Comments

Because balance flux is a maximum at the equilibrium line, maximum speed is expected in the central reach of a glacier, with an upglacier decrease in the accumulation area and a downglacier decrease in the ablation area. Observations confirm this general behavior [e.g., Meier and Tangborn (1965)], but commonly it is overprinted with complex variations caused by longitudinal variations of bed slope and channel cross-section shape [Nye (1959)], bends in valleys [Meier *et al.* (1974)], and confluence of tributaries [Collins (1970)]. Wherever it has been observed, the longitudinal strain rate has been found to have distinct variations in the vertical, lateral, and longitudinal directions [e.g., Raymond (1971)]. These conditions suggest the presence of significant longitudinal stress gradients, so the shear stress distribution and the corresponding velocity distribution at any cross section cannot be computed from local parameters.

Once longitudinal variations become significant, the flow becomes three dimensional with vertical, lateral, and longitudinal variations all acting together. As a start to understanding longitudinal variations, these have been analyzed by a plane strain rate approximation which neglects lateral gradients. Nye (1967) found a solution for a wedgelike glacier terminus, assuming perfectly plastic ice rheology. Solutions for more general geometry or complete longitudinal profiles are possible assuming linear rheology by analysis of the biharmonic equation [e.g., Shumskiy (1967)]. However, the main focus of effort has been on approximate methods in which the total unbalanced longitudinal force on a column is estimated and used to correct the statically estimated basal shear stress [Eq. (25)].

B. Formalism for Describing Longitudinal Stress Gradients

In Eq. (25) the corrections to basal shear stress from unbalanced longitudinal forces are expressed by $h\langle\partial\tau_{xx}/\partial x\rangle_h$ and $h\langle\partial p/\partial x\rangle_h$. The first of these terms involves the longitudinal deviatoric stress τ_{xx} which can be related to the longitudinal strain rate $\partial u/\partial x$ through the flow law [Eq. (13)]. The second involves the pressure p which cannot be expressed in terms of strain rate components directly. Budd (1970) shows how $\partial p/\partial x$ can be so expressed by differentiation of Eq. (11b) with respect to x followed by integration with respect to y. For planar deformation ($\tau_{zz} = 0$) and homogeneous density ($\partial \rho/\partial x = 0$), this gives

$$\left\langle \frac{\partial p}{\partial x} \right\rangle_h = - \left\langle \frac{\partial \tau_{xx}}{\partial x} \right\rangle_h + \frac{1}{h} \int_{y_s}^{y_b} \int_{y_s}^{y'} \frac{\partial^2 \tau_{xy}}{\partial x^2} \, dy'' \, dy' - \frac{\partial \sigma_{yy}}{\partial x} \bigg|_{y_s} \tag{34}$$

when averaged over depth. If $h\langle \partial \tau_{xx}/\partial x \rangle$ is expressed in terms of $\partial h \langle \tau_{xx} \rangle /\partial x$, the basal shear stress may be expressed from Eqs. (25) and (34) in the notation of Budd (1968, 1970) as

$$\tau_{xy}|_{\text{bed}} = -\rho g \sin \alpha h - 2G + T$$

$$+ \left[\tau_{xy}|_{y_s} - \frac{\partial \sigma_{yy}}{\partial x} \bigg|_{y_s} - 2 \frac{\partial \tau_{xx}}{\partial x} \bigg|_{y_s} \frac{\partial y_s}{\partial x} + 2 \frac{\partial \tau_{xx}}{\partial x} \bigg|_{y_b} \frac{\partial y_b}{\partial x} \right], \tag{35}$$

where

$$G = (\partial/\partial x)h\langle \tau_{xx} \rangle_h \tag{36}$$

and

$$T = \int_{y_s}^{y_b} \int_{y_s}^{y'} \frac{\partial^2 \tau_{xy}}{\partial x^2} \, dy'' \, dy'. \tag{37}$$

As long as the surface and bed slopes do not deviate strongly from the slope of the x axis, the terms in the square brackets may be neglected.

Budd (1968) argues that the relative importance of each of the three principal terms on the right-hand side of Eq. (35) depends on the wavelength λ of the longitudinal variations. For λ very large compared to depth h, the geometry approaches that of the planar slab model and G and T are negligible. Budd claims that G is larger than T except when λ is small. This introduces the concept of three distinct longitudinal scales: large λ for which both G and T are negligible, intermediate λ for which T is negligible, and small λ for which all terms must be kept. Budd (1968) suggests that the transition from large to intermediate longitudinal scale occurs at about $20h$ and the transition from intermediate to small scale occurs at $3h$ to $4h$ on ice sheets. Normally a glacier will show longitudinal fluctuations in surface and bed slope, and corresponding fluctuations of depth with a spectrum of wavelengths spanning the short, intermediate, and large scales. In this case, Budd (1968) argues that, if Eq. (35) is averaged longitudinally, the same concept of large, intermediate, and small scales applies, depending on the averaging length.

In order to extend this approach to account for the effects of valley sides, Budd and Jenssen (1975) and Budd (1975) suppose the combined effects of lateral and intermediate scale longitudinal gradients can be expressed as

$$\tau_{xy}|_{\text{bed}} = -f\rho gh \sin \alpha - 2G, \tag{38}$$

where f is the shape factor [Eq. (26)]. This seems to be a reasonable approach. However, the form

$$\tau_{xy}|_{\text{bed}} = f[-\rho gh \sin \alpha - 2G] \tag{39}$$

could also be considered. The latter form is more consistent with the concept behind the shape factor which describes the partitioning between $\partial\tau_{xy}/\partial y$ and $\partial\tau_{xz}/\partial z$ of an effective longitudinal force density irrespective of whether force density comes directly from a component of gravity or indirectly from a longitudinal stress gradient. Without further analysis or testing, both formulations are suspect.

To make Eqs. (35), (38), or (39) useful it is necessary to evaluate the various terms on the right-hand side. So far this has been considered only for the intermediate scale where it is supposed that T may be neglected. The first term is determined easily from geometry. The term G can be related to the longitudinal strain rate variations through the flow law, which in the form of Eq. (12) gives

$$\tau_{xx} = 2\eta(\dot{\varepsilon})d_{xx}. \tag{40}$$

In practice this has been used in two approximations.

Robin (1967) and Budd (1968) estimated $\eta(\dot{\varepsilon})$ from the surface longitudinal strain rate d_{xx} to find

$$G = (\partial/\partial x)hB|d_{xx}|^{1/n} \operatorname{sgn}(d_{xx}). \tag{41}$$

This would be valid for power flow law if d_{xx} and d_{yy} are the dominant contributions to $\dot{\varepsilon}$ and $|d_{xx}|^{1/n}$ at the surface is representative of the average over depth. This equation was used to study the relationship between surface slope and longitudinal strain rate variations to bed topography in ice sheets, assuming that the shear stress is constant along the bed. It was concluded that slope fluctuations are largely supported by longitudinal gradients (see Chapter 1).

Budd and Jenssen (1975) and Budd (1975) allow η in Eq. (40) to be a separate flow parameter or function η^*, in which case

$$G = 2(\partial/\partial x)h\eta^*\langle d_{xx}\rangle_h. \tag{42}$$

They then also used Eq. (19) to relate basal shear stress to surface velocity and eq. (21) to relate surface velocity u_s, depth-averaged velocity, and depth-averaged d_{xx} to write Eq. (38) as

$$\left[\frac{n+1}{2A}\frac{u_s - u_b}{h}\right]^{1/n} = +f\rho g h \sin \alpha + 4 \frac{\partial}{\partial x}\left\{\frac{h\eta^*}{n+2}\left[(n+1)\frac{\partial u_s}{\partial x} + \frac{\partial u_b}{\partial x}\right]\right\}.$$
(43)

With choices for the flow law parameters A and n, the parameter or function η^*, and some basal boundary condition, Eq. (43) can be integrated along the length of a glacier to determine distribution of velocity. In actual calculations, η^* has been chosen as a constant parameter [Budd (1975)], or as a function of longitudinal strain rate corresponding to a hyperbolic sine-type flow law [Budd and Jenssen (1975)]. These choices lead to the inconsistency that the surface parallel shear deformation and the longitudinal straining are controlled by different flow laws. Although some of the assumptions are uncertain, these methods provide a basis for practical numerical modeling of glacier adjustment to climate and internal perturbations.

C. Example of Theoretical Effects of Longitudinal Stress Gradients

The scale of longitudinal fluctuations for which longitudinal stress variations produce significant influence on the flow can be found by examining the velocity changes caused by placing harmonic perturbations of various wavelengths on the upper surface of a planar, parallel-sided slab of thickness h_0. "Laminar flow" theory predicts that velocity variations are in phase with thickness variations at long wavelengths and with slope variations at short wavelengths, and the amplitude of velocity variations increases with decreasing perturbation wavelength for fixed perturbation amplitude (Fig. 9). Intuitively these predictions should be valid for very long wavelengths. However, over an intermediate range of wavelengths longitudinal variations of velocity would be reduced by compensating longitudinal stress variations and longitudinal variations of velocity would be completely suppressed at very short wavelengths. What are the actual scales that define long, intermediate, and short?

For a linear fluid this can be investigated rigorously by a relatively simple analysis of the biharmonic equation to find the results shown in Fig. 9. Langdon and Raymond (1978) used a numerical method valid for $n \geq 1$ based on the assumptions that the depth profile of velocity has the same shape as in laminar flow, that there is no slip, and that deformation is planar. With these assumptions and the requirement of incompressibility, the velocity field is determined completely by surface veloc-

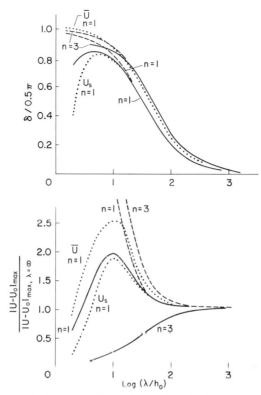

Fig. 9. Phase δ and relative amplitude of deviation of velocity u from datum velocity u_0 caused by harmonic perturbations of wavelength λ on the upper surface of a planar slab of thickness h_0 and slope α_0 of 5.7°. When $\delta = 0(\pi/2)$ maximum depth (slope) and velocity are in phase. Terms u_s and $\bar{u}$ represent surface- and depth-averaged velocity, respectively; relative amplitude is the same for both "laminar flow" theory (dashed line) and numerical calculations (solid line) which assume a specific shape for the depth profile. Dotted lines indicate analytical solution. [After Langdon and Raymond (1978).]

ity $u_s(x)$ and its longitudinal distribution is chosen based on a variational principle equivalent to the relevant differential equations governing the motion. The good agreement between the assumed laminar flow shape and observations from various glaciers indicates the assumption is reasonable (Fig. 5).

For a linear fluid ($n = 1$) the effect of longitudinal stress variations is apparent for λ less than about $100h_0$ and dominant for λ less than about $10h_0$. For a nonlinear material ($n = 3$), the effect of longitudinal stress gradients is apparent for λ less than several hundred h_0 and dramatic for λ less than $100h_0$. The longitudinal stress gradients are of such dominating importance that no increase in the velocity fluctuations associated with slope fluctuations occurs for decreasing wavelength when $n = 3$.

The large range of longitudinal stress gradients for nonlinear material can be explained in terms of the vertical variation of effective viscosity. The amplitude of velocity fluctuation predicted from "laminar flow" theory is controlled by an effective viscosity near the base of the glacier where the shear stress is high. In contrast the amplitude of the longitudinal strain rate is controlled by an effective viscosity averaged over the full depth. For $n = 1$ the two controlling viscosities are the same, the unique viscosity of the fluid. For $n > 1$ the second is larger than the first. Because shearing dominates near the base, there is a relatively low effective viscosity, strain-rate-softened layer near the bottom. The relatively high effective viscosity of the overlying ice acts to suppress longitudinal variations of velocity.

It is important to recognize that the quantitative aspects of the conclusions reached in the above example may not apply to other circumstances, especially for the nonlinear flow law where a different pattern of flow can force a different pattern of effective viscosity. For example, the scale of fluctuations for which longitudinal stress gradients are significant may be different for a case where the bottom has undulating topography and undulations on the surface are steady-state features associated with them. Also, effects from valley sides may be important in determining the effective range of longitudinal stress gradients. However, the example supports the idea of Budd (1968) that longitudinal stress variations are important even when longitudinal fluctuations occur on a scale of many ice depths.

D. Observations on Valley Glaciers

The longitudinal variation of velocity observed on Variegated Glacier [Bindschadler *et al.* (1977)] has been analyzed to determine whether the complex velocity variation (Fig. 10b) could be understood in terms of the geometry of the glacier (Fig. 10a). Figure 10c shows the actual surface slope averaged for three length scales and an effective slope represented by the product $f \sin \alpha$ needed in Eqs. (26) and (19) to get the measured surface velocity from the measured ice depth assuming Glen's law. The surface slope averaged over a longitudinal distance of about one glacier depth shows strong longitudinal variations which are incompatible with the smoothly varying effective slope needed to account for the observed speed variation. The surface slope averaged on a longer longitudinal scale of 8–16 depths is considerably more consistant in smoothness with the needed effective slope, but there is a systematic displacement. The dis-

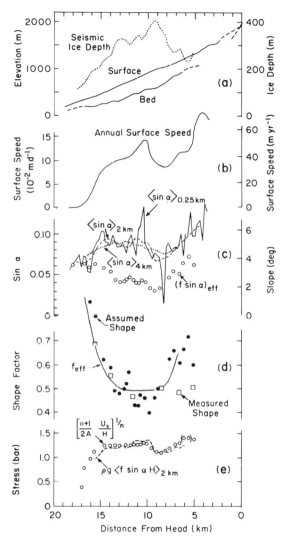

Fig. 10. Longitudinal profiles of (a) centerline surface elevation, bed elevation, and ice depth; (b) annually averaged surface velocity; (c) surface slope measured with various longitudinal length scales and effective slope times shape factor $[((n + 1)u_s/2A)^{1/n}(\rho g h^{(n+1)/n})^{-1}]$; (d) shape factor estimated assuming no slip from geometry of measured sections and from assumed parabolic shape elsewhere and an effective shape factor $(f \sin \alpha)_{\text{eff}}/\langle \sin \alpha \rangle_{2\text{km}}$; and (e) distribution of base stress determined from Eq. (19) and first term of Eq. (38) smoothed by a 2-km running average.

placement can be explained by the shape factor f. This is tested in Fig. 10d which shows the shape factor needed to correctly predict surface velocity and compares it to shape factors deduced theoretically from the geometry of the cross sections. The good agreement in Fig. 10 shows the longitudinal profile of surface speed can be explained in terms of Eqs. (19) and (26) as long as the surface slope is averaged on a length scale about one order of magnitude larger than the ice depth.

The longitudinal variation of base shear stress can be investigated in several ways. One approach is to use measured surface velocity and ice depth in Eq. (19) and solve for base stress. For the correct prediction of surface velocity in the modeling approach of Budd and Jenssen (1975), this is the best estimate for base shear stress. This can be compared to the terms on the right-hand side of Eq. (38). Figure 10e shows the local values of basal shear stress from Eq. (19) and smooth curves representing base stress from Eq. (19) and the first right-hand side term of Eq. (38) averaged over about ten glacier depths, which corresponds to the intermediate scale appropriate to Eq. (38). These are in good agreement, as would be expected from the foregoing discussion. The slight differences between the two curves would presumably be explained by the term $2G$. Because highly accurate velocity measurements are required to calculate the longitudinal strain rate and its gradient needed for evaluation of G by Eq. (41) or (42), a test of this is very difficult. Thus far it has not been possible to actually verify the applicability of Eq. (38) to valley glaciers [R. A. Bindschadler (unpublished)]. In this particular case, the longitudinally averaged base stress can be computed reasonably well from only the first term, even at the intermediate scale.

Figure 10 also shows that the local base stress does not deviate significantly from the large-scale average because it varies slowly. The surface velocity at a given place therefore can be calculated from Eq. (19) using a longitudinally averaged form of Eq. (26). These concepts may be expressed as

$$\tau_{xy}|_{\text{bed}} \approx -\rho g \langle fh \sin \alpha \rangle \tag{44}$$

and

$$(u_s - u_b) = [2A/(n + 1)](\rho g \langle fh \sin \alpha \rangle)^n h. \tag{45}$$

In this section we have dealt with the most thoroughly analyzed example of a longitudinal profile of velocity in a valley glacier. However, Meier *et al.* (1974) have reached a similar conclusion about the behavior of Blue Glacier where they found the internal deformation of the ice was controlled by a nearly constant effective slope and not by the local surface slope. The importance of a large-scale average slope was suggested in earlier work by Robin (1967) and Budd (1968) on ice sheets (see Chapter 1). The lack of strong response of surface velocity to local slope fluctuations is also consistent with the results of numerical calculation discussed in Section VIII.C. This, we can expect, is a general feature of glacier behavior.

IX. FLOW LAW OF GLACIER ICE INFERRED FROM DEFORMATION MEASUREMENTS

The profiles of velocity versus depth are broadly consistent with a power-type flow law with an exponent n about equal to 3 (Fig. 5). In order to determine the flow law parameters more carefully, the approach has been to plot $\dot{\varepsilon}$ versus τ (or alternatively $\tau/2\dot{\varepsilon}$ versus $\dot{\varepsilon}$) with the implicit assumption that Eq. (13a) holds. The results from borehole measurements (Fig. 4) are discussed here. There are two main problems.

The first concerns determination of $\dot{\varepsilon}$ versus depth in a borehole. The two components of tilting in a borehole cannot uniquely determine the 9 gradients of velocity needed to calculate $\dot{\varepsilon}$. The primary concern has been to determine longitudinal strain rate, which contributes directly to $\dot{\varepsilon}$ and also affects the tilting of a borehole where it is not normal to the surface [Nye (1957); Shreve and Sharp (1970)]. An example of the problem is given by Gerrard et al. (1952), who neglected a surface longitudinal strain rate of about 0.14 yr^{-1} in their analysis. This gives a value of surface longitudinal stress τ_0 between 0.5 and 1.0 bar (depending on whether their flow law interpretation or Glen's law is assumed). This corresponds to $\tau_0/|\tau_{xy}|$ of about 0.8–1.4, making softening of the ice caused by longitudinal strain rate highly significant (Fig. 6). This will influence the shape of the profile and the overall strain rate, affecting conclusions about both A and n. The possible importance of the longitudinal strain rate in this case was one of the motivations which led to the important analysis of Nye (1957). The problem can be circumvented somewhat by using additional information obtained from arrays of boreholes [Raymond (1973)] or downward extrapolation of surface measurements [Paterson and Savage (1963)]. However, in spite of these efforts, useful information has come only from the lower parts of holes where the bed parallel shear strain rate dominates and can be directly determined from the hole tilting rate. Very close to the bed local deformations around bumps can introduce unknown contributions to $\dot{\varepsilon}$.

The second and probably more serious problem concerns the determination of τ versus depth in a borehole. This has been done by using Eq. (26) to estimate τ_{xy}. Then ζ in Eq. (13a) is given by d_{xy}/τ_{xy} and the remaining components of deviatoric stress can be determined from estimates of the corresponding deformation rate components. This is a reasonable approach but depends on the validity of the estimate of τ_{xy}. In this regard there has been a tendency to locate boreholes at inflections on the surface with the hope of minimizing effects of longitudinal variations. These are places where the gradient of local surface slope is zero, but also local slope reaches its extreme values with maximum deviation from the mean. This becomes significant when it is realized that the results given by Paterson and Savage (1963), Shreve and Sharp (1970), and Raymond (1973) were based on analyses in which slope was estimated on a scale of about one glacier depth, which is much too short in view of the observations discussed above. The choice of location and method of analysis combine to maximize the error in estimating shear stress versus depth, especially in the deep ice where the useful tilt data are obtained. The error could be reduced by using a large-scale surface slope for estimating stress from Eq. (26). A further uncertainty in the calculation of shear stress from Eq. (26) lies in the choice of the shape factor (f), which is sensitive to the particular basal boundary condition. Only Raymond (1973) took account of this, whereas other analyses used shape factors based on hydraulic radius or ignored shape factor altogether.

The above factors may be partly responsible for the apparent differences in flow behavior observed at different locations (Fig. 4). When the borehole data are reinterpreted using current concepts about large-scale surface slope and shape factors, the various borehole tilting profiles are quite consistent with one another and also the experimental law of Glen (1955), $A = 0.148 \text{ bar}^{-4.2}a^{-1}$, $n = 4.2$. (See Table I.) This is remarkable in view of the structural complexity of glacier ice in comparison to the experimental samples of Glen (1955). Any major differences that do exist are perhaps on a local scale and are averaged out at the effective scale of the borehole measurements.

The values of n deduced from borehole experiments are in the range 3–4, which is somewhat less than the value of 4.2 given by Glen (1955). The reinterpretation of Table I does not affect the value of n deduced from boreholes; however, a more sophisticated assessment of depth dependence of stress might lead to some adjustment of estimates of n. At the very bottom of boreholes it is common to find a very high apparent value of n [Kamb and Shreve (1966)]. This may arise from several possible causes such as depth-dependent structure giving softer ice near the base or strain rate softening of the basal ice associated with motion around bumps in the bed [Kamb (1970)].

TABLE 1

Reinterpretation of Borehole Experiments[a]

Source	h_i	h_m	Original analysis					Alternative analysis				
			sin α	f	n	A	Δ	sin α	f	n	A	Δ
Mathews (1959)	495	482	0.035[b]	1.0	2.8	0.111	0.9	0.061[e]	0.5[c]	2.8	0.15	1.0
Paterson and Savage (1963)	209	184	0.11[b]	0.62[c]	4[i]	0.044[i]	0.7	0.089[f]	0.62[g]	4.	0.21	1.1
	322	321	0.061[b]	0.58[c]	5.2	0.050[j]	0.7	0.054[f]	0.50[d]			1.1
Kamb and Shreve (1966)		116	0.22[b]	0.75			0.8					
Shreve and Sharp (1970)	260	180	0.080[b]	0.56[c]	3.3	0.550	1.5	0.12[f]	0.61[h]	3.3	0.20	1.1
Raymond (1973)	316	310	0.068[b]	0.50[d]	3.6	0.076	0.8	0.054[f]	0.50[d]	3.6	0.17	1.0

[a] Δ is a factor by which estimated shear stress τ_{xy} must be multiplied to give agreement at 1 bar with Glen's law ($A = 0.148$ bar^{-n} yr^{-1}, $n = 4.2$); h_i and h_m are depths in meters of ice and deepest measurement, respectively; A is in units of bar^{-n} per year and n are flow law parameters [Eqs. (13)].

[b] Slope averaged over approximately one glacier depth.

[c] Based on hydraulic radius.

[d] Based on analysis of transverse variation of velocity.

[e] 3-km average.

[f] 2-km average.

[g] Assumed same as original because of modest slip velocity of 0.2–0.3 of surface velocity (Savage and Paterson, 1963).

[h] From Nye (1965), assuming no slip, as found later from borehole measurements [Engelhardt et al. (1978)].

[i] Parameters from Shreve and Sharp (1970).

[j] Parameters from Kamb (1970).

Assuming no basal slip, the observed variation of surface velocity with depth and slope (Fig. 2b) can also be used to determine flow parameters through Eq. (45) [Budd and Jenssen (1975)]. The resulting flow law (Fig. 4) gives deformation rates in agreement with borehole and experimental results for 1-bar shear stress. However, the apparent value of n is lower than would be expected from them. This discrepancy might possibly represent a systematic effect from sliding depending on ice depth and base stress, but other systematic effects might also be present. As yet there is no viable explanation for this discrepancy.

X. SLIDING BEHAVIOR OF GLACIERS

A. General Description

When the base of a glacier is below freezing, no slip is found [Goldthwait (1960)]. When temperate basal conditions exist, the contribution from sliding may vary from values small in comparison with internal deformation to exceptionally high values characteristic of icefalls, the terminal zone of tidewater glaciers, and glacier surges. Glaciers on steep mountain slopes may occasionally enter an unstable sliding regime which leads to acceleration and fall as a catastrophic glacier ice avalanche [Lliboutry (1968)]. Actual speeds of sliding span an enormous range far exceeding that which is thought possible by internal deformation.

Measurements in locations where natural subglacial air-filled cavities exist have given the most detailed measurements of basal sliding. Such cavities have been reached through direct connection from the margin, through tunnels from the rock below the ice, or through the ice itself. In these cavities, sliding is rather easily measured from displacements of markers in the basal ice, but these measurements are restricted to places where the ice is thin and the surface is relatively steep. The most detailed record has been obtained from cavities beneath Glacier d'Argentiére [Vivian and Bocquet (1973)] near the top of a steep icefall where the ice thickness is about 100 m. The sliding velocity is quite high with an average value of about 0.03 m h^{-1} (260 m yr^{-1}). An outstanding feature is that sliding ceases entirely for periods of hours and at other times is 2–3 times the average rate. Similar jerky slip was found beneath an icefall [Kamb and LaChapelle (1964)] and in cavities near the margin of a glacier [Theakstone (1967)]. On the other hand, McCall (1952) and D. N. Peterson (unpublished) have found smooth rates of motion for ice in close proximity to the ice-bed contact. In all of these cases involving measurements in natural cavities, the slip velocity is very close to that at the surface, so sliding is the major contribution to the motion.

Measurements in boreholes permit the investigation of sliding at more typical locations of deep ice and moderate to low slope. Standard geodetic survey of the surface and inclinometer survey of the profile beneath the surface provides one method [e.g., Savage and Paterson (1963)]. A major problem with it is that the standard thermal drilling technique may be halted well above the actual bed by dirt, so only upper limits are obtained reliably. Also, the tilting in boreholes cannot be measured accurately enough to allow short-time-scale determination of slip rate; only averages over several months to a year or more are possible. In order to use boreholes better, Harrison and Kamb (1973) have used combined thermal and mechanical drilling and debris bailing methods to penetrate the basal ice and expose the bottom. Sliding motion can be measured accurately by observing the bottom with a camera fixed in the hole. The main difficulty arises from turbidity in the water and debris which accumulates on the bottom of the hole, thus obscuring features of the underlying bed. These methods have shown that in some cases sliding may contribute only a small percentage to the total motion. The best examples come from Blue Glacier, Washington, in a zone where surface slope is about 12° and depth is about 120–130 m [Englehardt *et al.* (1978)].

An indirect indication of sliding is given by variations of surface velocity which cannot be explained by changes in ice mass distribution and the consequent change in stress and internal deformation rate. Strong diurnal variations and dramatic increases over times as short as one day have been found in some cases [Iken (1973, 1977)] (Fig. 11). It is common for velocity to be higher in summer than winter. Peak speeds are usually

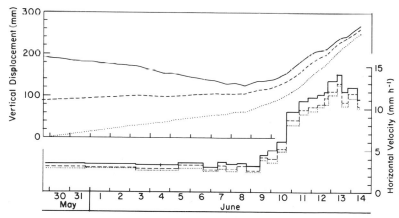

Fig. 11. Variation of horizontal velocity and vertical displacement measured over a two-week interval during early summer on Unteraargletscher [Iken (1977)]. Results from three different markers frozen into the surface are shown.

reached in the early part of the melt season [e.g., Hodge (1974); Iken (1977)].

These observations show that sliding rate cannot be explained entirely by the geometry of the glacier and local topography of the bed (roughness) because these are constant over short time scales. Conditions at the bed must be changeable. It is thought that the principle factor is water, which enters the glacier in variable amounts. The principal hydrological control of sliding velocity has, however, not been determined.

Seasonal variation of velocity is not in phase with either water discharge at the terminus or rate of water input from the surface (Fig. 12). Since water input from melting and rain and water discharge at the terminus are usually not in phase, water is stored on various time scales. On South Cascade Glacier, seasonal storage was found to be a maximum in the early part of the melt season [Tangborn *et al.* (1975)]. Because the timing of this roughly fits the timing of the seasonal peak in velocity seen

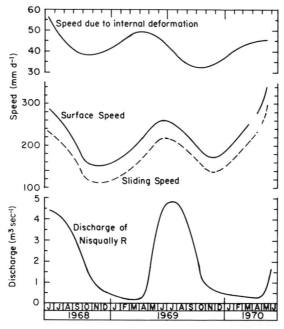

Fig. 12. Surface speed, estimated contributions from internal deformation and sliding, and water discharge of the Nisqually Glacier versus time. Although the velocity from internal deformation is uncertain in magnitude, the phasing of variations is known well from variation of load determined from thickness changes and mass balance. Note that surface velocity and inferred sliding rate changes are not in phase either with load or with water discharge at the terminus. [From Hodge (1974).]

on Nisqually Glacier in a similar environment, Hodge (1974) suggested that the storage may be a principal parameter affecting sliding. More direct indication of this was found by Iken (1977) on glaciers in the Alps. In one case, sudden increases in velocity were correlated with upward swelling of the glacier surface, possibly indicative of lifting of the sole and basal water storage (Fig. 11). Iken (1977) found, however, that the maximum speed was more closely coincident with the maximum rate of uplift than with the uplift itself.

An important effect from pressure in a basal hydraulic system is suggested when diurnal variations in speed are present. On this time scale water level in moulins and velocity have been found to be approximately in phase [Iken (1973)]. Although measurement of actual basal water pressure involves several difficult problems, some indication of expected values have been obtained [Mathews (1964); Hodge (1976); Röthlisberger (1976); Englehardt et al. (1978)]. It can range from zero in places of steep and moderate slope (e.g., where natural air-filled cavities exist) to $\frac{1}{2}-\frac{2}{3}$ of the ice overburden stress. High average values have been found during times of increasing water input such as during the early part of the melt season or periods of exceptional ablation or rainfall, but it also can increase when input and discharge are low. Substantial fluctuations of the order of $\frac{1}{3}-\frac{1}{2}$ of the overburden stress can occur on a short time scale of a day or less; sometimes these are related to diurnal or other variations of water input, but other times there is no apparent correlation or pattern.

Although the amount and pressure of water apparently has a major effect on sliding, observations are still inadequate to establish any quantitative phenomenological relationships.

B. Physical Processes

Irregularities of the bed projecting upward into the ice prevent sliding by rigid translation of the ice. Weertman (1957) proposed two physical processes by which ice could slip past such obstacles. The first is regelation and could occur if the contact between ice and rock were everywhere at the melting point. Because melting point is depressed by pressure in the liquid, high-pressure uphill sides of bumps are colder than low-pressure downhill sides. The consequent heat flow produces melting on the uphill sides and freezing on the downhill sides. Water is transferred in a layer microns thick between the ice and the rock under the pressure gradient from the uphill to downhill sides. The second process is a creep response to the local variations of pressure causing the ice to move over bumps or in a zigzag course around them. Based on a simple analysis, Weertman (1957) was able to show quite clearly that the regelation process should be most effective for small-scale obstacles and that the creep process is more effective for large obstacles.

Evidence for the action of these two processes and their relative importance at different spatial scales has been found by Kamb and LaChapelle (1964) in a tunnel to the bed of Blue Glacier Icefall. They found evidence for a thin layer of water between the ice and rock, a layer of structurally distinct ice in association with bumps of longitudinal scale of about 0.1 m, indicative of regelation, and warping of foliation on a larger longitudinal scale of about 10 m, indicative of accommodation of the moving glacier sole to large-scale bumps by creep. The subglacial precipitates and exsolution features observable on recently exposed glaciated surfaces [Hallet (1976a)] also demonstrate the role of regelation in the sliding process around short scale features.

Although regelation and creep are simple in concept, the physics involved is quite complex. Recent experiments and analyses of the regelation process reveal a number of important effects [Drake and Shreve (1973); Morris (1976); Chadbourne et al. (1975)]. The depression of melting point by impurities in the liquid phase slows the rate because impurities tend to collect on the downhill sides of bumps, lower the temperature there, and retard the heat flow. The existence of subglacial precipitates shows this effect is sometimes important to glacier sliding [Hallet (1976a)]. The temperature difference across the thin water layer between ice and rock may be significant [Nye (1973a)]. Liquid water and possibly gas bubbles in the basal ice complicate its thermal behavior [Harrison (1972); Raymond (1976)]. Permeability of the ice [Robin (1976)] or the bed [Boulton (1975)] to water flow also are possible complications. The local creep deformation around bumps may be complicated by transient creep effects as the ice moves through the spatially varying stress field [Lliboutry (1968)]. Local stresses may be so large that the usual flow law does not hold [Weertman (1969)], and the mechanical properties of basal ice might be different than typical glacier ice because of different texture, water content, and impurity content [Vivian and Bocquet (1973)].

The existence of rock particles on the interface and in the overlying ice can complicate both the regelation and creep processes and introduce the new element of rock-to-rock friction into the problem. The motion of such particles and the forces acting on them are not only relevant to the speed of sliding, they are the key to understanding erosion by abrasion, basal transport of debris, and subglacial deposition [Boulton (1974)]. The importance of rock-to-rock friction is shown by abundant finely ground rock flour discharged in streams from glaciers. The energy used up in the grinding process may be a significant fraction of the energy lost in glacier motion [Kamb et al. (1976)]. In some places the rock fragments at the contact may be rather dispersed, which is characteristic of observations at the base of rapidly sliding icefalls, but evidence from boreholes [Englehardt *et al.* (1978)] and the existence of fluted moraines shows that a layer of active subsole drift may exist between the glacier sole and bedrock in many places.

Lliboutry (1968) has proposed that the important effect of water on sliding may be associated with a third process involving separation of the ice from the bed. This will affect the boundary conditions important to the other processes; the principal feature is that the sole of the glacier is smoothed because some bed roughness elements no longer contact the ice. The existence of natural air-filled cavities beneath thin ice suggest analogous water-filled cavities could exist under deep ice as long as the water pressure were great enough. Opening and closing of cavities in response to changing hydraulic conditions could presumably account for variations in sliding velocity. This idea is compatible, at least qualitatively, with much of the existing data on velocity variations and their relationship to water input.

The geometry and pressure of the subglacial drainage system is a crucial factor in the process of separation. Water may be transported in permeable bed material, passages between the bed and glacier sole, or in conduits within the ice. The latter two are complicated by the fact that they can change with time by creep response to pressure differences between the ice and water or by melting due to viscous dissipation and advection in the flowing water [Röthlisberger (1972)]. As a result, configuration of the basal water system and its pressure distribution evolve in time, depending on the history of water input and basal stress conditions.

XI. MATHEMATICAL THEORY OF SLIDING

A. Structural Model of Bed

Theories of glacier sliding assume that clean ice rests on a fixed undeformable rock surface. This is a major simplification in view of the general presence of particles in motion at or near the contact and possible occurrence of subsole drift or unconsolidated sediments. However, where particle concentration is not high, as has been found in icefalls, it may be a reasonable assumption. In any case, this view represents the first step toward a practical theory.

A second important assumption made in most theoretical analyses is that the glacier sole is free of local shear stress. This is true where the ice is separated from the bed by a gas- or water-filled cavity. It is also expected even where there is contact, because the thin layer of water associated with regelation would prevent friction. This is reasonable but not entirely certain because the water layer may be locally pinched out [Frank (1967)] or frozen [Robin (1976)]. Assuming no friction, the mechanical equilibration of the glacier depends on the presence of rock bumps which project upward into the glacier sole, the average base stress required being produced by normal stress acting on the sloped surfaces of the irregular glacier sole. If the local normal compressive stress acting on the glacier sole is P_0, mechanical equilibrium requires

$$\tau_b = \frac{1}{a} \int_a p_0 \hat{\mathbf{n}} \cdot \hat{\mathbf{x}} \frac{dA}{|\hat{\mathbf{n}} \cdot \hat{\mathbf{z}}|}, \qquad (46a)$$

$$\sigma_b = \frac{1}{a} \int_a p_0 \hat{\mathbf{n}} \cdot \hat{\mathbf{z}} \frac{dA}{|\hat{\mathbf{n}} \cdot \hat{\mathbf{z}}|} = \langle p_0 \rangle_a, \qquad (46b)$$

where $\hat{\mathbf{n}}$ is the unit vector normal to the local bed, $\hat{\mathbf{x}}$ a unit vector down-slope parallel to the mean bed, $\hat{\mathbf{z}}$ normal to the mean bed, and a an averaging area on a plane parallel to the mean bed large compared to the bed parallel scale of the bumps. The average value of P_0 is equal to the compressive stress σ_b needed to support the weight of the overlying ice against accelerations perpendicular to the mean bed. In order to balance the component of weight parallel to the mean bed, P_0 must deviate from its mean with relatively high pressure on the uphill sides of bumps and low pressure on the downhill sides. If the basal shear stress τ_b is large or the bed is smooth ($|\mathbf{n} \cdot \mathbf{x}|$ very small), the amplitude of the fluctuations must be correspondingly large. For overall stability against gravity it is necessary that some of the bed surface slope uphill, which is probable unless the mean slope is very steep or the bed is very smooth [Kamb (1970)].

B. Sliding with No Separation

Nye (1969, 1970) and Kamb (1970) have analyzed the above model assuming that the glacier sole and bed surface are in contact everywhere and that the slope fluctuations of the bed are small. Their approach is to consider an area of the bed large enough to be statistically representative of the topography but small enough so that the sliding velocity u_b is constant over it. This involves a major implicit assumption that the scale of spatial variation of u_b is longer than the relevant scale of bed topography. A local bed topography is defined in the area as the deviation of the actual bed elevation from a mean bed defined in the area by a smoothing operation. The essential features of the local bed topography are expressed in terms of a spectral decomposition by Fourier transform methods. For a definite value of u_b there must be a certain combination of melting, refreezing, and bending of the stream lines, assuming the ice remains in contact with the rock. The problem is to find the detailed distribution of normal stress P_0 on the bottom of the ice, which will produce the required heat flow and local deformations. The average base shear stress is then determined by Eq. (46a). To calculate effects from regelation, Nye and Kamb used the simplest possible theory, in which temperature of the contact is affected only by pressure, heat is transported only by conduction, and liquid water flows only along the contact. Creep effects were calculated, assuming ice behaves as a linear fluid of viscosity η. These assumptions enabled them to find an essentially exact mathematical solution. Although the analysis can be carried out for an arbitrary spectral content of the bed topography, many of the results have been expressed assuming that the bed is nondimensional in the sense that the amplitude to wavelength ratio of the topographic waves is independent of wavelength. This model of a bed is sometimes referred to as white roughness [Kamb (1970)].

An important finding of the analysis is a characteristic wavelength

$$\lambda_* = 2\pi(\eta/\Gamma)^{1/2}, \tag{47}$$

where Γ is a constant arising from the treatment of regelation depending on latent heat of fusion, thermal conductivities of ice and bed, and depression of the melting temperature by pressure; λ_* has a numerical value of roughly $\frac{1}{2}$ m, given viscosity valid for an effective stress of about 1 bar and standard values of thermal parameters. Kamb (1970) calls λ_* the transition scale. Ice slides past bumps of scale smaller than λ_* mainly by regelation, but, for bumps of scale larger than λ_*, creep is dominant. Nye (1970) has shown that 90% of the drag arises from topography in the spectral range $\lambda_*/13 < \lambda < 13\lambda_*$, assuming white roughness. Most of the drag comes from bumps of scale close to λ_*, which is a feature originally deduced by Weertman (1957). The value of λ_* fixes the scale of the topography, which is of primary concern for understanding sliding; features with lengths from millimeters to meters are most important in the sliding process.

The relationship between base stress and sliding velocity derived by Nye (1969) and Kamb (1970) has been expressed by Lliboutry (1975) in a simplified notation as

$$\tau_b = u_b(\Gamma\eta)^{1/2}m_*^2. \tag{48}$$

The quantity m_*^2 depends on the spectral power density of the bed, and can be described as the mean quadratic slope in the direction of slip of the local bedrock filtered in a way which emphasizes those spectral components near the transition scale λ_*. Although m_*^2 is a measure of the roughness of the bed at the most important wavelengths, it is important to notice that it depends not only on the local bed topography but also on the theory through λ_* and the filter.

Nye (1969) also introduced the description of the bed in terms of its autocorrelation function, which allows the drag to be expressed as an integral over the bed rather than as a summation of spectral components. Nye (1970) has given careful attention to the effect of the scale of averaging area and the smoothing operation used to define the mean bed within it, and has shown that, as long as its dimensions are large compared to λ_* and the smoothing represents a running mean on a similar or larger scale, there is little quantitative effect on the predictions.

It is important to consider the nonlinear creep behavior of ice. Based on his simplified model of the bed and scale analysis, Weertman (1957) deduced a sliding law of the form $u_b \sim \tau_b^{(n+1)/2}$, where n is the power in the flow law. Kamb (1970) and Lliboutry (1975) attempted to extend the linear theory based on spectral decomposition of the bed to include nonlinear creep by approximate methods.

With nonlinear properties the effective viscosity depends on strain rate and both depend on position. Kamb assumes that the patterns of local motion and strain rate for the linear and nonlinear cases are approximately the same for a given sliding velocity over a given bed. Based on that assumption and the flow law [Eq. (14)], the strain rate distribution for the linear solution provides a distribution of effective viscosity from which the stress distribution for the nonlinear case may be estimated. The effective viscosity distribution shows pronounced spatial variation with a minimum at about $\lambda_*/2\pi$ above the bed. Kamb attempts to account for this spatial variation by assuming that the stress needed to drive the spectral component of the motion at wavelength π is determined by the effective viscosity at a distance $\lambda/2\pi$ from the bed, where strain rate for that component is maximum. Lliboutry (1975) made an additional simplification by assuming that the controlling effective viscosity is the minimum value for all spectral components. Lliboutry's approach is likely to underestimate the appropriate effective viscosity for wavelengths substantially different from λ_*, but the mathematical development turns out to be much simpler than the more rigorous approach of Kamb. Both approaches give results in the form

$$\lambda_* = (2\pi/K_\lambda)(A\Gamma)^{-1/2}(m_*/\tau_b)^{(n-1)/2}, \tag{49}$$

$$u = (K_u/m_*)(A/\Gamma)^{1/2}(\tau_b/m_*)^{(n+1)/2}, \tag{50}$$

for the case of white roughness. Here A and n are flow law parameters in Eq. (13). In Lliboutry's treatment K_λ and K_u are the same constant, but in

Kamb's treatment they are different and each depends slightly on roughness. The numerical differences for typical values of u_b, τ_b, and m_* are quite small [Lliboutry (1975)]. In Eq. (49), λ_* has the same interpretation as transition scale in the linear theory [Eq. (47)], but it is not constant; rather it varies with roughness and stress because they affect the effective viscosity distribution through the nonlinear flow law. Both treatments reproduce the original stress dependence deduced by Weertman (1957). However, when the roughness is not white, a different stress dependence is possible. If short wavelengths were absent, the stress dependence would approach $u_b \sim \tau_b{}^n$. Kamb (1970) suggested this spectral characteristic could be common as a result of preferential smoothing at the small scale by glacier abrasion and he modeled it using a "truncated white roughness."

Since bed roughness, slip velocity, and basal shear stress have not been measured simultaneously at any one place, definitive field testing of the sliding theories is not possible. As a start Kamb (1970) has taken eight examples where both u_b and τ_b have been estimated and calculated the roughness needed in the theory to explain these values. Based on his observations of recently deglaciated beds, Kamb indicates that the roughness required by the theories is lower than expected by a factor of 2–4, especially for the examples with sliding velocity $u_b > 20$ m yr ^{1}. Measurements of topography by Hallet (1976b) suggest the discrepancy may be even larger. Although the bed roughnesses might actually be unexpectedly low for some of the sliding glaciers, there may be a real discrepancy, which could be explained in several ways. Kamb (1970) considered the possibility of truncated white roughness; because short wavelengths are absent the remaining long wavelengths have more realistic roughness when used to explain the field examples. Another possible explanation would be ice-bed separation, which results in an effective smoothing of the bed. This was known to exist or was suspected in several of the cases. Identification of transition wavelength is potentially the most effective way to quantitatively test the theories. The predicted transition wavelengths are in the range 0.1–0.5 m. These are reasonable in light of the observations of Kamb and LaChapelle (1964), but the observations are still too limited to make a quantitative test on this basis.

One of the principal assumptions of the model is that the ice–rock interface is locally perfectly slippery because of the water layer associated with the regelation process. Morland (1976) has studied the effects that friction might have on the sliding process over a sinusoidal bed assuming linear rheology. Obviously the effect of friction is to reduce the sliding velocity. Morland finds also that the existence of friction influences the normal stress distribution. This alters the conditions under which local separation of the ice can occur, as discussed below. It is also an important consideration with regard to glacier erosion by quarrying and plucking [Morland and Morris (1977)].

Another simplification of the theories described above is that they neglect any possible effect of impurities on the regelation process. Hallet (1976a) has modified the theory of Nye (1969) to include an effect from solutes in the water layer associated with regelation. Because regelation rate is slowed, both the transition wavelength λ_* and sliding velocity are lowered, assuming the bed is not smooth at small wavelengths. A fully quantitative treatment of the impurity concentration distribution for a general bed and its effect on sliding is prevented by the complex interaction of kinetic effects at the liquid–solid surfaces, diffusion and advection in the water, and other factors. Hallet argues that the effect can be substantial, especially when the bed is composed of calcareous rocks.

C. Sliding with Separation

The normal stress distribution P_0 on the bottom of the ice needed to maintain contact with the bed is given by the theories of Nye (1969, 1970) and Kamb (1970). Since the interface cannot support a tensile stress, separation can occur where P_0 drops below a certain minimum (P_w). P_w has the absolute lower limit of triple-point pressure. However, if air has access to the bed, it is atmospheric pressure, and if pressurized water has access to the bed, it can be considerably higher. In this latter case, P_w is identified with the pressure in the basal hydraulic system, which transports water to the glacier terminus. P_w is controlled in quite a different manner than is the pressure P_0 in the very thin layer between ice and rock associated with local transport of water in the regelation process [Nye (1973b)].

The most likely locations for such separation are the downhill sides of bumps where P_0 is below the mean σ_b by an amount given by $\langle(\sigma_b - P_0)^2\rangle^{1/2}$ typically, or by about $2\langle(\sigma_b - P_0)^{1/2}\rangle^2$ in the extreme [Kamb (1970)]. Consequently, separation should start in some locations when $(\sigma_b - P_w) \leq 2\langle(\sigma_b - P_0)^2\rangle^{1/2}$ and be extensively developed when $(\sigma_b - P_w) \leq \langle(\sigma_b - P_9)^2\rangle^{1/2}$. The quantity $\sigma_b - P_w$ represents an effective normal stress N_e at the base of the glacier. The overburden stress (σ_b) is given approximately by Eq. (16b) as

$$\sigma_b = \rho_i g h \cos \alpha + p_a. \tag{51}$$

The root-mean-square deviation of P_0 from its mean is given by the sliding theory of Kamb (1970) or Nye (1970) as

$$\langle(\sigma_b - P_0)^2\rangle^{1/2} = (2/\pi)^{1/2}\tau_b/m_* \tag{52}$$

for white roughness and linear rheology. It is determined primarily by the condition for static equilibrium [Eq. (46a)]. Kamb (1970) finds a relation close to this for nonlinear rheology, but the right-hand side is divided by a factor between 1 and 1.5, depending on roughness and n in the flow law. The third influence on the initiation of separation comes from P_w. In terms of these parameters separation becomes likely as overburden or roughness decrease or as water pressure or shear stress increase. For reasonable values of roughness (e.g., $m_* = 0.2$), Kamb (1970) shows that some separation is likely when surface slope exceeds about 15°, even when P_w is zero, and when P_w exceeds one-half σ_b on typical surface slopes.

Separation of the ice from the bed can be expected to increase sliding velocity since contact is lost with some of the roughness elements of the bed and the ice sole is correspondingly smoothed. Weertmann (1964) and Kamb (1970) have attempted to account for this within the context of their theories by using a reduced roughness or spectral model. Kamb (1970) argues that separation will be most prevalent at wavelengths close to the transition value for which the normal stress fluctuations are largest. The result will be to smooth the base at this and shorter scales, in effect producing a configuration somewhat like "truncated white roughness."

Lliboutry (1968) suggested that the effect of separation on sliding is best described in terms of a function which gives the fractional area of separation. If the ceilings of cavities were roughly parallel, this would correspond to the area of bed in shadow when illuminated from upstream at the inclination of the ceilings. The shadowing function represents a new aspect of the bed description which is not determined by the spectral power density [Benoist and Lliboutry (1978)]. The slope of a cavity roof is determined by its relative velocity normal and tangential to the mean bed, and this will be affected by the difference between overburden and cavity pressures. Thus, once separation exists, the effective normal stress N_e enters as an important variable affecting the sliding relationship.

Various approximate relationships between u_b, τ_b, and N_e have been derived by Lliboutry (1968, 1978) using different models of the bed such as a single sine wave, several superimposed sine waves, or a nondimensional distribution of hemispherical knobs. When described with N_e held constant, these include the possibilities of a maximum τ_b and corresponding u_b for which stable motion is possible, a multiple-valued u_b versus τ_b with unstable ranges of velocity, and single-valued monotonic relationships between u_b and τ_b. It is difficult to evaluate these theoretical results because a relationship between u_b, τ_b, and N_e is not known from field measurements, and in any case even the qualitative theoretical behavior is quite sensitive to properties of the bed topography, which are not easily determined.

The principal hydrological parameter affecting the amount of separation is basal water pressure P_w. Its theoretical calculation is complicated by a lack of a definite picture of how water arrives at the bed and how it is transported along the bed. For typical conditions about 10^{-2} m yr^{-1} is melted at or near the bed by mechanical dissipation and geothermal heat. To this may be added some water which percolates through the vein network from the surface, but this is also probably quite small [Raymond and Harrison (1975)]. Most of the water produced at the surface, which is on the order of 1 m yr^{-1}, probably reaches the bed at isolated points. Motion of water along the bed may occur in conduits melted upward into the glacier sole [Röthlisberger (1972)] or in linear depressions in the bed [Nye (1973b)], or it may occur in a network of hydraulicly interconnected cavities on the downhill sides of bumps [Weertman (1972)]. Observationally it is known from tracing experiments that conduits exist beneath the ablation areas of glaciers [e.g., Behrens *et al.* (1975)]. Drainage of water arriving homogeneously over the bed and the very existence of short-term velocity variations and their explanation in terms of separation implies the existence of interconnected systems covering wide areas of the bed, at least transiently. Thus, all of these modes of transport apparently play a role, but their relative importance and how that changes with time in response to variations in water input are unknown.

By analyzing the balance between creep closure and thermal opening, Röthlisberger (1972) developed a mathematical theory of pressure in conduits assuming steady-state discharge. This has been extended by Nye (1976) to include nonsteady conditions. Principal conclusions about steady conditions are: first, that pressure decreases with increasing discharge, and second, that conduit pressure could be typically one-half or more of the ice overburden but would reach the ice overburden in only very unusual circumstances. The first indicates a tendency for large conduits to capture flow from small ones and thus concentrate flow into trunk conduits. The second is compatible with observations described above and, in view of the discussion of Eq. (52), it indicates that conduit pressures could reach levels adequate to inject water behind bedrock bumps, which would tend to divert flow into a layer and cause separation. This latter possibility is especially likely in transient conditions, when water input increases so abruptly that conduit sizes cannot adjust in size, in which case discharge and pressure increase together and pressure can reach high values.

The pressure in cavities has been estimated by Weertman (1972), assuming that it is given by the average normal compressive stress on the downstream sides of bumps predicted by the theory of sliding in the absence of separation; the resulting pressure is smaller than the overburden by an amount given by Eq. (52). The implication is that the pressure in a cavity is controlled by the motion of the roof, which is reasonable if the cavity is hydraulically isolated but is less certain if the cavity is part of an hydraulically interconnected system in which pressure is affected by remote boundary conditions.

An important factor in the geometry of the basal hydraulic system and separation of ice from the bed is transfer of water between conduits and an interconnected network of cavities. Weertman (1972) shows that in some circumstances stress redistribution associated with a conduit may cause a band of above-average normal compressive stress along either edge. This effect could inhibit the transfer of water, but it could be counteracted by effects of normal stress fluctuations around irregularities on the bed.

XII. ADVANCE AND RETREAT OF GLACIERS

A. General Comments

The elevation of the surface of a glacier may change in response to mass balance or vertical motion. When this happens at the terminus, the glacier advances or retreats. In a steady state the mass balance and vertical motion cancel one another to produce no net change. However, the rate of balance is variable as a result of weather and changing climate. Internal processes may alter the motion independently of any direct climatic influence. Consequently, glaciers are in a continual state of adjustment.

From a glacier dynamics point of view, the principal problem is to calculate the motion. Direct calculation of the detailed distribution of velocity in the complete volume and over the surface is impractical and unnecessary. The problem has therefore been formulated in terms of the one-dimensional equation of continuity representing an integration over the area of a cross section [Eq. (4)]. From this point of view the principal goal is to determine the dynamic flux Q_v in terms of appropriate parameters of the glacier. While it is not possible to do this accurately, some initial steps in this direction are possible and some useful models have been developed [e.g., Nye (1960); Budd and Jenssen (1975)]. In branched glaciers, ice fields, or ice sheets, the flow system cannot be conveniently thought of as one dimensional. For these cases two-dimensional models based on the vertically integrated equation of continuity have been attempted [e.g., Rasmussen and Campbell (1973)].

B. Response to Changing Climate

To analyze this problem Nye (1960) assumed the flux through a cross section at longitudinal position X is determined by depth h and local slope

α depending on cross-section shape, local basal conditions, and other parameters. This is represented as $Q_v(X, h, \alpha)$. If only small perturbations of h, α, and the corresponding Q_v from a steady datum state are allowed, the equation of continuity can be expressed in the approximate linearized form

$$\partial Q_v{}'/\partial t = C_0(\partial Q_b{}'/\partial X - \partial Q_v{}'/\partial X) + D_0\, \partial^2 Q_v{}'/\partial X^2, \qquad (53)$$

where prime values represent small perturbations from the datum state. The quantities $C_0(X)$ and $D_0(X)$ are, respectively, $(\partial Q_v/\partial h)/\bar{\rho}W$ and $(\partial Q_v/\partial \alpha)/\bar{\rho}W$, evaluated in the datum state and give the change in volume flux per unit width caused by changes in depth and slope.

The theory predicts that perturbations of the ice thickness or flux should propagate downglacier as kinematic waves with speed approximately equal to C_0. For a parabolic channel, constant density, and no slip, the speed is about $\frac{2}{3}(n + \frac{5}{2})$ times the mean ice velocity from Eqs. (29), (32) and (33). For n equal to 3 to 4, this factor is about 4, but it might be modified by sliding. This kind of theoretical behavior has been confirmed by observation of waves on several glaciers [e.g., Reynaud (1977)]. The dependence of ice flux on surface slope α tends to even out perturbations and strongly modify the propagation of the waves by the introduction of a diffusive effect with coefficient equal to D_0. From Eq. (33) this would be $(Q_v/\bar{\rho}W)(n/\sin \alpha)$, which shows diffusion becomes especially important when the slope is small. Damping by diffusion is probably one reason that kinematic waves are not more commonly observed on glaciers. The distributions $C_0(X)$ and $D_0(X)$ can be found from measurements of the ice velocity and flux on a glacier [Nye (1963)], and these and Eq. (53) determine the theoretical response of the glacier to small variations in climate.

The response of a simple model glacier to a sudden step change in climate was used by Nye to illustrate important characteristics of glacier response. There is an initial short period during which the thickness changes unstably at an accelerating rate in the lower reaches of a glacier, where there is compressive flow. This phase lasts for a time about equal to that required for a kinematic wave to travel from the equilibrium line to the terminus, characteristically about several decades in a typical mountain glacier. The rate of thickness change then decreases, but the adjustment continues for several centuries. The actual time scales can be quite different from glacier to glacier, depending on length, slope, and rate of motion. Nye also examined other types of climate change history, including short-duration pulses and harmonic variations. All of these show that advance and retreat of different glaciers may be asynchronous, even though they may be responding to the same climate.

The inverse problem, in which past mass balance variations are de-

duced from a history of advance and retreat of glacier margins, is potentially important for climatic interpretation of the geological record and
practical hydrological deductions in remote regions where continuous
direct measurement of mass balance is not possible. Nye (1965a) finds the
mass balance perturbation for a given year can be estimated theoretically
from measurements of terminal position over a relatively short period extending back over an interval similar in length to the initial phase of
response for a step change in climate. Tests of this theory show that on a
year-to-year basis the predictions are not good, but the general trends of
10-yr running mean values are in reasonable agreement [Nye (1965a)].

The main difficulty with the above theory is that it neglects longitudinal
stress gradients which suppress longitudinal variations of velocity in
response to variations in local slope. It is the longitudinal variation of
velocity from places of thick to thin ice and steep to shallow slope which
results in wave propagation and diffusion. Intuitively, one expects perturbations of very short longitudinal scale to move passively on the glacier
surface at the ice velocity whereas perturbations of very long length ought
to conform to the theory. Langdon and Raymond (1978) show that the
long-scale behavior exists only for perturbations which are many times
the ice depth, which for typical mountain glaciers can be as long or longer
than the total glacier length. Kinematic propagation and diffusion of ef
fects will, in general, be slower than predicted by neglecting longitudinal
stress gradients, thus suggesting the initial short period of accelerating
response and the longer period of continued adjustment would be extended. However, the action of longitudinal forces introduces a new short
time scale element into the behavior; effects can be propagated longitudinally by forces without the time lag required for propagation of topo-
graphic effects by mass redistribution.

Several recent numerical models provide a potential for more realistic
calculations of glacier adjustment to climate by eliminating the restriction
that changes be small and by attempting to account for longitudinal stress
gradients. R. A. Bindschadler (unpublished) bases his model of Q_v primarily on Eqs. (45), (31), and (4) and attempts to account for longitudinal
gradients by using a large-scale slope with a secondary contribution from
local slope which provides a diffusive effect needed for stability in the numerical treatment of the equation of continuity. Budd and Jenssen (1975)
calculate center-line velocity distribution by integration of the longitudinal force equation for intermediate scale [Eq. (43)] and from this evaluate Q_v in a way similar to Eqs. (31). This gives additional effects from
longitudinal gradients beyond those taken into account by the simpler ar-
tifice of slope averaging and has a somewhat more rigorous mechanical
foundation. This more complex treatment may not have any advantage in

accuracy for typical flow situations, but, in some cases of especially complex longitudinal profile or during surges, the longitudinal stress gradients may be of such overwhelming importance that their explicit calculation is essential [Budd (1975)].

All of the models are subject to the uncertainty about the basal boundary condition which at this stage cannot be parametrized. Without additional understanding about this, reasonable predictions will be possible only when sliding makes a negligible contribution to the motion.

C. Internally Controlled Fluctuations

Seasonal and shorter-period fluctuations of velocity commonly found on glaciers give an example of changes in Q_v not associated with changes in depth and slope. Because of the short time scale, seasonal fluctuations of normal amplitude do not lead to major changes in geometry. Surge behavior represents a case in which velocity changes of an order of magnitude or more occur over a cycle period of many years [Meier and Post (1969)]. Because of the large amplitude and the relatively long time scale, this does lead to substantial alterations in geometry. Possibly there is a continuous spectrum of behavior between these extremes. Meier and Post (1969) attempt to classify surge-type glaciers into several classes, depending on the speeds attained during surges and other factors. The main characteristic of all of the cases is that the balance flux is transported by a pulsating dynamic flux. This represents a major difficulty for interpretation of glacier fluctuations in terms of climate.

The very fast surge-type glaciers are the most spectacular manifestations of this behavior [Meier and Post (1969)]. The main feature is a period of months to several years of exceptionally rapid motion with speeds up to several kilometers per year during which a large mass is transported downglacier from a reservoir area to a receiving area. During a surge the dynamic flux far exceeds the balance flux, and the reservoir area is depleted with a consequent drop in elevation while the receiving area experiences a corresponding increase in elevation. The surface elevation changes can reach up to 100 m. If the receiving area extends to the terminus, the ice margin can advance up to the order of 10 km. Such surges are separated by intervals of one to several decades of relative quiescence in which rates of motion are more normal. During a quiescent period the dynamic flux is much less than the balance flux, and the reservoir area slowly recharges while the receiving area is slowly depleted. During a surge the total displacement may be a significant fraction of the total glacier length and exceed the displacement accumulated over the much longer period of quiescence. This illustrates the high relative speed of the surge and its importance in achieving a long time scale balance between motion and mass balance. Table II describes characteristics of some typical surge-type glaciers. It shows glaciers of a variety of lengths and slopes may surge.

Surge-type glaciers comprise only a small percentage of all glaciers. Their distribution within glaciated regions is remarkably nonuniform. In some mountain regions they are absent, but in others they are relatively common [e.g., Post (1969); Dolgushin and Osipova (1975)]. Their distribution indicates that special environmental circumstances are necessary for their existence, but these remain obscure. Surge-type glaciers exist with temperate and subpolar thermal regimes in maritime and continental environments, are not characterized by any particular geometrical shape, and do not display any consistent relationship to bedrock type, altitude, or orientation. Although many surging glaciers are associated with fault-related valleys, there is little evidence that surges are initiated by seismic activity.

Budd (1975) suggested that a critical parameter may be the flux per unit width times slope which gives the average rate of mechanical dissipation of heat by potential energy loss. If this is high enough, then a glacier may be forced to flow in a distinct exceptionally fast mode. Three classes of behavior could then be expected for a given geometry of cross section and longitudinal slope: low balance flux transported continuously by a normal mode of flow (normal behavior); very high balance flux transported continuously by the fast mode of flow (continuous fast behavior); intermediate balance flux which is too large to be transported by the normal mode of flow but too small to feed continuous fast motion, transported by oscillations between the slow and fast modes (surge-type behavior). Budd's (1975) comparison of slope, velocity, and flux of normal, surge-type, and continuously fast-moving glaciers supports this suggestion, but much more observation is necessary before any definite pattern can be established.

A number of theoretical explanations for glacier surges have been proposed involving changes in the mechanical properties of the ice [Jonas and Müller (1969); Nielson (1969)], nonunique thickness profiles caused by longitudinal stress gradients [Robin (1969)], wave propagation effects [Palmer (1972)], and basal or internal temperature variations with time [Clarke (1976)] or position [Schytt (1969)]. Some of these might contribute to the wide range of internally controlled fluctuations of glaciers. However, based on current knowledge, the very fast surges are probably due to rapid sliding. Since some of the very fast surge-type glaciers are temperate [Bindschadler *et al.* (1976)], thermal triggering of sliding by a transition from subfreezing to melting conditions at the bed is not an essential

TABLE 2

Characteristics of Some Surge-Type Glaciers in Western North America[a]

Glacier	Mountain range and location	Area total (km²)	Length Total (km)	Length Surging part (km)	Slope Average over surging part (°)	Cycle period (yr)	Surge duration (yr)	Surge velocity Maximum annual rate (km yr⁻¹)	Surge displacement (km)
Bering	Chugach–St. Elias, Alaska	5800	200	>153	0.7	~30	~3	?	9.7
Klutlan	Icefield, Yukon Territory	1072	55	40	1.3	~30	~3	>3.2	6.5
Walsh	St. Elias, Yukon Territory	830	89	86	1.0	~50	~4	>5.6	11.5
Muldrow	Alaska, Alaska	393	63	46	2.2	~50	~2	6.6	6.6
Variegated	St. Elias, Alaska	49	20	19	4.2	~20	~2	>5	>5
Tyeen	Fairweather, Alaska	11	7	7	13.6	~20	~3	>1.5	>2.4

[a] From Meier and Post (1969).

aspect to the process. Initiation of a surge apparently depends on an interaction of changing patterns of basal stress distribution and hydraulic conditions.

Lliboutry (1969) describes an easily visualized explanation of two distinct modes of flow and surge behavior based on a double-valued relationship between slip velocity and base stress at fixed effective normal stress (Fig. 13). The unstable range of velocity in which sliding velocity decreases with base stress is deduced theoretically as a consequence of bed separation from small-scale bumps. From this point of view, a surge is initiated when basal shear stress rises to a critical value, or alternatively if a change in effective normal stress through the action of subglacial water drops the critical base shear stress to the existing one. Although these ideas are illustrative and represent an attractive explanation, they are based on a tentative analysis of the immensely complex problem of sliding in the presence of bed separation.

Another explanation offered for surge behavior is based on a positive feedback between water lubrication and sliding rate which arises because increased rate of potential energy loss associated with increased sliding yields increased basal melting and increased water lubrication. Weertman (1969) has proposed that a water layer could build up to separate the glacier sole from obstacles of the size most effective in producing drag with a consequent major increase in sliding speed. Without any particular physical model in mind, Budd (1975) assumes the local friction is reduced according to the local rate of energy dissipation and a lubrication parameter in a way which holds constant the average basal stress needed for overall mechanical equilibrium. This basal boundary condition is incorporated into the flow model of Budd and Jenssen (1975). With a range of choices of lubrication parameter and viscosity parameter which determines longitudinal interactions [Eq. (43)], Budd was able to simulate two distinct modes of flow and the range of normal, surge-type, and continuous-fast behavior, depending on the balance flux.

A major problem with the theories of Weertman (1969) and Budd (1975)

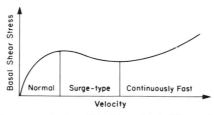

Fig. 13. Schematic representation of double-valued sliding velocity similar to one proposed by Lliboutry (1968).

is that they do not give careful attention to the effect of water penetrating from the glacier surface to the bed. Except during the very fast motion of a surge itself, the water production rate at the surface is orders of magnitude larger than that at the base. The short period and seasonal variation of velocity typical of temperate glaciers indicates surface water penetrates to the bed and affects an area large enough to influence the sliding rate and overall motion. This same kind of behavior is found on the surge-type Variegated Glacier, indicating the basal hydraulic system carrying water from the surface is a major element in the glacier's dynamic behavior during its quiescent phase [Bindschadler *et al.* (1976, 1977)]. In this case large, progressive, year-to-year increases in summer velocity are accompanied by relatively small changes in winter velocity, indicating these hydraulic effects are probably of major importance in the surge behavior [Bindschadler *et al.* (1978)]. Distinct seasonal activity, reported from the surge-type Medvezhy Glacier [Dolgushin and Osipova (1975)] also supports this conclusion.

In addition to input of water to the bed, it is essential to consider how water drains from the bed. If the outflow were blocked by some process, water would accumulate, flood the bed, and lead to fast motion. Robin and Weertman (1973) give one scheme by which this might happen cyclicly. It is based on a model of glacier adjustment between surges in which mass balance is assumed to dominate [$\partial Q_v/\partial X$ negligible in Eq. (4)], and on the notion that water flows along the base of the glacier in a network of cavities with pressure determined by the hypothesis of Weertman (1972) discussed in Section X.C. By this hypothesis, pressure increases with decreasing shear stress. As the reservoir area thickens and the receiving area is depleted during the quiescent phase, a zone of increasingly negative basal shear stress gradient develops in between. It is proposed that eventually the downglacier water flow is blocked which initiates a surge. Since the hypothesis about cavity pressure becomes uncertain when cavities are part of an interconnected drainage system, ice motion is likely to be a significant factor in the glacier adjustment between surges, and basal shear stress gradient is difficult to calculate because of lateral and longitudinal stress gradients, this intriguing model is quite speculative at present. If water flows along the bed in conduits, a surge might be initiated by constriction of the conduits as a result of either the changing patterns of basal stress or slip rate. A positive feedback seems possible because rapid sliding over a bumpy bed would tend to obstruct basal conduits unless they were incised downward into the bed [Nye (1973b)].

The role of increasing basal shear stress in the reservoir area might be important. Since the amplitude of normal stress fluctuations along the bed climbs rapidly with increasing basal shear stress [Eq. (52)], the rising shear stress promotes a tendency for bed separation and consequent instability, without any dramatic alterations in the basal hydraulic system and effective normal stress. This is the fundamental concept behind the behavior visualized in Fig. 13. Whatever the actual quantitative effects, it may be an important feature of the surge initiation.

Although some of the above theoretical ideas can simulate general features of surge behavior, field data from surge-type glaciers are still inadequate to test whether any of them explain the actual physical mechanisms. It seems clear from the existing observations that surge behavior is controlled by the evolution of the glacier surface and the consequent changes of basal stress as a result of the direct mechanical effect at any one location and an indirect effect of the overall pattern through its effect on the basal hydraulic system. The key to understanding surge behavior lies in an expanded observational knowledge of the sliding process and the behavior of water at the bases of glaciers. This is also essential for better understanding of response to climate of all glaciers with temperature bases.

ACKNOWLEDGMENTS

The author is grateful to Robert Bindschadler, Almut Iken, Louis Lliboutry, Stan Paterson, Hans Weertman, and the editor, who made valuable comments on an early version of this chapter.

REFERENCES

The following special abbreviations are used:

CRREL   U.S. Army Cold Regions Research and Engineering Laboratory, Hanover, New Hampshire
IAHS 104   International Association of Hydrological Sciences Publication 104. (Similarly for other numbers.)
IASH 86   International Association of Scientific Hydrology Publication 86. (Similarly for other numbers.) (IASH is the former name of IAHS.)
SIPRE   U.S. Army Snow, Ice, and Permafrost Research Establishment, Wilmette, Illinois. (The former name for CRREL.)

Agassiz, L. (1847). "Nouvelles Etudes et Expériences sur les Glaciers Actuels, leur Structure, leur Progression et leur Action Physique sur le Sol." Masson, Paris.
Allen, C. R., Kamb, W. B., Meier, M. F., and Sharp, R. P. (1960). *J. Geol.* **68**, 601–625.
Anderson, D. L., and Benson, C. S. (1963). *In* "Ice and Snow" (W. D. Kingery, ed.), pp. 391–411. MIT Press, Cambridge, Massachusetts.
Atherton, D. (1963). *J. Glaciol.* **4**, 547–557.
Baker, R. W. (1978). *J. Glaciol.* **21**, 485–500.
Barnes, P., Tabor, D., and Walker, J. C. F. (1971). *Proc. R. Soc. London Ser. A* **324**, 127–155.

Behrens, H., Bergmann, H., Moser, H., Ambach, W., and Jochum, O. (1975). *J. Glaciol.* **14,** 375–382.
Benoist, J., and Lliboutry, L. (1978). *Ann. Geophys.* **34,** 163–175.
Bindschadler, R. A. (1978). Ph.D. Thesis, University of Washington, Seattle (unpublished).
Bindschadler, R. A., Harrison, W., Raymond, C., and Gantet, C. (1976). *J. Glaciol.* **16,** 251–259.
Bindschadler, R. A., Harrison, W. D., Raymond, C. F., and Crosson, R. (1977). *J. Glaciol.* **18,** 181–194.
Bindschadler, R. A., Raymond, C. F., and Harrison, W. D. (1978). *Dokl. Acad. Sci. USSR, Glaciol. Sect.* **32,** 109–117 and 224–229.
Bjornsson, H. (1974). *Jökull* **24,** 1–26.
Blümcke, A., and Finsterwalder, S. (1905). *Sitzber. Bayer. Ak. Wiss. Math.-Nat. Kl.* **35,** No. 1, 109–131.
Boulton, G. S. (1974). *In* "Glacial Geomorphology" (D. R. Coates, ed.), pp. 41–87. State University of New York, Binghamton.
Boulton, G. S. (1975). *In* "Ice Ages Ancient and Modern" (A. E. Wright and F. Moseley, eds.), pp. 7–42. Seel House Press, Liverpool.
Budd, W. F. (1968). *IASH* **79,** 58–77.
Budd, W. F. (1970). *J. Glaciol.* **9,** 19–27.
Budd, W. F. (1975). *J. Glaciol.* **14,** 3–21.
Budd, W. F., and Jenssen, D. (1975). *IAHS* **104,** 257–291.
Budd, W. F., and Radok, U. (1971). *Rep. Prog. Phys.* **34,** 1–70.
Budd, W. F., Jenssen, D., and Radok, V. (1971). *Aust. Natl. Antarct. Res. Exped. Publ.* **120,** 1–178.
Byers, B. A., Chalupnik, J. D., and Raymond, C. F. (1973). *EOS, Trans. Am. Geophys. Union* **54,** 452 (abstr. only).
Carbonnell, M., and Bauer, A. (1968). *Medd. Groenl.* **173,** 1–78.
Chadbourne, B. D., Cole, R. M., Tootill, S., and Walford, M. E. R. (1975). *J. Glaciol.* **14,** 287–292.
Clarke, G. K. C. (1976). *J. Glaciol.* **16,** 231–250.
Colbeck, S. C., and Evans, R. J. (1973). *J. Glaciol.* **12,** 71–86.
Collins, I. F. (1970). *J. Glaciol.* **9,** 169–194.
Deeley, R. M. (1895). *Geol. Mag.* [4] **2,** 408–415.
Deeley, R. M., and Parr, P. H. (1913). *Philos. Mag. J. Sci.* **26,** 85–111.
Deeley, R. M., and Parr, P. H. (1914). *Philos. Mag. J. Sci.* **27,** 153–176.
Dolgushin, L. D., and Osipova, G. B. (1975). *IAHS* **104,** 292–304.
Drake, L. D., and Shreve, R. L. (1973). *Proc. R. Soc. London, Ser. A* **332,** 51–83.
Duval, P. (1977). *IAHS* **118,** 29–33.
Embleton, C., and King, C. A. M. (1968). "Glacial and Periglacial Geomorphology." St. Martin's Press, New York.
Engelhardt, H. F., Harrison, W. D., and Kamb, W. B. (1978). *J. Glaciol.* **20,** 469–508.
Flint, R. B. (1971). "Glacial and Quaternary Geology." Wiley, New York.
Forbes, J. D. (1846). *Philos. Trans. R. Soc. London* **136,** 143–210.
Frank, F. C. (1967). *Philos. Mag.* [8] **16,** 1267–1274.
Gerrard, J. A. F., Perutz, M. F., and Roch, A. (1952). *Proc. R. Soc. London Ser. A* **213,** 546–558.
Glen, J. W. (1955). *Proc. R. Soc. London Ser. A* **288,** 519–538.
Glen, J. W., and Lewis, W. F. (1961). *J. Glaciol.* **3,** 1109–1122.
Glen, J. W., and Perutz, M. F. (1954). *J. Glaciol.* **2,** 397–403.

Goldthwait, R. P. (1960). *SIPRE Tech. Rep.* **39**.

Hallet, B. (1976a). *J. Glaciol.* **17**, 209–221.

Hallet, B. (1976b). *EOS, Trans. AM. Geophys. Union* **57**, 325 (abstr. only).

Harrison, W. B., and Kamb, B. (1973). *J. Glaciol.* **12**, 129–137.

Harrison, W. D. (1972) *J. Glaciol.* **61**, 15–29.

Higashi, A. (1969). *Phys. Ice, Proc. Int. Symp., 3rd, 1968* pp. 197–211.

Hodge, S. M. (1974) *J. Glaciol.* **13**, 349–369.

Hodge, S. M. (1976) *J. Glaciol.* **16**, 205–218.

Hoinkes, H. C. (1964). *Res. Geophys.* **2**, 391–424.

Hooke, R. L., Dahlin, B. B., and Kauper, M. T. (1972). *J. Glaciol.* **11**, 327–336.

Iken, A. (1973). *Z. Gletscherkd. Glazialgeol.* **9**, 207–219.

Iken, A. (1977). *Z. Gletscherkd. Glazialgeol.* **13**, 23–25.

Jonas, J. J., and Müller, F. (1969). *Can. J. Earth Sci.* **6**, 963–968.

Kamb, W. B. (1959). *J. Geophys. Res.* **64**, 1891–1909.

Kamb, W. B. (1964). *Science* **146**, 353–365.

Kamb, W. B. (1970). *Rev Geophys, Space Phys.* **8**, 673–728.

Kamb, W. B. (1972). *In* "Flow and Fracture of Rocks" (H. C. Heard, I. Y. Borg, N. L. Carter, and C. B. Raleigh, eds.), pp. 211–241. Am. Geophys. Union, Washington, D.C.

Kamb, W. B., and LaChapelle, E. R. (1964). *J. Glaciol.* **5**, 159–172.

Kamb, W. B., and Shreve, R. L. (1963). *EOS, Trans. Am. Geophys. Union* **44**, 103 (abstr. only).

Kamb, W. B., and Shreve, R. L. (1966). *EOS, Trans. Am. Geophys. Union* **47**, 190 (abstr. only).

Kamb, W B., Pollard, D., and Johnson, C. B. (1976). *EOS, Trans. Am. Geophys. Union* **57**, 325 (abstr. only).

Ketcham, W. M., and Hobbs, P. V. (1969). *Philos. Mag.* [8] **19**, 1161–1173.

Langdon, J., and Raymond, C. F. (1978). *Dokl Acad Sci, USSR, Glaciol. Sect.* **32**, 123–133 and 233–239.

Lliboutry, L. (1964). "Traité de glaciologie," Vol. 1. Masson, Paris.

Lliboutry, L. (1965). "Traité de Glaciologie," Vol. 2. Masson, Paris.

Lliboutry, L. (1968). *J. Glaciol.* **7**, 21–58.

Lliboutry, L. (1969). *Can. J. Earth Sci.* **6**, 943–953.

Lliboutry, L. (1971a). *Adv. Hydrosci.* **7**, 81–167.

Lliboutry, L. (1971b). *J. Glaciol.* **10**, 15–29.

Lliboutry, L. (1975). *Ann. Geophys.* **32**, 207–226.

Lliboutry, L. (1978). *Ann. Geophys.* **34**, 147–162.

Lliboutry, L., Arnao, B. M., Pantri, A., and Schneider, B. (1977). *J. Glaciol.* **18**, 239–254.

McCall, J. G. (1952). *J. Glaciol.* **2**, 122–131.

Mathews, W. H. (1959). *J. Glaciol.* **3**, 448–454.

Mathews, W. H. (1964). *J. Glaciol.* **5**, 235–240.

Matsuda, M., Wakahama, G., and Budd, W. F. (1976). *Teion Kagaku, Butsuri-Hen* **34**, 163–171.

Mayo, L. R., Meier, M. F., and Tangborn, W. V. (1972). *J. Glaciol.* **11**, 3–14.

Meier, M. F. (1960). *U.S., Geol. Surv., Prof. Pap.* **351**, 1–70.

Meier, M. F. (1972). *IAHS* **107**, 353–370.

Meier, M. F., and Post, A. (1969). *Can. J. Earth Sci.* **6**, 806–817.

Meier, M. F., and Tangborn, W. V. (1965). *J. Glaciol.* **5**, 547–566.

Meier, M. F., Kamb, W. B., Allen, C. R., and Sharp, R. P. (1974). *J. Glaciol.* **13**, 187–212.

Morland, L. W. (1976). *J. Glaciol.* **17**, 463–477.

Morland, L. W., and Morris, E. M. (1977). *J. Glaciol.* **18**, 67–75.
Morris, E. M. (1976). *J. Glaciol.* **17**, 79–98.
Nielsen, L. E. (1969). *Can. J. Earth Sci.* **6**, 955–961.
Nye, J. F. (1951). *Proc. R. Soc. London, Ser. A* **207**, 554–572.
Nye, J. F. (1952). *J. Glaciol.* **2**, 82–93.
Nye, J. F. (1953). *Proc. R. Soc. London Ser. A* **219**, 477–489.
Nye, J. F. (1957). *Proc. R. Soc. London Ser. A* **239**, 113–133.
Nye, J. F. (1959). *J. Glaciol.* **3**, 387–408.
Nye, J. F. (1960). *Proc. R. Soc. London Ser. A* **256**, 559–584.
Nye, J. F. (1963). *Proc. R. Soc. London Ser. A* **275**, 87–112.
Nye, J. F. (1965a). *J. Glaciol.* **5**, 589–607.
Nye, J. F. (1965b). *J. Glaciol.* **5**, 661–690.
Nye, J. F. (1967). *J. Glaciol.* **6**, 695–716.
Nye, J. F. (1969). *Proc. R. Soc. London Ser. A* **311**, 445–467.
Nye, J. F. (1970). *Proc. R. Soc. London Ser. A* **315**, 381–403.
Nye, J. F. (1973a). *In* "Physics and Chemistry of Ice" (E. Whalley, S. J. Jones, and L. W. Gold, eds.), pp. 387–399. Royal Society of Canada, Ottawa.
Nye, J. F. (1973b). *IASH* **95**, 189–194.
Nye, J. F. (1976). *J. Glaciol.* **17**, 181–207.
Nye, J. F., and Frank, F. C. (1973). *IASH* **95**, 157–161.
Palmer, A. C. (1972). *J. Glaciol.* **11**, 65–72.
Paterson, W. S. B. (1969). "The Physics of Glaciers." Pergamon, Oxford.
Paterson, W. S. B., and Savage, J. C. (1963). *J. Geophys. Res.* **68**, 4537–4543.
Peterson, D. N. (1970). *Inst. Polar Stud., Rep.* **36**, 1–161 (unpublished).
Post, A. (1969). *J. Glaciol.* **8**, 229–240.
Post, A., and Mayo, L. R. (1971). *U.S., Geol. Surv., Atlas* **HA-455.**
Rasmussen, L. A., and Campbell, W. J. (1973). *J. Glaciol.* **12**, 361–374.
Raymond, C. F. (1971). *J. Glaciol.* **10**, 55–84.
Raymond, C. F. (1973). *J. Glaciol.* **12**, 19–44.
Raymond, C. F. (1974). *J. Glaciol.* **13**, 141–143.
Raymond, C. F. (1976). *J. Glaciol.* **16**, 159–171.
Raymond, C. F. (1978). *In* "Rockslides and Avalanches" (B. Voight ed.), Vol. 1, pp. 793–833. Am. Elsevier, New York.
Raymond, C. F., and Harrison, W. D. (1975). *J. Glaciol.* **14**, 213–233.
Reynaud, L. (1973). *J. Glaciol.* **12**, 251–258.
Reynaud, L. (1977). *Z. Gletscherkd. Glazialgeol.* **13**, 155–166.
Rigsby, G. P. (1958). *J. Glaciol.* **3**, 273–278.
Rigsby, G. P. (1960). *J. Glaciol.* **3**, 589–606.
Rigsby, G. P. (1968). *J. Glaciol.* **7**, 233–251.
Robin, G. de Q. (1967). *Nature (London)* **215**, 1029–1032.
Robin, G. de Q. (1969). *Can. J. Earth Sci.* **6**, 919–928.
Robin, G. de Q. (1976). *J. Glaciol.* **16**, 183–196.
Robin, G. de Q., and Weertman, J. (1973). *J. Glaciol.* **12**, 3–18.
Röthlisberger, H. (1972). *J. Glaciol.* **11**, 177–203.
Röthlisberger, H. (1976). *J. Glaciol.* **16**, 309–310 (abstr. only).
Savage, J. C., and Paterson, W. S. B. (1963). *J. Geophys. Res.* **68**, 4521–4536.
Schytt, V. (1969). *Can. J. Earth Sci.* **6**, 867–873.
Shreve, R. L., and Sharp, R. P. (1970). *J. Glaciol.* **9**, 65–86.
Shumskiy, P. A. (1967). *Phys. Snow Ice, Conf., Proc., 1966* Vol. 1, Part 1, pp. 371–384.
Shumskiy, P. A. (1969). "Dinamicheskaya Glastsiologiya I." Itogi Nauki, Moscow.

Swithinbank, C. W. (1963). *Science* **141**, 523–524.

Tangborn, W. V., Krimmel, R. M., and Meier, M. F. (1975). *IAHS* **104**, 185–196.

Theakstone, W. H. (1967). *J. Glaciol.* **6**, 805–816.

Tyndall, J., F. R. S. (1896). "The Glaciers of the Alps." Longmans, Green, New York.

Untersteiner, N., and Nye, J. F. (1968). *J. Glaciol.* **7**, 205–214.

Vallon, M., Petit, J. R., and Fabre, B. (1976). *J. Glaciol.* **17**, 13–28.

Vivian, R. (1975). "Les Glaciers des Alpes Occidentales." Imprimerie Allier, Grenoble.

Vivian, R., and Bocquet, G. (1973). *J. Glaciol.* **12**, 439–451.

Weertman, J. (1957). *J. Glaciol.* **3**, 33–38.

Weertman, J. (1964). *J. Glaciol.* **5**, 287–303.

Weertman, J. (1969). *Can. J. Earth Sci.* **6**, 929–942.

Weertman, J. (1970). *Rev. Geophys. Space Phys.* **8**, 145–168.

Weertman, J. (1972). *Rev. Geophys. Space Phys.* **10**, 287–333.

Weertman, J. (1973a). *IASH* **95**, 139–145.

Weertman, J. (1973b). *In* "Physics and Chemistry of Ice" (E. Whalley, S. J. Jones, and L. W. Gold, eds.), pp. 320–337. Royal Society of Canada, Ottawa.

3 SEA ICE GROWTH, DRIFT, AND DECAY

W. D. Hibler III

U.S. Army Cold Regions Research and Engineering Laboratory
Hanover, New Hampshire

I. INTRODUCTION

Large areas of the polar oceans are covered with a thin, variable thickness layer of ice formed from freezing seawater. The growth, drift, and decay of this ice cover are intrinsically related to both the dynamic and thermodynamic variations in the atmosphere and ocean. In particular, rates of growth and decay of the ice depend on the thermal characteristics of the atmosphere and ocean, as well as on the distribution of ice thicknesses. The thicknesses are, in turn, modified by the ice transport and deformation patterns which are driven by atmospheric and oceanic stresses. In addition the sea ice itself can substantially change the sea–air heat and momentum exchanges and hence modify the behavior of the atmosphere and ocean.

DYNAMICS OF SNOW AND ICE MASSES

Determining the dominant mechanisms of this complex coupled system has been the goal of considerable empirical and theoretical research. In the empirical studies, various sea ice characteristics have been classified and correlated with atmosphere and oceanic variations. Such studies have identified the general characteristics of sea ice, and determined the general relationship of such ice characteristics to atmosphere and ocean variables. In the theoretical studies, attempts have been made to deduce the behavior of sea ice from a quantitative treatment of various physical processes. In such studies numerical models have helped identify the effect of different mechanisms on the behavior of sea ice.

In the following three sections progress in understanding the growth drift and decay of sea ice is reviewed. Section II outlines the general characteristics of sea ice, emphasizing empirical data studies, both to portray an overall view of the sea ice problem and to identify areas where knowledge of the physical processes is most essential. Section III examines the physics of sea ice and reviews a variety of observational, theoretical, and numerical studies that yield insight into the various physical mechanisms. Section IV discusses large-scale numerical simulation models for sea ice. Finally, in the concluding remarks some comments on the role of sea ice in climate are presented.

II. GENERAL CHARACTERISTICS OF SEA ICE

In examining the general characteristics of sea ice, most of the emphasis in this paper will be placed on arctic sea ice because it has been examined in more detail. However, where appropriate, certain aspects of antarctic sea ice will also be discussed.

A. Temporal and Spatial Variations in Ice Cover Extent

Probably the most obvious manifestation of sea ice growth and decay is the variation in the ice concentration and extent. This effect is particularly pronounced in the Antarctic where the ice cover is reduced by as much as 75% in the summer. Figure 1 shows estimated maximum and minimum extents of sea ice for the Arctic and Antarctic. The average seasonal variations of ice conditions in different portions of the Arctic are reviewed in more detail in Vowinckel and Orvig (1970). A notable feature of ice located in the Arctic Basin is the small variation in its boundary. This feature is caused by the landlocked nature of the Arctic Basin, which restricts ice expansion in winter, and contrasts with the Antarctic where the ice is unconstrained. Another salient feature of the arctic ice cover is

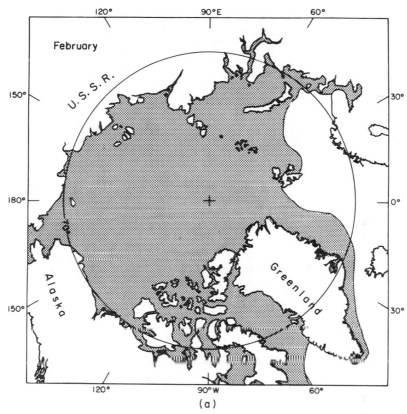

Fig. 1.   Arctic [(a) and (b)] and antarctic [(c) and (d)] ice extents based on 75% concentration limits. The arctic extents are climatological means from H. O. Pub. No. 705, U.S. Naval Oceanographic Office. The antarctic extents are for 1975 as compiled by Navy Fleet Weather Facility, Suitland, Md.

the presence of ice off the East Coast of Greenland, even during times of minimum ice extent. This is largely caused by ice drift patterns, which make this area the primary locus for the ice flow out of the Arctic Basin into the North Atlantic.

With regard to the magnitude of sea ice coverage, Nazarov (1963) estimates that antarctic sea ice at its maximum extent covers about $25 \cdot 10^6$ km², while Walsh and Johnson (1979) estimate the maximum arctic ice cover to be about $15 \cdot 10^6$ km². Walsh and Johnson (1979) also find an average seasonal variation of 50% in the Arctic with the minimum extent in August and maximum extent in February. For the Antarctic, Treshnikov (1967) estimates a 75% variation with minimum extent occurring in March and the maximum extent in September.

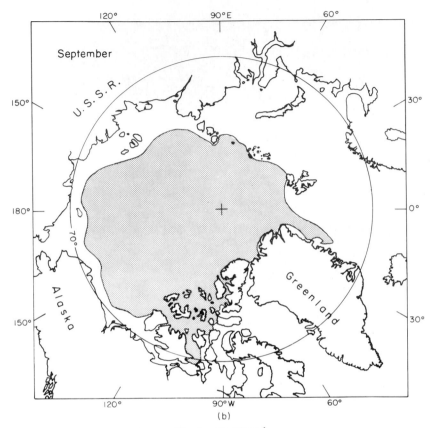

Fig. 1. *Continued*

The extent of the ice can vary substantially from year to year. Using arctic data from 1968 to 1976, Kukla (1978) finds a substantial increase in ice coverage between 1971 and 1973 followed by a gradual decrease through 1975. In a more extensive study Walsh and Johnson (1979) find the average extent of arctic ice to have generally increased over the last 25 yr (see Fig. 2). While detailed analyses are not complete, this trend seems to agree with decreases in arctic surface temperatures between 1954 and 1975 [Walsh (1977)]. Although based on considerably fewer data, similar conclusions about correlations between temperatures and ice extent were reached by Budd (1975) for the antarctic ice cover.

Empirical studies of the extent of ice in the North Atlantic have been made by several authors [Vinje (1976); Skov (1970); Sanderson (1975)]. The studies by Vinje emphasized the variation of the ice edge location from year to year. Vinje plotted the extreme extent of the ice for each

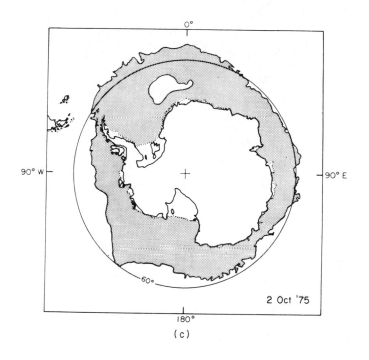

(c)

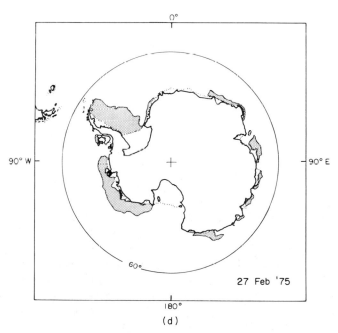

(d)

Fig. 1. *Continued*

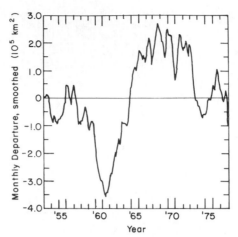

Fig. 2. Time series of the departures from monthly means of the area covered by arctic sea ice. The record has been smoothed with a 24-month running mean. [From Walsh and Johnson (1979).]

month from 1966 to 1974 along with the 1975 ice border. The combined results for February and August are reproduced in Fig. 3. A particularly prominent feature of this figure is the ice-free region to the west of Spitsbergen. This region is caused by the warm West Spitsbergen current that is the main source of heat entering the Arctic Basin [Aagaard and Greisman (1975)]. This current combines with the cold East Greenland current passing southward to create the gyrelike structure especially apparent in the maximum February ice extent in Fig. 3.

The primary mechanisms suggested for the variations in the extent of ice on the Greenland Sea have been the atmospheric wind and temperature [Sanderson (1975)] and the oceanic heat transport [Skov (1970)]. The correlations by Skov (1970) tend to indicate that northward oceanic heat transport is the primary factor responsible for year to year variations. However, for short-term (over a month or so) variations, wind and temperature effects appear critical [Sanderson (1975)]. Sanderson also examines correlations between different regions from 1966 to 1974 and suggests that anticorrelations (i.e., one region receding while another expands) may be due to the development of certain long-wave patterns in the upper level westerlies. This type of explanation has also been advanced by Kelly (1978), who suggests that eastern arctic ice extent may correlate with the position of the Icelandic low.

Recent studies of ice extents using empirical orthogonal functions have also found this type of anticorrelation [Walsh and Johnson (1979)]. Walsh and Johnson compiled and analyzed arctic sea ice data spanning the 25-yr

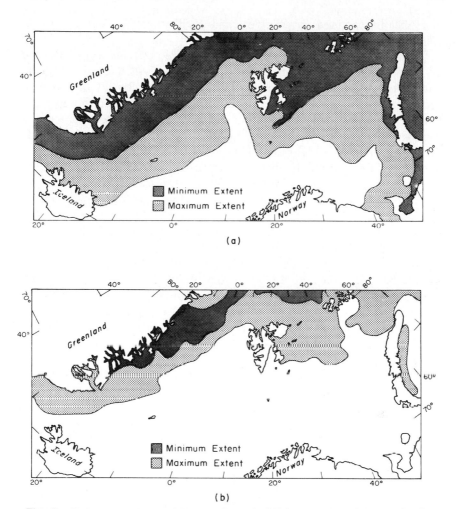

Fig. 3. Extreme sea ice conditions at the end of February (a), and August (b), from 1966 to 1975. [Redrawn from Vinje (1976).] Maximum and minimum extremes (for 1966–1974) are denoted by solid lines. The border is defined by a concentration of $\frac{3}{8}$ or less ice coverage.

period 1953–1977, reducing the data to digitized monthly grids covering the polar cap. The dominant fluctuation exhibited by these data was an asymmetric spatial change in ice extent with the ice in the North Atlantic having an anomalous change in extent opposite in sign to the anomalous change over the remainder of the polar cap. Similar asymmetric changes have been reported on antarctic sea ice (although only a few years' data have been examined) by Ackley and Keliher (1976) and Zwally and Gloersen (1977).

Another aspect of the ice extent is the degree of temporal persistence. To examine this effect, Lemke *et al.* (1980) has divided ice coverage data into 36 equal longitudinal segments, and analyzed monthly data over the 11-yr period between January 1966 and December 1976. In these data the strongest annual changes in extent were found in the regions west and east of Greenland while the smallest changes were found in the Beaufort Sea. The ice anomaly variance also exhibited a variation similar to the annual cycle, but was only about one-third as large. The persistence of the data was analyzed using a red noise model (first-order Markov process). Most of the segments were found to fit this model with typical persistence times of about 3 months. Deviations from the independent red noise model were attributed to significant amounts of zonal ice advection.

While primarily limited to summer variation, the ice extent off the North Slope of the U.S. and Canada has been studied by a number of investigators [e.g., Barnett (1979); Walsh (1979); Rogers (1978)]. As in the North Atlantic, this ice edge can vary substantially from year to year. In a bad navigation summer (i.e., a large amount of ice) the ice can remain right up to the coast. During a relatively ice-free summer, on the other hand, the ocean can be relatively ice free for several hundred kilometers off the coast.

Barnett (1979), in an attempt to understand the historical variation of the North Slope summer ice edge, compiled statistics going back about 20 yr. On the basis of these statistics he constructed a historical severity index. A component of this index that correlates well with the overall index is the distance to the ice edge from Point Barrow. This parameter is plotted as a time series in Fig. 4. Probably the salient characteristic of Barnett's study was the discovery of a quasi-periodic oscillation with about a 5-yr frequency. Barnett found this severity index correlated well with atmospheric pressure changes in the central basin in April. These pressure changes, in turn, were related to the location of the Siberian high. However, Barnett was not able to draw any conclusions on the cause of this correlation.

To help determine the physical mechanisms responsible for North Slope ice conditions, Rogers (1978), Wendler (1973), and Walsh (1970) have done empirical studies. Wendler argued that winds were the dominant factor in determining the variation of the near-shore concentration. In particular, he showed ice conditions tended to depend on surface wind directions derived from 5-day, mean sea level pressure maps. Rogers studied the interannual variability of the ice extent in the Beaufort Sea. He found that air temperature, in the form of thawing degree days, was

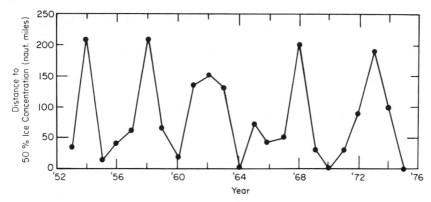

Fig. 4. Record of ice conditions off the North Slope of Alaska. [Drawn from data in Barnett (1979).] The ordinate is the distance in nautical miles from Point Barrow northward to the boundary of 50% ice concentration on September 15.

the parameter that correlated most highly with the summertime ice margin distance. Rogers also found that mild summers were often associated with pressure anomalies that resulted in more southerly surface winds advecting warmer air. In Walsh's study, the emphasis was on the predictability of the North Slope ice extent. To study this Walsh carried out empirical orthogonal function expansions for ice extent, surface atmospheric pressure, and surface air temperatures. The time series of the coefficients of the dominant eigenvectors were then correlated. While the long-range temperature-based predictions of summer ice extent showed some accuracy, Walsh generally found the predictability to be limited to one or two months, with persistence being the dominant factor.

B. Ice Drift Characteristics

The general pattern of the ice cover circulation in the Arctic Basin has been known for some time. The general characteristics (shown in Fig. 5) consist of a gyre in the Beaufort Sea and a transpolar drift stream. The transpolar drift stream carries ice from the East Siberian Sea (and even from the Bering Strait), passing it across the North Pole and down the East Coast of Greenland. Because of the recirculating character of the Beaufort gyre, it is possible for ice that survives the summer melt to stay in the gyre for a large number of years [e.g. Koerner (1973)]. Consequently some of the thickest multiyear ice is found in this region.

Drifting stations in the Arctic typically have been found to drift at a rate of about 6 km day^{-1} [Reed and Campbell (1962); Dunbar and Wittmann (1963)], but drift rates can vary from day to day and from year to

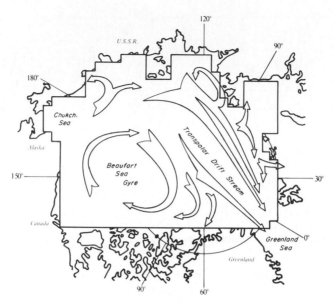

Fig. 5. The pattern of mean ice drift in the Arctic Ocean. [Adapted from Gordienko (1958).]

year. One-day drifts of over 20 km have been observed [Thorndike and Colony (1979)]. It is also possible for the ice to remain effectively motionless for several weeks. Such behavior was documented by Thorndike and Colony (1979) in the spring of 1976 in the Beaufort Sea. However, except for ice very near shore, this motionless behavior appears to be the exception rather than the rule.

To give some idea of the general drift rates and variability of overall drift patterns from year to year, a number of mean monthly drifting station positions are plotted in Fig. 6. Two notable features illustrated by this figure are the presence of substantial drift (even in winter) parallel to the Alaskan coast and a relatively small component of motion perpendicular to the Canadian Archipelago near the pole. A feature of drift tracks not shown in this figure is the meandering nature of the drift, with week-long reversals in direction being common. Additional perspective on the drift rates in the gyre may be gained by noting that it took the ice island T-3 about 10 yr to make a full circuit [Dunbar and Wittmann (1963)]. As drifting stations approach the Greenland–Spitsbergen passage, drift rates begin to increase and meandering decreases. This phenomenon continues after the ice exits from the Basin with the drift rates being as large as 25 km day $^{-1}$ in the East Greenland region [Vinje (1976)].

While valuable for determining the general characteristics of ice drift,

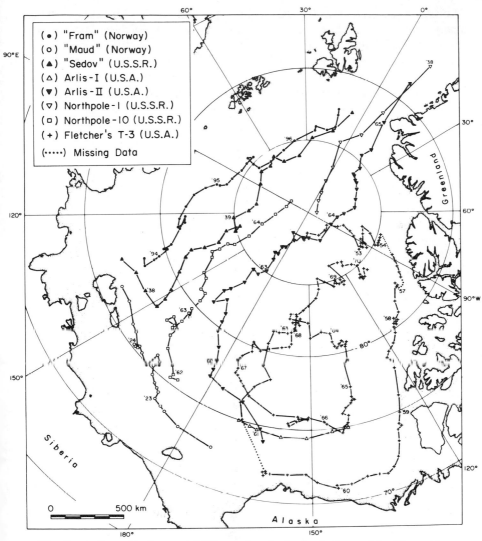

Fig. 6. Drift tracks of several drifting stations in the Arctic Ocean. Monthly positions are plotted with the year of each January 1st position listed. [From Hastings (1971).]

the ice station data are not accurate enough to allow detailed examination of the temporal and spatial variations of ice drift. More recent measurements [Thorndike and Colony (1979)] using satellite navigation equipment and laser surveying equipment [Hibler *et al.* (1974a)] have supplied a more complete description of the temporal and spatial variations in ice drift. These data show that while the ice velocity has the greatest variations at low frequencies (less than about $\frac{1}{5}$ cycle day^{-1}) significant velocity perturbations do occur at higher frequencies. These high-frequency motions are dominated by a peak at about 2 cycles day $^{-1}$. This peak is prominent during the summer months, but is substantially suppressed during the winter months. Hunkins (1967) and McPhee (1978) have explained these 12-h oscillations as inertial motions arising from the balance between inertia and the Coriolis force. Typical seasonal velocity spectra (Fig. 7) clearly show such oscillations in the summer and (to a much lesser degree) in winter.

In addition to drift, the pack ice regularly undergoes deformation. Measurements of deformation on scales of 10–20 km [Hibler *et al.* (1974a)] and on scales of 100–500 km [Thorndike (1973); Thorndike and Colony (1979)] show typical magnitudes of the strain rate to be about 1% day $^{-1}$. These strain rates generally have little correlation with the ice velocity. A typical month-long record of such measurements is shown in Fig. 8. This figure also shows how the drift speed can fluctuate widely over several days.

On scales smaller than about 20 km the strain rates become very erratic since motion will often be due to a single lead. Because of the paucity of flaws in the pack ice (especially in winter) across which motion can occur, it becomes difficult to adequately define the deformation except statistically [Nye (1973a)]. Typical observed deviations of the ice velocity field from a linear spatial variation have varied from 0.05 km day^{-1} [Hibler *et al.* (1974a)] to about 1.0 km day^{-1} [Thorndike and Colony (1978)].

Less is known about drift in the Antarctic, although some information has been gathered from ships caught in the ice. A summary of drift data for the Weddell Sea based on ship tracks and iceberg motion is given

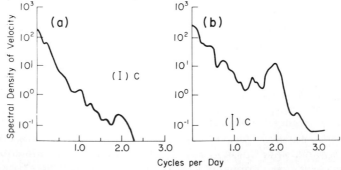

Fig. 7. Seasonal spectra of ice velocity measurements for winter (a) and summer (b). [From Thorndike and Colony (1979).] The ice drift data were obtained by satellite navigation procedures in the Beaufort Sea in the vicinity of 74° N 145° W.

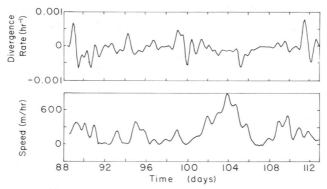

Fig. 8. Deformation rate and drift speed in the vicinity of an ice station located at approximately 75° N and 148° W in April 1972. The deformation rates were estimated from a strain array approximately 20 km in diameter. [Reproduced from Hibler *et al.* (1974a), by permission of the International Glaciological Society)]

in Ackley (1980). These data show average drift rates of about 4 km day $^{-1}$, which are generally commensurate with arctic values.

A particularly important facet of the ice drift in the Arctic is the formation of a relatively discontinuous shear zone off the North Slope of Alaska and the Canadian Arctic. Basically the winter drift rates very near shore are often substantial in this region. This introduces a discontinuity in the velocity fields very near the coast. Analysis of this shear zone region suggests the shearing nature is sporadic and may be preceded by a reduction in onshore stress caused by offshore motion. A March 1973 Landsat study by Hibler *et al.* (1974b) off Barrow, for example, showed the development of strong shearing motion to be initially accompanied by an offshore motion of the pack. Initial and final location of a series of ice points over a one-day interval from the study are shown in Fig. 9.

Drift measurements taken further to the east in the same year indicate a shear zone was also well defined there [Reimnitz *et al.* (1978)]. In particular Reimnitz *et al.* (1978) find a relatively persistent shearing motion off Prudhoe Bay from the middle of March to mid-May. This motion tended to occur on the seaward side of a heavily ridged and partially grounded zone approximately 20 km wide. This zone, which they termed a "stamukhi" zone, tends to straddle the 10–30-m depth range. An important observation of both studies [Hibler (1974); Reimnitz *et al.* (1978)] was that the seaward ice next to the shear zone often moves westward (3–10 km day $^{-1}$) at a faster rate than the ice farther from the coast. Thus the ice motion takes on the characteristics of a large rotating cohesive wheel with slippage at the boundary.

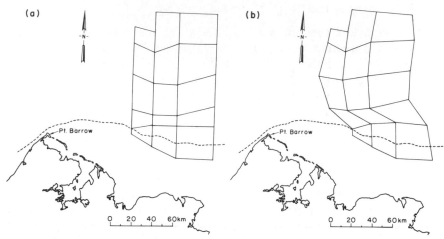

Fig. 9. Observed ice motion over a one-day interval during shear activity in the near-shore region: (a) 19 March 1973; (b) 20 March 1973. The observations were made by LANDSAT imagery and the dashed line marks the boundary between the shorefast ice and the pack ice. [From Hibler *et al.* (1974b).]

More recent measurements, however, indicate that a well-defined shear zone does not always exist. During the springs of 1975–1976 and 1976–1977, detailed measurements of the near-shore ice regions up to about 30 km from shore were made near Prudhoe Bay [Weeks *et al.* (1977); Tucker *et al.* (1979a)]. These measurements show a near-shore ice cover that undergoes few large motions (typically one large deformation event per month). During such an event the pack ice may move several kilometers away from the near-shore ice for a day or so and then return. However, in the absence of such large events, the ice did exhibit high-frequency fluctuations of about 10 m.

Explanations of this complex near-shore behavior emphasize the ability of grounded ice to anchor the near-shore ice [Kovacs and Mellor (1974)]. Another factor is the ''shading'' effect caused by boundary irregularities [Reimnitz *et al.* (1978)]. This shading effect can cause the zone of shearing to be relatively regular even though the coastline is irregular. However, in terms of the time evolution of the near-shore ice, both Kovacs and Mellor (1974) and Reimnitz *et al.* (1978) emphasize that strengthening of the near-shore ice due to ridging and general ice buildup can cause the shear effects to move farther seaward. Because of this complex evolution with time, it seems likely that the near-shore ice velocities are strongly affected by the time and magnitude of earlier ice drift events.

C. Ice Thickness and Ridging Characteristics

Due to almost constant motion, formation, and decay, sea ice has a highly variable thickness, which is significantly affected by the amount of ridging. The thin ice plays a particularly important role because of its low strength and high growth rates. Probably the most valuable data for estimating ice thickness characteristics are submarine sonar profiles of the underside of the pack ice [Swithinbank (1972); Williams *et al.* (1975); Le-Shack *et al.* (1971); Kozo and Tucker (1974); Wadhams (1977, 1980a); Wadhams and Horne (1978)]. Because about 90% of the ice thickness is underwater, these profiles can be used to estimate the distribution and spatial variability of thickness. While considerable data have been taken, few have been analyzed in detail. However, in the Arctic basin a sufficient amount of data have been analyzed to infer the general shape of the thickness distribution. The most complete analysis of submarine sonar data to date has been carried out by Wadhams (1980a) using data taken by the HMS Sovereign near the North Pole. These data consisted of almost 4000 km of sonar profiles taken from 2 October to 5 November 1976. Also, along part of the track simultaneous laser profile data were taken, thus allowing comparison of top and bottom roughness characteristics.

Some typical sonar thickness distributions are shown in Fig. 10. A salient characteristic of the ice thickness distribution illustrated by Fig. 10a is the marked bimodal shape. Basically the ice can be thought of as a combination of a thin ice distribution (up to about 2 m) plus a thick ice distribution. The thin ice mode consists of young first-year ice while the thicker mode is composed of level multiyear ice and ridged ice (both first-year and multiyear). In the early fall, the thin ice consists mostly of ice formed after the cessation of summer melt conditions. However, ice can be added to the thin mode throughout the winter by the creation of leads via diverging deformation. The presence of substantial amounts of ice less than 1 m thick is especially apparent in the data in Fig. 10 taken by the USS Gurnard in April 1976 in the Beaufort Sea [e.g., Wadhams and Horne (1978)].

While the unridged young first-year ice is normally quite level, it should be noted that the undeformed thicker multiyear ice is not. The thicker multiyear ice tends to have a characteristic undulating surface which varies on about a 30-m horizontal scale. Detailed measurements of the thickness and roughness variation of "level" multiyear ice [Hibler *et al.* (1972a); Ackley *et al.* (1976)] show typical standard deviations of thickness of about 0.6 m. The mean thicknesses were typically of the order of 2.5 m. The reason for this variability is the presence of summer melt ponds which create low spots (sometimes completely melting

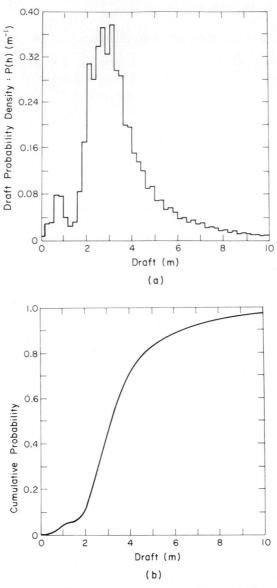

Fig. 10. Distributions of ice draft taken from two 200-km sections of submarine track intersecting at 72°50′ N and 144°17′ W in April 1976. [Data from Wadhams and Horne (1978).] (a) The probability density function of ice draft; (b) The cumulative probability.

through the ice). Concomitant with this thickness variation is a salinity and density variation [Ackley *et al.* (1976)] with the thicker ice being less salty and less dense.

Very little methodical analysis of the large-scale spatial variations of thickness has been done. However, L. A. LeShack (private communication) has analyzed selected portions of submarine tracks over the Arctic Basin. By averaging the mean thickness of the ice-covered portion of different submarine tracks (regardless of season) approximate contours of mean ice thickness for part of the Arctic Basin can be constructed (see Fig. 11). The dominant characteristic of this observed variation is a high

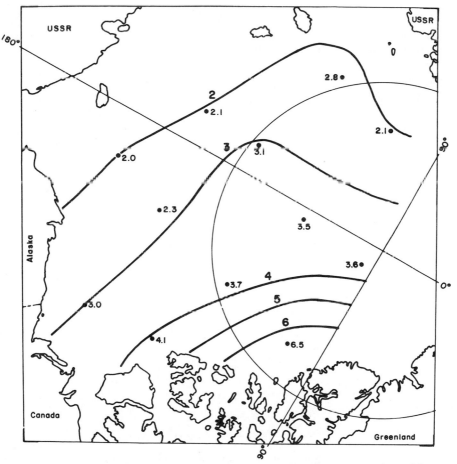

Fig. 11.   Contours (in meters) of observed mean ice thickness values estimated from submarine sonar data. [LeShack (private communication).]

thickness off the Canadian Archipelago and a lesser thickness off the Siberian Coast and North Slope.

Because they dominate the thick portion of the thickness distribution, ridges are particularly important. While normally accounting for about 25% of the ice mass, ridges can comprise more than 50% of the ice in heavily ridged regions [Williams *et al.* (1975)]. Field observations [Kovacs (1972); Weeks *et al.* (1971)] show the cross sections of ridge keels and sails to be approximately triangular in shape. For multiyear ridges the keel depth is consistently about 3.2 times the sail height. For first-year ridges the ratio tends to be larger (~4.5) and more variable [Kovacs and Mellor (1974)].

In an attempt to classify pressure ridges, statistical distribution models have been developed [Hibler *et al.* (1972b); Wadhams (1976)] and used to examine the spatial and temporal variability of ridges [Hibler *et al.* (1974c); Tucker *et al.* (1979b); Wadhams (1978)]. Specifically, Hibler *et al.* (1972b) derived height and spacing distributions using general assumptions of randomness and assuming ridge keels and sails to have triangular cross sections. Later modifications to these distributions based on the finite spacing between ridges have been made [Lowry and Wadhams (1979)]. These distributions have been found to fit ridge keels well. They also work well for ridge sail heights except at higher ridge heights [Tucker *et al.* (1979b); Wadhams (1976)] where a simpler height distribution proposed by Wadhams (1976) performs better.

Studies of the spatial and temporal variations of ridging in the West

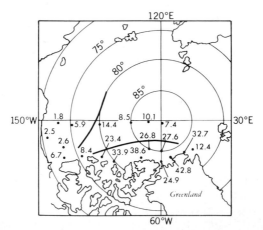

Fig. 12. Ridging intensity observations and contours obtained from laser profilometer measurements taken during February 1973. The ridging intensity is approximately proportional to the volume of deformed ice per unit area and thus has units of meters. [From Hibler *et al.* (1974c).]

Arctic Basin [Hibler *et al.* (1974c)] generally show the spatial variability of ridging to correlate well with the thickness contours shown previously. Figure 12, for example, shows ridging intensity contours for February 1973 based on laser profiles of the topside of the arctic ice cover. In addition to these basin wide variations, near-shore ridge studies [Wadhams (1976); Tucker *et al.* (1979b)] off the North Slope of Alaska and Canada show a buildup of ridging near the coast which decays several hundred kilometers off shore. Field studies show this region to be highly deformed with substantial amounts of rubble pile-up [Kovacs (1976)]. While these relative geographical ridge characteristics remain more or less constant, the magnitude of the ridging can vary from year to year, especially near shore. In particular, Hibler *et al.* (1974c) found the ridging intensity (approximately equivalent to the mean thickness of ridged ice) off the Canadian Archipelago to be twice as large in the winter of 1972 as in 1971. Other regions, however, exhibited less change. Also, Tucker *et al.* (1979b) find the near-shore ridging intensity off Prudhoe Bay to be substantially less in the winter of 1978 than in 1977.

D. Mass Balance Characteristics

A useful way to analyze the various aspects of sea-ice growth and decay is to consider the ice cover to be in a seasonal equilibrium. With this assumption the magnitude of the various source and sink terms can then be estimated. Such studies have been made from thermodynamic considerations [Untersteiner (1961)] and by using data taken while crossing the ice by dogsled over a period of a year and a half [Koerner (1973)]. To supplement these ground data, Koerner used a spatter of submarine data, and drift, deformation, and thermodynamic estimates.

The ice balance B of the Arctic Ocean can be expressed by

$$B = C - A - G, \tag{1}$$

where C and A are the accumulation and ablation of ice and G is the amount of ice that drifts out of the Arctic Ocean. Among the average statistics estimated by Koerner (1973) is a total ice growth of about 1.1 m of ice per year, which is dissipated by about 0.6 m of ablation, and an equivalent of 0.5 m per year of outflow. While most of this ablation occurs due to summer melt, small amounts can also occur in winter [Untersteiner (1961)]. In order to identify the sources of summer melt more precisely, Koerner studied a localized test area in some detail. He found the greatest ablation occurs in new ridges and, to a lesser degree, in melt ponds (these areas had melt rates 1.7–2.6 times as large as level multiyear ice). Old multiyear hummocks or ridges, on the other hand, ablated more slowly (by a factor of 0.6) than level multiyear ice.

Koerner also found that about 20% of the ice production ends up in ridges, that annual snow accumulation was typically 0.1 m or less and decreased toward the center of the Arctic Basin, and that multiyear ice covered 70–80% of the area surveyed. To examine ice thickness characteristics, Koerner made a variety of ice thickness measurements in both first-year and multiyear ice. For first-year ice he found a maximum thickness of 2.04 m at the end of the growth season. For multiyear ice he found a wide range of thicknesses that, while they peaked at 2.5 m, demonstrated no well-defined thickness mode. Also, he was unable to discern any significant old ice thickness differences between the Pacific Gyre and the transpolar drift (see Fig. 5).

One of the more important features of the mass balance is outflow of ice from the Basin (felt to be dominated by outflow through the Greenland Spitsbergen passage). Estimates of this parameter by other authors [e.g., Vowinckel and Orvig (1970)] are generally lower than Koerner's, but are usually in the neighborhood of 0.1 Sverdrup ($Sv = 10^6 m^3 sec^{-1}$). While this is not large in terms of oceanic transport, it does represent a substantial heat flux because of the latent heat of ice. Estimates of the heat budget of the Arctic Ocean [Aagaard and Greisman (1975)] indicate that this flux accounts for about one-third of the total oceanic advective heat budget. This ice outflow is also undoubtedly an important factor in the ice edge conditions in the Greenland Sea.

Due to the paucity of data, less is known about the mass balance characteristics of the antarctic pack ice. However, several aspects of the mass balance of antarctic pack ice in the Weddell Sea have been discussed by Ackley (1980), who concluded that summer ablation inside the edge of the pack ice is much smaller in the Antarctic than in the Arctic. Also, in a theoretical study Gordon and Taylor (1975) argue that the effects of wind stress can be particularly effective in expanding the antarctic ice cover. Based on these studies it is possible that the decay of the ice cover in the Antarctic may be carried out predominantly at the ice edge, with ice drift transporting ice to this region to be melted.

E. Physical Properties of Sea Ice

Since it is formed from sea water, sea ice tends to be a complex mixture of saltwater (brine) and salt-free ice. This has profound ramifications for the thermal and mechanical properties of sea ice, and greatly complicates its physical properties. While a review of the complex fine structure is outside the scope of this paper, it is useful to give a brief discussion of sea-ice salinity characteristics. These salinity characteristics are particularly relevant to the thermodynamic behavior, as will be discussed later.

For more comprehensive reviews, interested readers are referred to
Weeks (1968) and Weeks and Gow (1978) for general physical properties,
to Schwarz and Weeks (1977) for small-scale mechanical properties, and
to Kohnen (1976) for some typical field measurements of the physical
properties.

As sea ice freezes, salt is entrapped in the ice in the form of liquid
pockets of brine. As the freezing continues or melt begins, the amount of
brine entrapped is gradually reduced. The primary brine drainage mechanisms are expulsion and flushing [Untersteiner (1968)] and gravity
drainage [Lake and Lewis (1970)]. Brine expulsion is primarily a thermal
process, while flushing occurs when surface meltwater percolates through
the ice. Gravity drainage refers to the movement of the heavier brine due
to the gravitational force. These drainage processes tend to create characteristic variations of salinity (salinity profiles) versus depth with generally
lower salinity at the top surface of the ice.

While the mechanisms of brine drainage are not well understood, Cox
and Weeks (1974) have demonstrated that, at least in the Arctic, the vertically averaged salinity is closely correlated with the thickness and thermal
character of sea ice. They made a variety of salinity measurements in multiyear ice during the spring of 1972, and combined these with other existing data. Figure 13 shows a plot of salinity versus thickness for cold ice

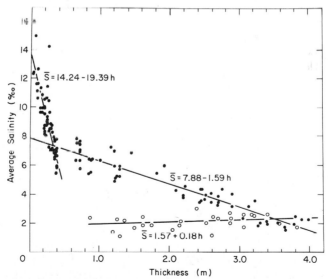

Fig. 13. Average salinity of sea ice as a function of ice thickness for cold sea ice sampled during the growth season, and warm sea ice sampled during or at the end of the melt
season: (solid circles) cold sea ice, (open circles) warm sea ice. [Redrawn from Cox and
Weeks (1974) by permission of the International Glaciological Society.]

and warm ice. As can be seen, the average salinity drops rapidly as the thickness approaches 0.4 m, and then decreases more slowly for cold ice. On the basis of laboratory experiments, Cox and Weeks (1975) suggest the transition at about 0.4 m is due to a change in the dominant brine drainage mechanisms from expulsion to gravity drainage. For warm ice the situation is different and, as shown in Fig. 13, there are lower average salinities and little change with thickness.

After ice survives a melt season its salinity profile takes on a characteristic appearance. This is made possible by a "steady-state" seasonal growth cycle of old ice, which consists of ablation in the summer and ice replacement in the winter. The multiyear salinity profile normally assumed by researchers was developed by Schwarzacher (1959) on the basis of 40 cores in multiyear ice. This profile is shown in Fig. 14. However, while such a standard profile appears reasonable for hummocked ice, the study by Cox and Weeks indicates that different profiles can occur in low spots of multiyear ice where melt ponds have formed in summer (Fig. 14). This fact again emphasizes that the characteristics of old "level" ice are not horizontally homogeneous.

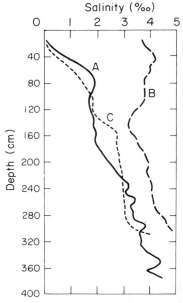

Fig. 14. Average salinity profiles for multiyear ice. Curves A and B are the average hummock and depression salinity profiles, respectively, obtained by Cox and Weeks (1974). Curve C is the multiyear ice average salinity profile determined by Schwarzacher (1959). [Reproduced from Cox and Weeks (1974) by permission of the International Glaciological Society.]

III. PHYSICS OF SEA ICE GROWTH, DRIFT, AND DECAY

Considerable progress has been made over the last several years in understanding and parametrizing the essential physics of sea ice growth, drift, and decay. Perhaps the most intensive effort has been the Arctic Ice Dynamics Joint Experiment (AIDJEX), which focused on the development of models for the simulation of sea ice dynamics. The basic approach in this program was to try to deduce large-scale behavior from a quantitative treatment of small-scale processes [e.g., Rothrock (1975a, 1980)]. Perhaps the most important contribution of this and other recent research has been the formulation of a framework for a complete model of sea ice dynamics. While considerable work remains for understanding the essential physics, this framework provides a convenient focus for coherent examination of physical processes in sea ice. Consequently, the highlights of recent process and model studies are reviewed here to identify the physics of sea ice. Both quantitative process studies and empirical-numerical studies are included.

A concept essential to understanding the large-scale physics of sea ice is that the ice dynamics and the rates of ice growth and decay are intrinsically related. For example, winter rates of freezing depend on the distribution of ice thicknesses, which in turn depends on the ice transport patterns. The ice transport, on the other hand, is modified by the ice thickness distribution (especially the thin-ice percentage, which largely determines the amount of stress the ice can transmit).

This coupled system is discussed in three parts. First, the concept of an ice thickness distribution is examined and the manner in which this distribution evolves due to dynamics and thermodynamics is discussed. This ice thickness distribution concept is then used to help relate the thermal and dynamic processes examined in the next two parts.

A. Ice Thickness Distribution

A sea ice cover typically contains a variety of ice thicknesses that are constantly evolving in response to deformation, advection, growth, and decay. Ice deformation creates thick ice through ridging and creates open water and thin ice through the formation of leads. Growth will decrease the amount of thin ice by converting it into thicker ice, while melt can create thin ice and open water by gradually reducing the ice thickness.

Because of its low strength and high growth rate, many of the dynamic and thermal characteristics of sea ice are dominated by the thin ice. Consequently, for many purposes, the ice thickness distribution may be approximately characterized by breaking it into two parts: thick and thin.

The idea here is that if the thin-ice component is monitored, then uncertainties in the remaining ice thicknesses may not be critical. Within this two-level approach the ice cover is broken down into an area A (often called the compactness) that is covered by thick ice and a remaining area $(1 - A)$ that is covered by thin ice. The area evolves according to a continuity equation [e.g., Nikiforov (1957); Hibler (1979); Parkinson and Washington (1979)]

$$\partial A/\partial t = -\nabla \cdot (\mathbf{u}A) + S_A, \tag{2}$$

where $\mathbf{u}$ is the ice velocity and S_A is a thermodynamic source and/or sink term. In addition, to ensure conservation of mass, a continuity equation for the ice mass per unit area is often used in conjunction with Eq. (2):

$$\partial m/\partial t = -\nabla \cdot (\mathbf{u}m) + S_m, \tag{3}$$

where S_m is a thermodynamic term.

Treatment of the thermodynamic term S_A in Eq. (2) varies with investigators, and can make a significant difference in the physics "compactness" represents. In ice-forecasting applications [e.g., Nikiforov (1957); Doronin (1970)], the S_A term is often omitted and A is simply taken as the area covered by ice (as opposed to open water). In longer-term seasonal simulations [Hibler (1979); Parkinson and Washington (1979)], S_A has been used to cause the thin-ice fraction to disappear in proportion to the ice growth rates. Under these formulations, $(1 - A)$ represents the combined fraction of thin ice and open water.

While such two-level approaches have proven useful for identifying the relative roles of thin and thick ice, their approximate treatment of the ice thickness distribution causes certain shortcomings. For example, Hibler (1979), in a two-level seasonal simulation of the arctic ice cover, found equilibrium thickness values to be smaller than observed. He attributed this, at least partially, to inadequate growth rate estimates for the thick ice (greater than about 0.5 m in thickness) due to a lack of detail in the ice thickness distribution.

A more general framework for dealing with an ice cover of variable thickness, which is continually modified by deformation, growth, and decay, has been developed by Thorndike et al. (1975). They introduced an areal ice thickness distribution $g(h)$, where $g(h)\,dh$ is the fraction of area covered by ice with thickness between h and dh. The equation developed for the dynamic–thermodynamic evolution of this distribution is

$$\partial g/\partial t = -\nabla \cdot (g\mathbf{u}) - \partial(gf)/\partial h + \psi, \tag{4}$$

where f is the growth rate for ice of thickness h and ψ a redistribution function that describes the creation of open water and the transfer of ice from one thickness to another by rafting and ridging. In general, f is considered to be a function of $g(h)$, and ψ a function of both $g(h)$ and the rate of deformation. Except for the last two terms, Eq. (4) is a normal continuity equation for g. The second term on the right-hand side can also be considered a "continuity" requirement in thickness space since it represents transfer of ice from one thickness category to another by the growth rates. The redistribution function can be thought of as a general mechanical source and sink function. The basic requirements on ψ are that it renormalize the g distribution to unity due to changes in area, which leads to the equation

$$\int_0^\infty \psi \, dh = \nabla \cdot \mathbf{u}, \tag{5}$$

and that it does not create or destroy ice but merely changes its distribution. This second requirement leads to the equation

$$\int_0^\infty h\psi \, dh = 0. \tag{6}$$

An important tacit assumption in the derivation of Eq. (4) is that the growth rate f is independent of the thermal history of the ice. This is somewhat troublesome for perennial multiyear ice, where heat stored through internal melting during the summer substantially affects later growth rates [e.g., Maykut and Untersteiner (1971); Semtner (1976)]. However, Semtner (1976) has shown that if some error in the seasonal variations of thickness is tolerated, reasonable mean annual thicknesses can be obtained from a history-independent thermodynamic ice model. Further discussion of these time-independent modifications will be given later in the thermodynamics section.

Characterizing ψ for use in Eq. (4) is basically a question of determining how the dynamics interact with the ice thickness distribution. The actual functional form of ψ is still a subject of debate. There is, however, general agreement that under converging conditions thin ice should be ridged (hence ψ transfers thin ice to thick ice categories) and that open water is created under diverging conditions. It also seems likely that both open water creation and ridging may occur simultaneously under deformation, especially under pure shear. Although progress has been made [e.g., Thorndike (1979)], the precise manner and magnitude of the transfer of ice from one category to another and the extent of redistribution under shear are still unclear.

An additional redistribution problem is how to treat ridged ice. Since ridges result from ice deformation, the amount of ridging, in principle, supplies a measure of the volume of deformed ice. How to characterize this deformed ice is, however, a subject of debate. One approach [e.g., Hibler *et al.* (1974c)] is to introduce a "ridging intensity" parameter defined as being proportional to the mean square ridge height times the ridge frequency. If ridges are randomly oriented and tend to the same angle of repose, the ridging intensity is a measure of the volume of deformed ice per unit area. Wadhams (1980a) and Bugden (1979) have discussed in more detail the limitations and advantages of using ridge statistics to characterize deformed ice. The basic problem with this approach is that it is difficult in practice to distinguish undeformed ice from small ridges in sonar and laser profiles.

The extent to which the thickness distribution is modified by deviations from the average large-scale strain rate is also questionable. Thorndike and Colony (1978), for example, have suggested that creation of thin ice due to nonlinear spatial fluctuations of the ice deformation field may be commensurate with mean strain effects. In addition, McPhee (1978) has suggested that inertial oscillations of the ice cover may create significant amounts of thin ice.

B. Thermal Processes

1. General Characteristics

Many aspects of the thermal processes responsible for sea-ice growth and decay can be identified by examining semiempirical studies of ice breakup and formation of relatively motionless lake and sea ice [e.g., Langleben (1971, 1972); Zubov (1943); Bilello (1961, 1979)]. Observations of fast ice at the border of the Arctic Ocean indicate that radiation is primarily responsible for breakup. In particular, once the initial stages of breakup (with concomitant melt of the snow cover and formation of melt ponds) have passed, the remaining decay of a stationary ice cover is almost all due to the short-wave radiation incident on the surface [Langleben (1972)]. This relatively simple situation occurs because (1) the outgoing long-wave radiation from the earth is very nearly balanced by the incoming long-wave radiation from the clouds during the summer season [Vowinckel and Orvig (1970)] and (2) the turbulent energy fluxes of sensible and latent heat are generally negligible in summer [Langleben (1968)]. However, at the initial stages of breakup and decay, the sensible and latent heat fluxes are important. In particular, once the air temperature reaches 0°C, very rapid melting of the surface snow cover takes place. This melting forms melt ponds [Langleben (1971)], which reduce the albedo and greatly enhance the rate of ice melt. After only a few weeks, drainage canals and vertical melt holes develop, and the characteristic appearance of a summer ice cover evolves with melt ponds and surrounding smooth hummocks.

Once these melt ponds have formed, the remaining decay is dominated by the radiation absorbed by the water. Zubov (1943), for example, has constructed an empirical rule for the breakup of fast ice, using only the radiation absorbed by the water and assuming this energy is expended solely in decreasing the horizontal dimensions of the floes. Although approximate, this approach has found wide acceptance in the literature. The model has been improved by Langleben (1972), who showed that including the radiation absorption by the ice can reduce the decay time up to 30%, in better agreement with observations.

Most empirical models of growth parametrize all the heat budget components (except conduction) by the air temperature. Stefan's law, in which the ice is assumed to be of constant conductivity with a top temperature equal to the air temperature, is the simplest approach [e.g., Pounder (1965)]. In this approach a steady state is assumed, so the temperature gradient in the ice is linear. Since the bottom is at freezing, the freezing rate is therefore proportional to the air temperature and inversely proportional to the ice thickness. Integration of the freezing rate leads to a thickness proportional to the freezing degree days. In practice, Stefan-like laws often yield very good agreement with observations [Bilello (1961)] but require various empirical parameters. That of Anderson (1961) is probably the most complete such empirical law for the Arctic.

2.   Thermodynamic Models of Level Ice

Due to the complexities of its physical properties, the thermodynamic characteristics of sea ice interact in a complex way with the various components of the heat budget. In order to better understand this complex thermodynamic structure, Untersteiner (1964) carried out a series of numerical calculations of the heat conduction in sea ice. This approach was subsequently refined and combined with heat budget components to form a time-dependent thermodynamic model for level multiyear ice [Maykut and Untersteiner (1971)]. Using this model, Maykut and Untersteiner (1971) carried out a variety of calculations that yielded considerable insight into the growth and decay of sea ice.

In this model, the greatest complexity lies in the treatment of heat conduction through the ice–snow system. In particular, the ice density, specific heat, and thermal conductivity are all functions of salinity and temperature (the dependence on temperature is also due indirectly to the salinity). These dependencies are caused by salt trapped in brine pockets that are in phase equilibrium with the surrounding ice. This equilibrium is maintained by volume changes in the brine pockets. A rise in temperature causes the ice surrounding the pocket to melt, diluting the brine and raising its freezing point to the new temperature. Because of the latent heat involved in this internal melting, the brine pockets act as a thermal reservoir, retarding the heating or the cooling of the ice. Since the brine has a smaller conductivity and a greater specific heat than ice, these parameters change with temperature.

To account for these complex variations, Maykut and Untersteiner used semiempirical formulas developed by Untersteiner (1961) that describe the product of the density and heat capacity ρc, and the thermal conductivity k in terms of the salinity S and temperature T. For the salinity profile versus depth, they assumed a constant salinity profile as given by Schwarzacher (1959). In practice the salinity profile varies somewhat in time, but knowledge adequate to calculate this important profile as a function of the thermal history of the ice is lacking [Weeks (1968)].

To calculate the heat conduction, Maykut and Untersteiner used the thermal diffusion equation, allowing for short-wave radiation penetration:

$$(\rho c)_i \, \partial T / \partial t = k_i \, \partial^2 T / \partial z^2 + K_i I_0 \exp(-K_i z) \tag{7}$$

where the subscripts i denote the ice, T is the temperature, K_i is the short-wave extinction coefficient, and I_0 is the amount of short-wave radiation passing through the ice surface. I_0 is usually assumed to be 17% of the incident radiation (although other values were used in sensitivity studies). For the snow cover, a similar diffusion equation was used with fixed values of the parameters ($\rho c)_s$, k_s, and I_0 set equal to zero. These equations were solved as a time-dependent, initial-value problem. Incoming longwave radiation (F_{lw}), incoming solar radiation (F_{sr}), latent and sensible heat fluxes (F_l, F_s), and outgoing short-wave radiation (σT_0^4) were included as driving functions. These functions are combined with the conduction specified by Eq. (7) to form a surface heat budget, which is solved by iteration to determine a surface temperature. This is then used as a boundary condition for further integration of Eq. (7). Also included as driving functions were a specified snowfall rate versus time and an oceanic heat flux F_w into the ice from below. A constant heat flux from the ocean was chosen because a proper oceanic temperature profile and eddy diffusivity would have required an ocean model (actual values and variations of F_w are poorly known). A typical simulation (after a 38-yr integration to yield an equilibrium state) is shown in Fig. 15. Over an annual cycle, a bottom accretion of 45 cm occurs, balanced by a surface ablation of 40 cm and a bottom ablation of 5 cm. Note that the bottom ablation occurs in early fall after most of the surface ablation is over.

In practice, uncertainty about the magnitude and variability of the oce-

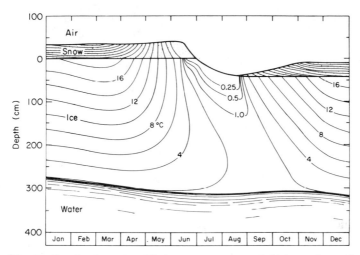

Fig. 15. Predicted values of equilibrium temperature and thickness for sea ice. [From Maykut and Untersteiner (1971).] Isotherms are labeled in negative degrees Celsius. In this simulation, an upward oceanic heat flux of 1.5 kcal/cm² yr is assumed, and 17% of the net short-wave radiation is allowed to pass through the ice surface (when it is snow-free). In addition the albedo of snow-free, melting ice is taken to be 0.64. The snow cover albedos are taken to vary seasonably, as specified by Marshunova (1961). Other heat budget input at the upper boundary was taken from Fletcher (1965). To distinguish between movements of the upper and lower boundary, they are plotted without regard for hydrostatic adjustment.

anic heat flux and snow cover creates substantial uncertainty in the simulated equilibrium thickness of sea ice. Some effects of these variables are illustrated in Figs. 16 and 17. In Fig. 16 the equilibrium thickness was calculated by Maykut and Untersteiner for different constant values of the oceanic heat flux. In Fig. 17, the magnitude, but not the temporal distribution, of the snow cover was modified.

In the case of the oceanic heat flux, the results are generally straightforward. Increasing the heat flux reduces the rate of accretion on the bottom of the ice and somewhat increases the bottom ablation. The processes at the upper boundary, on the other hand, are relatively insensitive to the ocean-ice heat flux. By reducing the heat flux to zero (as compared to the standard case of 1.5) the ice thickness almost doubles. However, Maykut and Untersteiner did not examine the effect of temporal variations in the heat flux. It is quite possible that the timing of the maximum fluxes could substantially affect the ice thickness.

The effects of varying the snow cover (Fig. 17) are more complex. While a thinner snow cover allows greater cooling of the ice during winter, it is also removed earlier in the spring, which decreases the average albedo and prolongs the period of ice ablation. The effect is fur-

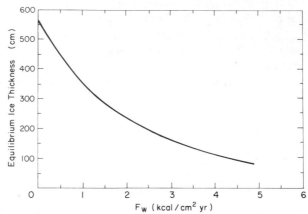

Fig. 16. Average equilibrium thickness of arctic sea ice as a function of oceanic heat flux. [From Maykut and Untersteiner (1971).] Other parameters are the same as the standard case shown in Fig. 15.

ther complicated by the internal melting due to the absorbed radiation once the ice is snow free. Because of these complex effects, the snow depth has unexpectedly small effects until 70 cm is reached. When this occurs the decrease in ice ablation tends to dominate and the ice thickens. Until 120 cm is reached, the snow cover in the model will totally disappear at some time in the summer.

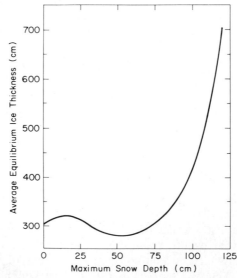

Fig. 17. Average equilibrium thickness of arctic sea ice as a function of maximum annual snow depth. [From Maykut and Untersteiner (1971).] Other parameters are the same as the standard case shown in Fig. 15.

More recent thermodynamic studies of level ice using the Maykut and Untersteiner approach have concentrated on young ice [Maykut (1978)]. For this type of ice Maykut argues that under cold conditions the conduction of heat to the surface can be reasonably approximated by a linear temperature distribution with a conductivity based on the salinity of the surface layer. Salinity profiles reported in the literature together with thickness-dependent surface albedos [Weller (1972)] were used for a variety of calculations. These calculations indicated that, although both the amount of absorbed short-wave radiation and emitted long-wave radiation depend on ice thickness, it is the turbulent fluxes that undergo the largest changes as the ice thickens. Because of this, the rate of heat exchange over thin ice appears to be extremely sensitive to the snow depth and atmosphere boundary layer temperature. Maykut also confirmed the importance of the heat exchange between thin ice and the atmosphere and ocean by estimating that winter heat input to the atmosphere through ice in the 0–40-cm range can be 1–2 orders of magnitude larger than from thicker multiyear ice.

While the Maykut and Untersteiner theory represents a comprehensive treatment of the thermodynamics of level ice, its time dependence and complicated differencing scheme make it difficult to use for practical problems. To remedy this, Semtner (1976) has proposed a simplified model that retains the most essential components of the thermodynamics. Besides reducing the dependence on empirical data, this study also helped identify the essential physics of the growth and decay. As part of this study, Semtner also examined a simpler time-independent thermodynamic model, and the case of sea ice which totally disappears in the summer.

Basically, Semtner argued that the key aspect of the thermodynamics of perennial level ice is the presence of internal melting in summer. In particular, sea ice can be thought of as a matrix of brine pockets surrounded by ice where melting can be accomplished internally by enlarging the brine pockets rather than externally by decreasing the thickness. The amount of internal melting is largely dictated by the percentage of short-wave radiation passing through the surface in summer. In practice, this internal melting effectively causes a thermal inertia, reducing ablation in the summer and retarding ice accretion in the fall. Because of this thermal inertia effect, the thickness of sea ice can be substantially larger than freshwater ice, given the same external factors.

Based on this "brine damping" concept, Semtner (1976) proposed a simple model in which the snow and ice conductivities were fixed. In this model the salinity profile does not have to be specified. To account for internal melting, an amount of penetrating radiation was stored in a heat reservoir without causing ablation. Energy from this reservoir was used to keep the temperature near the top of the ice from dropping below freezing in the fall. Using this simplified model, Semtner was able to reproduce Maykut and Untersteiner's results within a few percent.

For an even simpler time-independent model, Semtner proposed that a portion of the penetrating radiation I_0 be reflected away. The remainder of I_0 was applied as a surface energy flux. In addition, to compensate for the lack of internal melting, the conductivity was increased to allow greater winter freezing. In the simplest model, linear equilibrium temperature gradients are assumed in both the snow and ice. Since no heat is lost at the snow–ice interface, the heat flux is uniform in both snow and ice. The surface temperature is obtained by setting the surface heat balance equal to zero and solving numerically for the temperature. If the surface temperature is above the melting point, it is set at melting. A somewhat similar, but even simpler model has been used by Bryan *et al.* (1975) and Manabe *et al.* (1970), who reduce the snow to an equivalent ice thickness so that only ice has to be considered.

The results of both Semtner's time-dependent and time-independent models are compared to Maykut and Untersteiner's results in Fig. 18. This figure also shows the importance of the assumed penetration of radiation which causes internal melting. By allowing no radiation to pene-

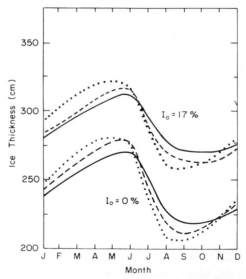

Fig. 18. Annual thickness cycles of three thermodynamic sea ice models for the cases of 0 and 17% penetrating radiation: (solid lines) Maykut and Untersteiner model, (dashed lines) three-layer Semtner model, (dotted lines) zero-layer Semtner model. The forcings (other than the penetrating radiation) are the same as discussed in Fig. 15. [Reproduced from Semtner (1976).]

trate, the internal melting is mitigated and the radiation instead is used to melt the ice, causing a reduced thickness. Note that while the time-independent model reproduces the mean thickness well, its amplitude and phase of the seasonal variation of thickness are somewhat different from the time-dependent models. As mentioned earlier, this simplest time-independent model has taken on special significance for numerical simulations of sea ice because the almost continual redistribution of a variable thickness ice cover makes it difficult to record the thermal history of a fixed ice thickness.

A final result of Semtner's study was the presence of multiyear cycles of ice thickness if the ice was allowed to disappear in summer. To investigate this case Semtner introduced a passive 30-m boundary layer (for heat storage) in the ocean. This allowed the model to be integrated through periods when the ice totally disappeared. The model then produced complex multiyear cycles that repeated themselves exactly. Semtner attributed these cycles to the differing conductivities of snow and ice in conjunction with open water. In particular he noted that following the appearance of open water in a given summer, the ocean does not refreeze until after the heavy snowfall season. With a thin cover of insulating snow the ice grows rapidly during the subsequent winter so that several years with the usual snow cover must pass before open water reappears.

It is also notable that complex thermal oscillations (albeit on a short time scale) have been found in laboratory freezing experiments with fixed thermal forcing [Welander (1977)]. In general the presence of such simulated and observed cycles emphasizes the highly nonlinear nature of the thermodynamics of sea ice.

3.   Thermal Processes Related to Spatial Thickness Variations

The growth and decay of sea ice can be substantially affected by spatial thickness variations [Rothrock (1980)]. Perhaps the most obvious example is the effect of open water on the adjacent pack ice. During melting conditions the radiation absorbed by leads can contribute to lateral melting by ablation at the edges of ice floes [e.g., Langleben (1972)]. This heat also can be carried under the ice where it will contribute to bottom ablation. If the ice cover is sufficiently disintegrated, some of the heat can be stored in the mixed layer thus delaying autumn freeze-up. Thin ice can also affect adjacent heat exchanges by modifying one or both of the planetary boundary layers. Substantial local modifications to the atmospheric boundary layer, for example, have been documented by Andreas et al. (1979).

Many of these open-water related processes are particularly pronounced in the marginal ice zone (i.e., the ice area close to an open ocean margin). Submarine studies [Kozo and Tucker (1974); Wadhams *et al.* (1979)] show this region to have a particularly wide range of ice development, ranging from multiyear ice to small newly formed "pancake" ice floes (0.3–2 m across). In addition, wave action contributes both to the breakup of floes and wave-induced melting [Wadhams *et al.* (1979)]. More details on a variety of physical processes in the marginal ice zone are given in a review by Wadhams (1980b).

It is also probable that thick ice in the form of pressure ridges behaves differently than level ice. Koerner's (1973) ablation observations indicate that the upper surface of first-year ridges ablate much more rapidly than level ice. Rigby and Hanson (1976) observed up to 2 m of summer melt on the bottom of an old 10-m-thick ridge. They also observed losses of 4 m from a 5-m first-year ridge. Basically, the deep keel of the ridge allows ablation at the sides as well as the bottom.

As this brief discussion suggests, present knowledge about the thermal processes of a spatially variable ice cover is sketchy. What is clear, however, is that our relatively clear conception of the thermodynamics of level ice may be substantially modified by the spatially varying character of the ice cover.

C. Ice Drift and Deformation

Recent research has substantially improved our understanding of the nature of ice drift. Basically, sea ice forms a complex three-dimensional continuum that moves in response to wind and water stresses. The internal stresses this continuum can sustain are substantial. Consequently, to understand the physics of ice drift and deformation requires understanding the nature of the ice interaction. Determining the nature of this interaction was the primary goal of the Arctic Ice Dynamics Joint Experiment (AIDJEX).

From research to date the general character of ice drift and deformation has emerged. It is generally agreed that the dynamics of this complex system on a large scale can be characterized using the following elements: (1) A *momentum balance* describing ice drift, including air and water stresses, Coriolis force, internal ice stress, inertial forces, and ocean current effects; (2) an *ice rheology* which relates (on some suitable continuum space and time scale) the ice stress to the ice deformation and strength; (3) an *ice strength* determined primarily as a function of the ice thickness distribution.

In discussing these components it will be assumed that a description of sea ice as a two-dimensional continuum is possible. While such an approximation appears reasonable, the fact that pack ice consists of large individual pieces causes substantial fluctuations in the large-scale motion. Results from studies of the kinematics of pack ice [e.g., Hibler *et al.* (1974a); Nye (1975); Thorndike and Colony (1978)] exhibit substantial deviations from a spatially smooth velocity field. These deviations tend to scale with the magnitude of the ice deformation. In practice, however, typical fluctuations are small enough that on scales larger than about 20 km they do not mask the continuum motion. However, their presence means that continuum motion has to be considered a statistically defined quantity for sea ice.

It should also be noted that the physical structure and small-scale strength of sea ice has been deliberately excluded from the above outline. Such physical properties do play a significant role in local force problems [e.g., Tryde (1977); Croasdale (1978)]. However, for large-scale modeling, the contribution of such small-scale characteristics to the ice interaction and drift is felt to be of lesser importance. Schwarz and Weeks (1977) have reviewed the small-scale strength properties of sea ice in some detail.

1. Momentum Balance

The momentum balance describes the forces that determine drift and deformation of sea ice. Denoting the ice mass per unit area by m, the momentum balance in a two-dimensional Cartesian coordinate system is

$$mD_t\mathbf{u} = -mf\mathbf{k} \times \mathbf{u} + \tau_a + \tau_w + \mathbf{F} - mg\,\nabla\mathbf{H}, \qquad (8)$$

where $D_t(\partial/\partial t + \mathbf{u} \cdot \nabla)$ is the substantial time derivative, $\mathbf{k}$ is a unit vector normal to the surface, $\mathbf{u}$ is the ice velocity, f is the Coriolis parameter, g the acceleration due to gravity, τ_a and τ_w are the forces due to air and water stresses, $\mathbf{H}$ is the dynamic height of the sea surface, and $\mathbf{F}$ is the force due to internal ice stress. In this formulation, τ_w includes frictional drag due to the relative movement between the ice and the underlying ocean. Also $\mathbf{H}$ can, in principle, include the variation of the sea surface height due to atmospheric pressure changes as well as due to geostrophic current balance.

With respect to the dominant components of this balance, early observations by Nansen (1902) indicated that much of the ice drift far from shore could be explained by wind variations. He established empirically that in such circumstances the ice drifts approximately 30° to the right of the wind with about one-fiftieth of the surface wind velocity. Zubov (1943) later made similar observations using geostrophic winds and the drift of ships locked in the ice. He noted that the ice drift effectively takes place parallel to the pressure isobars with a speed of about one one-hundredth of the geostrophic wind speed. Such simple rules are still used today for many operational ice forecasts.

To explain this wind-drift phenomenon, Nansen proposed that the acceleration and internal ice stress terms be neglected in the momentum balance and a constant current effect be added. He did not, however, specify how the water drag should be treated. Later studies of such equilibrium stress-free drift were made by Shuleikin (1938) and Reed and Campbell (1962), who proposed water drag models for the τ_w term. However, these models neglected steady-current effects. A boundary layer formulation similar to that of Reed and Campbell was later used by Campbell (1965) in a basin-wide sea ice circulation model.

Subsequent to this early work, Rothrock (1975b) presented a more unified discussion of the momentum balance in a study of the steady circulation of an incompressible arctic ice cover. Following the approach used in Russian ice-modeling work [e.g., Doronin (1970); Egorov (1971)], Rothrock (1975b) proposed linear forms based on simple Ekmann layer arguments for τ_a and τ_w:

$$\tau_a = C_A (\cos \theta_a \mathbf{U}_g + \mathbf{k} \times \mathbf{U}_g \sin \theta_a), \tag{9}$$

$$\tau_w = C_W(\cos \theta_w(\mathbf{U}_w - \mathbf{u}) + \mathbf{k} \times (\mathbf{U}_w - \mathbf{u}) \sin \theta_w), \tag{10}$$

where C_A and C_W are constants, and θ_a and θ_w are constant turning angles in the air and water. In the air stress [Eq. (9)] the ice velocity has been neglected, with an error of a few percent, in comparison to the geostrophic wind. The geostrophic wind $\mathbf{U}_g$ is related to the gradient of the atmospheric pressure $(1/\rho f)\mathbf{k} \times \nabla p$, and the current $\mathbf{U}_w$ is taken to be in geostrophic balance and is consequently related to the gradient of the dynamic height $gf^{-1}\mathbf{k} \times \nabla \mathbf{H}$.

Basically, in this formulation the essential causes of the ice motion can be thought of as the geostrophic wind above the atmospheric boundary layer and the geostrophic ocean current beneath the oceanic boundary layer. These forces are transmitted to the ice via simple integral boundary layers with constant turning angles. In addition, the steady ocean current introduces a tilt $\nabla \mathbf{H}$ of the sea surface height that also affects the ice motion.

While this overall framework appears valid for sea ice, more recent measurements indicate that the linear boundary layer formulations are inadequate. In particular, measurements of the oceanographic boundary layer by McPhee and Smith (1976) and Shirawawa and Langleben (1976) show C_w to be dependent on the difference of ice speed and ocean current (Fig. 19). In practice, this dependence is approximated well by a quadratic

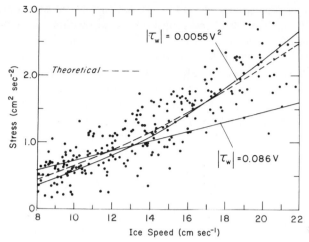

Fig. 19. Observed kinematic water stress (stress divided by water density) versus ice speed (relative to the geostrophic currents). The quadratic, linear, and theoretical curves are shown. [Modified from McPhee (1979a).]

water drag (Fig. 19). Similar results have been obtained for the atmospheric boundary layer [e.g., Brown (1979); Banke *et al.* (1976); Carsey (1979); Leavitt (1979)]. Within this quadratic approximation, the planetary boundary layers can be represented by Eqs. (9) and (10) with C_A and C_W given by

$$C_A = \rho_a C_a |U_g|, \tag{11}$$

$$C_W = \rho_w C_w |U_w - u|, \tag{12}$$

where C_a and C_w are dimensionless constants (with typical values of 0.0012 and 0.0055, respectively). To explain this quadratic stress dependence, McPhee (1979a) made use of Rosby number similarity arguments. Under this argument, changes in the ice velocity change the depth of the dynamic boundary layer. This change in depth makes the drag nonlinear. For small Rossby number variations, quadratic drag and constant turning angles can be used. However, for larger variations (which may be important at small velocities), changes in the turning angle and deviations from quadratic drag become important [McPhee (1979a)].

While simple, the quadratic formulations appear to give adequate wind and water stress estimates and have been used to successfully model stress-free (**F** equals zero) ice drift during the summer [McPhee (1979a)]. In general they form the framework used in many ice drift models to date. However, it should be noted that while such simple integral approxi-

mations may be adequate for surface stress estimates, their utility for estimating detailed heat flux characteristics is less clear. Considerable research remains to be done on the detailed structure of the planetary
boundary layers in the polar regions.

While wind and water stresses supply the key driving forces, other
components, especially the ice stress terms, are necessary to properly
model ice motion. This was qualitatively observed by Sverdrup (1928),
who noted that the drift angle right of the wind tended to decrease during
very compact ice conditions and increase when large numbers of leads
and polynyas existed. Also Nansen (1902) ascribed the small wind-ice
turning angle he observed (of less than 30°) to the effect of ice resistance.

Empirical estimates of the various components in the force balance
made using ice drift and wind and water stress measurements confirm the
importance of the internal ice stress. Analysis of the drift records
[Thorndike (1973)] shows the momentum advection terms are negligible
and, at least in winter conditions, the temporal acceleration is not usually
significant. Neglecting accelerations, Hunkins (1975) obtained force balances with the typical internal ice force terms comparable to wind and
water drag (Fig. 20). In Hunkin's study, the internal ice stress was treated
as a residual with the other parameters calculated from observations.

Under summer conditions with large amounts of open water, the internal ice stress term decreases substantially. However, the temporal
acceleration term then becomes more important. Studies of summer ice
drift by McPhee (1978) show the coupling of the acceleration with the
Coriolis force causes marked inertial oscillations in the ice. These oscillations cause the 12-h cycles in ice drift discussed earlier. In the absence of

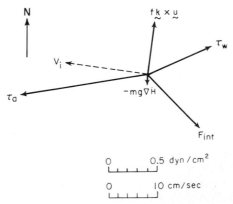

Fig. 20. An estimate of the force balance on sea ice for winter conditions based on
wind and water stress measurements. [From Hunkins (1975).] In this balance the force due
to internal ice stress is determined as a residual and the dashed line shows the ice velocity.

ice stress effects, McPhee found the momentum balance of the combined ice–oceanic boundary layer to be well approximated by

$$m \, d\mathbf{u}/dt = -mf\mathbf{k} \times \mathbf{u} + \tau_a. \qquad (13)$$

A particularly important feature of the McPhee study was that considering only ice with a passive drag by the ocean tended to overdamp the oscillations. However, by assuming that not just the ice cover but the whole boundary layer was undergoing oscillations, good agreement was found.

Another aspect to the momentum balance is the relative magnitude of the steady geostrophic current component of drift versus the wind-driven drift. Empirical model studies [e.g., Hibler and Tucker, (1979)] indicate that, in the central Arctic, steady current and tilt effects are normally quite small (several percent relative to the wind-driven components) for ice velocities averaged over a few days. However, current and tilt effects do appear significant for velocities averaged over years. Figure 21 shows the effect of currents and tilt on the simulated average drift of three drifting stations over a 2-yr period. The long-term significance of the cur-

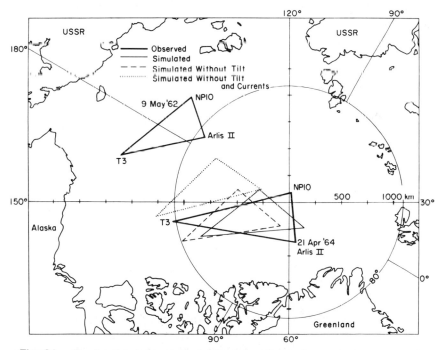

Fig. 21. Simulated and observed net drift ot three ice stations over a two-year period. Simulations are shown both with and without current effects. [Reproduced from Hibler and Tucker (1979) by permission of the International Glaciological Society.]

rent effects arises from their steady character. The wind effects, while
larger, are of a fluctuating character. Over a long time interval the fluctuations largely average out.

Both the inertial oscillations and steady current effects demonstrate the
extent to which ice drift is intrinsically coupled to the ocean. When the ice
cover is capable of sustaining high stress in the winter, it can be essentially
viewed as an independent entity, separate from the ocean and concentrated
in a thin layer. Under these conditions the ocean executes a damping drag
due to the ice motion and steady currents. However, when the ice loses
its capability to transmit stress the ice cover effectively becomes part of
the upper ocean with its inertial motion reflecting this fact. In addition, for
long-term simulations the effect of the ice motion and thermodynamics on
the average ocean currents must be considered. In particular the stress
transmitted to the ocean from the atmosphere, which affects the current
structure, can be modified by the ice interaction. Coon and Pritchard
(1980), for example, have shown that near shore the air-to-sea energy
transfer can be drastically changed by the ice cover because it acts as a
relatively cohesive block. Also, the change in the vertical air–sea heat exchange due to the ice cover can modify the thermohaline circulation of the
ocean. Overall, these interactions emphasize the importance of considering the ice and ocean as a "coupled system."

2. Sea Ice Rheology

Analysis of the momentum balance has shown the importance of
internal ice forces in ice drift. Because of this, the determination of a realistic constitutive law relating the ice stress to the deformation has been
critical to the understanding of ice drift and deformation. Glen (1970) has
outlined a framework for examining this constitutive law question with
particular clarity. Based on earlier glacier flow concepts (see Chapters 1
and 2), Glen suggested that a general constitutive law of the following
form might be applicable to sea ice:

$$\sigma_{ij} = 2\eta\dot{\varepsilon}_{ij} + [(\zeta - \eta)(\dot{\varepsilon}_{xx} + \dot{\varepsilon}_{yy}) - P]\delta_{ij}, \tag{14}$$

where σ_{ij} and $\dot{\varepsilon}_{ij}$ are the two-dimensional stress and strain rate tensors, η, ζ,
and P are general functions of the two invariants of the strain rate tensor,
and δ_{ij} equals one for i equal to j or zero for i not equal to j. These invariants can be taken as the principal values of the strain rate tensor or,
alternatively, as $(\dot{\varepsilon}_{xx} + \dot{\varepsilon}_{yy})$ and $(\dot{\varepsilon}_{xx}\dot{\varepsilon}_{xx} + 2\dot{\varepsilon}_{xy}\dot{\varepsilon}_{xy} + \dot{\varepsilon}_{yy}\dot{\varepsilon}_{yy})$. Within the
framework of this equation, η and ζ can be thought of as nonlinear bulk
and shear viscosities. To obtain an internal ice force for the momentum

balance, the stress tensor is differentiated, yielding

$$F_x = \partial(\sigma_{xx})/\partial x + \partial(\sigma_{xy})/\partial y, \tag{15a}$$

$$F_y = \partial(\sigma_{xy})/\partial x + \partial(\sigma_{yy})/\partial y, \tag{15b}$$

where F_x and F_y are the components of the force transmitted through the ice.

Particular forms of this law include both viscous and plastic models. In general, a "viscous" rheology is defined as one in which stress can only be sustained through nonrecoverable dissipation of energy by deformation. A "plastic" rheology, on the other hand, is one in which stress can be sustained through lack of deformation or elastic deformation where the energy is recoverable [e.g., Malvern (1969)]. Special cases of these rheologies often used in sea ice are a "linear viscous" rheology where the stress depends linearly on the strain rates and an "ideal rigid plastic" rheology where the stress state is either indeterminate or independent of the magnitude of strain rates. In the linear viscous case, η and ζ are taken to be linear in $\dot{\varepsilon}_{ij}$, and P is constant. This results in a stress state linearly dependent on the strain rates. In the rigid plastic case the stress state is taken to be fixed independent of the magnitude of the strain rate invariants provided the ratio of the invariants does not change. In this case η and ζ will be nonlinear functions of the invariants. In the rigid plastic case, however, η and ζ may not be well defined for zero strain rates. Under these conditions the ice is understood to move rigidly with the stress determined by external balances.

Some of the linear viscous and plastic rheologies proposed for sea ice are shown in Fig. 22, where the allowable stress states for two particular linear viscous cases of Eq. (14) are shown. These states are a Newtonian viscous fluid ($\eta = \text{const}$, $\zeta = P = 0$), and a fluid containing resistance

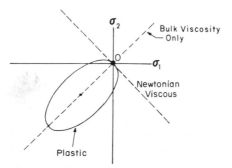

Fig. 22. Allowable stress states for a Newtonian viscous rheology, a linear viscous rheology with only a bulk viscosity, and an ideal rigid plastic rheology with an elliptical yield curve. The stress states are plotted as a function of the principal components of the two-dimensional stress tensor.

to dilation and compression but not to shear (ζ = const, $\eta = P = 0$). For an ideal plastic constitutive law the stress state lies either on or inside some fixed yield curve. In Fig. 22, an elliptical yield curve is shown. For a plastic rheology a rule is needed to uniquely relate the stress state to the strain rates. The most common rule [e.g., Malvern (1969)] is to take the ratio of the strain rates for a given stress state to be that of a vector normal to the surface (this is referred to as the normal flow rule). For the ellipse, the normal flow rule yields η, ζ, and P values of [e.g., Hibler (1977)]

$$P = P^*/2, \qquad \zeta = P^*/2\Delta, \qquad \eta = \zeta/e^2, \tag{16}$$

where Δ equals

$$(\dot{\varepsilon}_{xx}^2 + \dot{\varepsilon}_{yy}^2)[1 + (1/e^2)] + (4/e^2)\dot{\varepsilon}_{xy}^2 + 2\dot{\varepsilon}_{xx}\dot{\varepsilon}_{yy}[1 - 1/e^2)]^{1/2}$$

P^* is a constant and e is the ratio of the lengths of the principal axes of the ellipse. For rigid motion, the stress state will lie within the ellipse. Note that both linear viscous laws allow large amounts of tensile strength (i.e., positive stresses) whereas this particular plastic law allows only small tensile strengths.

In early efforts to approximate the ice interaction, viscous fluid concepts similar to those in meteorology were used. Laikhtman (1964), for example, proposed a linear Newtonian viscous model that was easy to deal with numerically and was subsequently used by a variety of authors [Egorov (1971); Doronin (1970)] to carry out empirical–numerical studies. Probably the most well-known application of this rheology was a calculation by Campbell (1965) of the steady circulation of the Arctic Basin ice cover using an idealized mean annual wind field [Felzenbaum (1958)]. These empirical models studies demonstrated that the Newtonian viscous rheology had some ability to simulate the effects of ice interaction. However, a particular defect of this rheology is that it has no resistance to convergence [e.g., Rothrock (1975a)].

Empirical studies using a more "general" linear viscous law containing both bulk and shear viscosities (and hence having compressive strength) have successfully simulated observed ice drift and deformation far from shore, both over short time intervals [Hibler (1974)] and over a seasonal cycle [Hibler and Tucker (1979)]. In particular, Hibler (1974) showed a linear viscous rheology could explain the observation that, in a low-pressure system, the ice pack often tends to converge in the winter and diverge in the summer. Best estimates of the viscosity magnitudes yielded viscosities (two dimensional) of the order of 10^{12} kg sec^{-1} with the bulk viscosity ζ about twice as large as the shear viscosity η. In a seasonal study Hibler and Tucker (1979) showed the "best-fit" viscous parameters to vary in a regular seasonal manner. This seasonal variation was commensurate with the physical notion that sea ice has less thin ice in winter (due to the higher freezing rates) and hence has higher strengths. The seasonal viscosity also tended to correctly simulate observed seasonal variations of the angle of ice drift relative to the wind.

While generally performing well far from shore, a major problem with the linear viscous rheology is that viscosities for reasonable simulations in the central pack are quite different from viscosities needed to simulate near-shore behavior [Rothrock (1975a)]. For example, viscosities of the order of 10^{11}–10^{12} kg sec $^{-1}$ [Hibler (1974); Hibler and Tucker (1979)] are needed to model pack ice behavior far from shore. Near shore, however, the effective viscosities may be as small as 10^8 kg sec^{-1} [Hibler et al. (1974b)] and in very small channels they can be as small as 10^5 kg sec^{-1} [Sodhi and Hibler (1979)]. Such variations suggest that inclusion of some type of nonlinear behavior is necessary.

An additional problem with linear viscous laws is that their physical basis is questionable. In particular, the local characteristics of sea ice appear to be nonlinear and "plastic" in nature. However, even though sea ice may not exhibit linear viscous characteristics on a small scale, Nye (1973b) has noted that such linear laws still might be appropriate for a time-averaged or spatially averaged response. A particular case of this concept has been proposed by Hibler (1977), who suggested that on appropriate time scales stochastic fluctuations in the deformation rate might cause the average stress–strain rate relation to be essentially linear. To examine this idea, Hibler used observed and simulated deformation records to simulate temporal variations of stress for an ideal plastic rheology. Comparison of the stress and strain rate records showed very good linear coherence for periods longer than a few days (and hence explainable by a linear viscous rheology with a bulk and shear viscosity). However, the best regression parameters depended on the magnitude of the deformation rate. This provides one explanation of why best-fit viscosities can vary widely in empirical studies.

To avoid the inherent deficiencies in linear viscous rheologies and to better describe the local real-time behavior of sea ice, Coon et al. (1974) and Rothrock (1975a,c) proposed a plastic rheology for sea ice. As discussed earlier, in a plastic constitutive law the stress state of the ice is presumed to be independent of the magnitudes of the stress and strain rate (e.g., Fig. 22). Such a law allows highly nonlinear behavior, which is helpful in explaining ice flow both near and far from shore. In addition, it contains a simple way to specify a low tensile strength concurrently with a high compressive strength.

To justify such a plastic law physically, Coon and Rothrock argued that such behavior is consistent with the local physics of sea ice. The essential argument for rate independence is that the work done in ice deformation is primarily due to ridging. The energy needed in ridging, however, would be expected to be independent of the rate of ridge building. This concept can be extended into a complete explanation (in terms of ridging and frictional losses) of all energy dissipated during plastic flow [Rothrock (1975c)]. This particular approach is discussed in the next section. With regard to small tensile strengths, it was argued that the presence (over a large scale) of many cracks prevented any effective cohesion under dilation. In addition, a plastic rheology appears to be consistent with observations of sea ice lead patterns which have been explained by rock mechanic failure criteria [e.g., Marko and Thomson (1977)].

While the concept of an ideal rigid plastic law is useful theoretically, it is difficult to implement in practice. In particular, to examine a plastic rheology numerically, some approximation must be made to describe the "rigid" behavior, i.e., the behavior when the ice is not flowing plastically. Two approaches have been suggested: the elastic-plastic approach [Pritchard (1975)], and the viscous plastic approach [Hibler (1979)]. In the elastic-plastic method the ice is assumed to behave elastically for certain strain states. The advantage of this is that it allows rigid behavior in the sense that the ice can sustain a variety of stresses while possibly undergoing no deformation. However, inclusion of the elasticity is inherently unwieldy and has the disadvantage of adding elastic waves to the behavior of the model [Colony and Rothrock (1979)].

In the viscous-plastic approach [Hibler (1979)], sea ice is considered to be a nonlinear viscous continuum characterized by both nonlinear bulk and shear viscosities and a pressure term [Eq. (14)]. These nonlinear viscosities are adjusted to give plastic flow for normal strain rates. To close the system, for very small deformation rates the ice is assumed to behave as a very stiff linear viscous fluid. Thus the rigid behavior is approximated by a state of very slow flow. This technique is computationally simple. Its disadvantage, however, is that ice that is actually stationary may be modeled as slowly creeping.

Simulations using plastic rheologies have reproduced many aspects of sea-ice dynamics. Near-shore simulations by Pritchard et al. (1977a) and Pritchard and Schwaegler (1976) have shown the capability of the elastic-plastic rheology to model both relatively stationary behavior under stress and intense shearing behavior near coasts. Also, Pritchard (1978) shows that for sufficiently high strengths, irregular boundaries can produce shaded regions of dead ice with leads forming farther off shore, in agreement with observations. With respect to seasonal simulations, Hibler (1979) has shown that in addition to producing near-shore shear zone effects, a plastic rheology yields geographical ice thickness buildup and ice outflow from the basin which agree well with observations.

Overall, these physical arguments and numerical studies demonstrate the appropriateness and utility of a plastic rheology for describing sea ice on the large scale. However, questions remain regarding the appropriate plastic yield curve and flow rule.

3.   Aggregate Ice Strength

A constitutive law relating stress to ice deformation provides a useful means of precisely describing the nature of the ice interaction. However, as input to the constitutive law, some measure of the ice strength is needed. In general, one intuitively expects the ice strength to depend on the aggregate ice thickness characteristics. Observations as early as Nansen's tended to support this. However, to quantify this concept it is helpful to identify some of the local mechanisms responsible for the stresses transferred during interaction of floes. In one of the first attempts at such a quantification, Coon (1974) (conference paper presented in 1972) proposed two controlling mechanisms dictating the compressive load that can be sustained by sea ice, namely, breaking of ice by bending resulting from rafting or ridging, and buckling of ice floes under a horizontal load. These compressive buckling loads depended on the thickness of the ice involved. In a closer examination of pressure ridging, Parmerter and Coon (1972) developed a kinematic model employing Coon's mechanism of breaking by bending but excluding buckling. After examination of the magnitude of the various forces, Parmerter and Coon concluded that on the average the horizontal stress was essentially equal to the increase in potential energy of the ridge per unit displacement (due to upward piling of ice and downward displacement of water).

Based on the Coon (1974) and the Parmerter and Coon (1972) concepts, Rothrock (1975c) subsequently proposed a unified approach to energy deformation. In essence Rothrock suggested the two-dimensional ice stress and strain rate might be related to the rate of work done on the ice through ridging. Variable thicknesses were included by means of the redistribution function [see Eqs. (4)–(6)]. Thus, the stress state in the ice during shearing might be explained by allowing ridging to occur under pure shear even though the divergence rate was zero. Similarly, under convergence, a variety of the thicknesses might be simultaneously deformed. By allowing, say, only the thinnest 10% of the ice cover to deform, the strength will tend to be primarily dependent on the young ice. Rothrock made some attempt to include frictional losses in ridging, but did not include frictional losses in shearing between floes (this particular mechanism will be referred to later as a "shear sink" of energy). Anisotropic effects also were not considered.

An important byproduct of Rothrock's work was that, assuming a plastic rheology, it related the plastic flow to the amount of work done by ridging under an arbitrary deformation state. Under this assumption, the way in which ridging occurs could in principle be used to determine both the strength and the plastic yield curve for arbitrary deformation. A feature of the work of Parmerter and Coon and of that of Rothrock is that the ice stresses would be expected to be rate independent, which is an important aspect of plastic laws.

In more formal terms, the basic assumption made by the above authors is to equate the rate of strain work ($\sigma_{ij}\dot{\varepsilon}_{ij}$) in the ice to the rate of mechanical production of potential energy per unit area ($\dot{R}_{pot}$) plus the rate of frictional loss per unit area ($\dot{R}_{fric}$). For isotropic materials, where the principal directions of stress and strain rate are aligned, this assumption yields the equation

$$\sigma_1\dot{\varepsilon}_1 + \sigma_2\dot{\varepsilon}_2 = \dot{R}_{pot} + \dot{R}_{fric}, \tag{17}$$

where $\dot{\varepsilon}_1$ and $\dot{\varepsilon}_2$ and σ_1 and σ_2 are the principal components of the strain rate and stress. Assuming isostatic balance, the potential energy per unit area of a block of ice of thickness h and density ρ_I floating in the water is [Rothrock (1975c)]

$$R_{pot} = \tfrac{1}{2}(\rho_I/\rho_w)(\rho_w - \rho_I)gh^2, \tag{18}$$

where ρ_w is the density of water and g the acceleration of gravity. Consequently, for an arbitrary distribution of ice thicknesses the potential energy rate $\dot{R}_{pot}$ is proportional to the rate of change of the mean square thickness.

To get some idea of the magnitude of stresses needed, it is useful to consider ice of a constant thickness $\bar{h}$ undergoing convergence. Equation (17) then yields

$$(\sigma_1 + \sigma_2)\nabla \cdot \mathbf{u} \cong -(\rho_I/\rho_w)(\rho_w - \rho_I)g\bar{h}^2\nabla \cdot \mathbf{u}, \tag{19}$$

where we have assumed $\partial \bar{h}^2/\partial t$ approximately equals $-2\bar{h}^2\nabla \cdot \mathbf{u}$. Substituting in typical numerical values for ρ_w and ρ_I,

$$\sigma_1 + \sigma_2 \cong -10^3\bar{h}2 \quad \text{N m}^{-1}, \tag{20}$$

where $\bar{h}$ is in meters and σ_1 and σ_2 are the principal components of the two-dimensional stress tensor. It is notable that such potential energy strengths are substantially less than buckling or crushing strengths for ice of similar thicknesses (see Table 1).

Whether the ridging potential energy is dominant and friction can be neglected is still an open question. However, regardless of how important friction is, the way in which ridging is parametrized can substantially change the ice strength. For example, Thorndike et al. (1975) assumed a

TABLE 1

Buckling and Crushing Strengths for Three Ice Thicknesses[2]

h(m)	P_{cr}(N m^{-1})	P_b(N m^{-1})
0.10	40×10^3	17×10^3
0.50	200×10^3	185×10^3
1.00	400×10^3	524×10^3

[a] After Rothrock (1975c).

redistribution that transfers ice into a fixed multiple of its thickness. In their initial study a multiple of 5 was assumed, which yielded strengths of the order of 10^3 N m^{-1}. A choice also must be made as to how much thick ice takes part in ridging. Clearly by varying these different parameters a wide range of strengths may be obtained. Using, for example, a redistribution based on the mean ridge height, Bugden (1979) obtains strengths about four times as large as those of Thorndike *et al.* (1975).

Other estimates of ice strengths have been made with numerical models. In these estimates appropriate driving forces are assumed together with a constitutive law. Ranges of strength that give reasonable agreement with observed drift and thickness characteristics are then determined. The difficulty with these studies is that wind and water drag coefficients have to be assumed. However, investigations of free drift [e.g., McPhee (1979b)] indicate that present treatments of wind and water stress yield reasonable drift estimates. Such rheological studies have been carried out both with basin-wide ice models [Hibler (1979)] and localized near-shore simulation [e.g., Pritchard (1978)].

In the basin-wide calculations, Hibler carried out a simulation for the entire Arctic Basin over an 8-yr period using a plastic rheology with an elliptical yield-curve. The plastic strength was taken to be proportional to the ice thickness. To obtain reasonable agreement between the net observed and predicted drift rates of ice stations over a 1-yr period, a proportionality constant of $5 \cdot 10^3$ N m^{-2} was used. For ice 3 m thick this yields typical compressive strengths (in the absence of open water) of about $1.5 \cdot 10^4$ N m^{-1}. Calculations were also performed with fatter and thinner ellipses and larger and smaller compressive strengths. The results of these sensitivity tests indicated that doubling the strengths (especially the shear strength) substantially reduced the simulated net drift and restricted flow out of the Basin. Halving the strengths yielded unrealistically large net drifts.

With regard to more localized plastic tests, Pritchard (1978) examined near-shore dynamical behavior in some detail over periods of several weeks in the winter of 1976 and spring of 1977. Pritchard found best-fit compressive strengths were of the order of $4 \cdot 10^4$ N m^{-1}. When the compressive strengths were an order of magnitude lower than this, the velocity became effectively free drift. On the other hand, for compressive strengths an order of magnitude higher, the ice velocity was almost entirely caused by the boundary motion. Pritchard also examined a yield surface containing greater shear strengths, which he argued yielded more physically reasonable results. This was primarily because a greater shear strength simulated the stress state without causing large traction along leads formed during the simulation. To justify such a large shear strength, Pritchard proposed adding a "shear sink" of energy.

In general, both the basin-wide and near-shore numerical tests indicate strengths on the order of 10^4-10^5 N m^{-1}, about an order of magnitude higher than theoretical estimates by Thorndike *et al.* (1975) based on Rothrock's energetics arguments. However, by modifying the redistribution of ice under ridging [e.g., Bugden (1979)], the theoretical strengths can be substantially increased.

IV. NUMERICAL SIMULATION OF SEA ICE GROWTH, DRIFT, AND DECAY

Using aspects of the physics discussed in the previous section, it is possible to numerically simulate certain characteristics of sea ice behavior. Such numerical work can be broadly divided into two categories: short-term simulations spanning a month or less, and seasonal simulations covering at least one annual cycle. The short-term simulations have generally been oriented toward ice-forecasting applications, whereas the seasonal simulations have been oriented toward simulating the role played by sea ice in the interaction between the atmosphere and ocean.

A. Short-Term Simulations

The simplest type of short-term ice simulations make use of semiempirical rules to estimate the ice drift from the local wind and current values. The most widely used rule for this purpose is that of Zubov (1943), which (as noted earlier) simply states that the ice drifts parallel to the isobars with about one-hundredth of the geostrophic wind speed. Ad hoc modifications to such rules include seasonally modifying the proportionality constant and turning angle between ice drift and wind, and spatially averaging the wind field in order to crudely take into account the effects of ice interaction. Sobczak (1977) has compared several of these empirical rules to observed ice drift in the Beaufort Sea.

More complete models employ a coupling between the ice thickness and dynamics together with some type of ice rheology to model the ice interaction. Examples of such models employed for short-term integrations are those of Udin and Ullerstig (1976) and Lepparante (1977) for the Baltic, Kulakov *et al.* (1979) and Ling *et al.* (1979) for parts of the Antarctic, and Neralla and Lui (1980), Doronin (1970), and Coon *et al.* (1974) for parts of the Arctic. The most sophisticated and complete of these models is that of Coon *et al.* (1974) as later refined by Pritchard (1978). This model makes use of an elastic-plastic rheology coupled to a variable ice thickness distribution. The numerical scheme is formulated for an irregular deforming Lagrangian grid which allows boundary irregularities to be taken into account. In practice this model has been used together with observed data buoy drift data to supply a moving boundary condition. Specific simulation results have been reported by Coon *et al.* (1976) and Pritchard *et al.* (1977b). A simplified fixed-grid model similar to that of Coon *et al.* (1974) has been used for near-shore studies by Hibler (1978). This model employs a two-level ice thickness distribution, and makes use of a viscous-plastic numerical scheme for modeling plastic flow. The other models mentioned above use Newtonian viscous rheologies [except for Ling *et al.* (1970), where a linear viscous rheology is used].

The emphasis in these short-term studies has been on forecasting ice drift. Overall, this work has demonstrated that with adequate input wind data [Denner and Ashim (1979)], short-term drift patterns can be simulated well if ice interaction effects are properly accounted for. However, the extent to which the dependence of the ice dynamics on the thickness characteristics can be accurately forecast has been less well demonstrated.

B. Seasonal Simulations

A class of model studies particularly relevant to this review are numerical simulations of spatially varying sea ice growth, drift, and decay over at least one seasonal cycle. Simulations of this nature have been performed both using observed input data [Washington *et al.* (1976); Parkinson and Washington (1979); Hibler (1979)] and as part of a global coupled atmosphere ocean model [Bryan *et al.* (1975); Manabe *et al.* (1979)]. Except for the simulation by Hibler (1979), however, these studies have been essentially thermodynamic in nature and include only crude parametrizations of the ice motion. In particular, Bryan *et al.* (1975) and Manabe *et al.* (1979) allow the sea ice to drift with the upper ocean until it reaches cutoff thickness, at which time the ice motion is totally stopped. Washington *et al.* (1976) allow no ice motion but do allow leads. In the Parkinson and Washington (1979) study, the ice velocity field obtained by assuming free drift is iteratively corrected to ensure the maintenance of a fixed fraction of leads. This iteration is, however, performed without regard for conservation of momentum, and in practice appears to stop all motion rather than modify the relative motion as occurs in reality.

In the Hibler (1979) simulation a more realistic treatment of the ice dynamics and its relation to thickness variations is made. In particular, Hibler uses a nonlinear viscous-plastic ice rheology to relate the ice deformation and thicknesses to the internal stresses in the ice cover. Using this approach, realistic ice velocities and velocity induced thickness effects are simulated. However, a relatively idealized thermodynamics, in the form of spatially invariant growth rates dependent on time and ice thickness, are used. Also, Hibler's simulation covers only the Arctic Basin, not the entire arctic region.

In the following section the results of these simulations are discussed. First, the global thermodynamic simulations using observed data by Washington *et al.* (1976) and Parkinson and Washington (1979) are examined. Second, the Arctic Basin dynamic-thermodynamic simulation by Hibler (1979) is discussed. Third, sea ice results from coupled atmosphere–ocean model studies by Bryan *et al.* (1975) and Manabe *et al.* (1979) are briefly reviewed.

1. Global Thermodynamic Simulations

The most straightforward simulation of the seasonal growth and decay of the polar ice covers is that of Washington *et al.* (1976). The idea in this work was to generalize Maykut and Untersteiner's (1971) level ice thermodynamic model to include spatial variations. To do this Washington *et al.* (1976) used observed climatological atmospheric data compiled by Crutcher and Meserve (1970) and by Jenne *et al.* (1971) in conjunction with Semtner's (1976) time-independent, level ice, thermodynamic sea-ice model to calculate the components of the surface energy balance. This energy balance supplies a boundary condition for the thermodynamic ice model, which yields the growth or decay rate of the ice–snow system. In the standard experiment, the oceanic portion of the system was modeled by a mixed layer 30 m in depth with a fixed upward oceanic heat flux of 1.5 kcal cm^{-2} yr^{-1}. In addition, to account crudely for deformation effects that tend to create thin ice, a lead parametrization suggested by Semtner (1976) was used in some of the sensitivity studies. This parametrization consisted of assuming a fixed minimum 2% fraction of leads. This fraction was allowed to increase slowly under melting conditions and decrease (to 2%) rapidly under freezing conditions.

The results of the standard simulation of Washington *et al.* (1976) for both the Arctic and Antarctic are shown in Fig. 23. This figure also shows estimated observed ice extents. In the Arctic the ice extent results are in good agreement with observations. However, while the magnitudes of the simulated arctic thicknesses are reasonable, the geographical variation is not in good agreement with observed estimates (see Fig. 11), which show a greater buildup of ice near the Canadian Archipelago. Also, notable in the arctic portion of the Washington *et al.* (1976) simulation is the summer retreat of ice in the basin with an August thickness of less than 1 m except near the pole.

In the case of the antarctic ice cover the retreat and advance of the ice cover in the standard experiment are in poor agreement with observation. In particular the ice extent and thickness are much too large. However, Washington *et al.* (1976) find that the inclusion of leads (see Fig. 24) improves the thermodynamic simulation of the ice cover extent. In particular, by including leads the summer extent is substantially reduced. The primary reason for this seems to be the increased melting due to short-

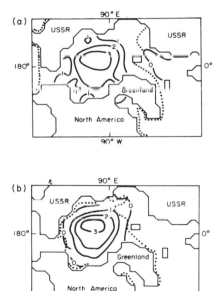

Fig. 23.   Simulated arctic [January (a), August (b)] and antarctic [March (c), August (d)] ice thickness contours obtained from thermodynamic considerations alone. [From Washington *et al.* (1976).] For the simulation a time-independent level ice thermodynamic model [Semtner (1976)] was used in conjunction with climatologically averaged atmospheric data. [Crutcher and Meserve (1970); Jenne *et al.* (1971).] The dotted line shows the estimated ice extent based on observations.

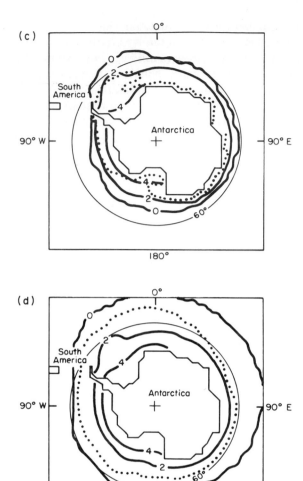

Fig. 23. *Continued.*

wave radiation absorption. Even better agreement is found by increasing
the oceanic heat flux to six times the amount used in the Arctic. Washington *et al.* (1976) argue that a reason for expecting the larger heat flux in the
Antarctic is that the surface layers in the Antarctic are less stratified than
in the Arctic [e.g., Gordon (1978)] and thus turbulent exchanges of heat
may occur more readily.

Following the Washington *et al.* (1976) approach with climatologically

averaged data, Parkinson and Washington (1979) have carried out seasonal simulations which, to a limited extent, include transport effects.
Normally Parkinson and Washington (1979) allow the ice to move in the
absence of internal ice stress effects. However, when the fraction of open
water reaches a minimum value, the divergence rate is iteratively corrected to maintain a fixed fraction of leads. Unfortunately this iteration is
performed without regard for conservation of momentum and appears to

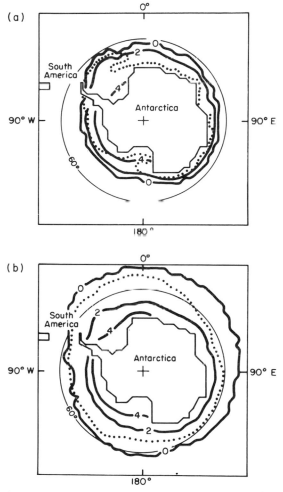

Fig. 24. Effects of including a fraction of leads and increased oceanic heat flux on the
thermodynamic simulation of the antarctic ice cover. [From Washington *et al.* (1976). In (a)
[March] and (b) [August] leads were included; in (c) [March] and (d) [August] both leads
and an increased oceanic heat flux (9 kcal cm^{-2} yr^{-1}) were used.

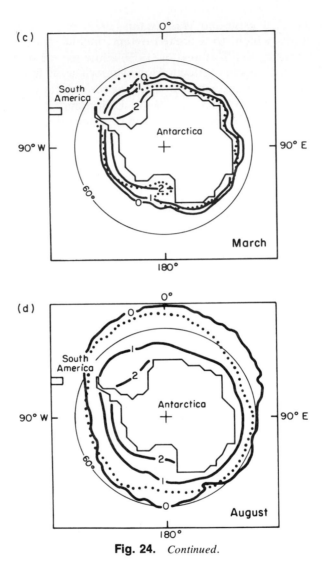

Fig. 24. *Continued.*

excessively damp the velocity field. A secondary problem is that the ice is not allowed to converge and increase in thickness by ridging, as occurs in reality. Because of these ad hoc ice interaction corrections, the arctic thickness contours of Parkinson and Washington (Fig. 25) differ little from those obtained by Washington *et al.* (1976) based on thermodynamic considerations alone, namely thicker ice near the pole with decreasing values as the coasts are approached. Similar statements also hold for the Antarc-

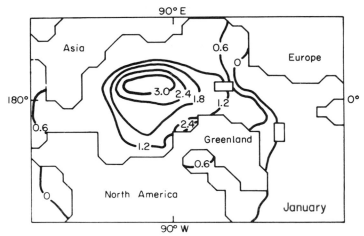

Fig. 25.   Simulated arctic ice thickness contours. [From Parkinson and Washington (1979).]

tic, although direct comparisons are difficult to make due to a different formulation of the oceanic heat flux. In particular, Parkinson and Washington (1979) assume an Antarctic Ocean heat flux of 25 W/m² (18.84 kcal cm⁻² yr⁻¹) compared to the values of Washington *et al.* (1976) of 1.5 and 9.0 kcal cm⁻² yr⁻¹.

Under summer melt conditions, however, the ad hoc velocity correction employed by Parkinson and Washington (1979) is needed less often. Consequently the summer transport effects are more reasonable and the simulated summer compactness (especially near shore) tends to be considerably less than that obtained from thermodynamics considerations alone, in better agreement with observed estimates. Also, the summer ice cover in the Arctic is more compact than that of the Antarctic. Parkinson and Washington (1979) were able to correctly simulate this phenomenon.

An interesting feature of the Parkinson and Washington simulation is the difference between simulated ice extent and the freezing air temperature isotherm. In particular, the ice was found generally to lag behind the freezing line, presumably due to the thermal inertia of the fixed-depth oceanic boundary layer. In the fall the freezing line expands southward faster than the ice and retreats more rapidly than the ice in the spring. This effect is in contrast to Stefan-law–like estimates of the ice thickness and indicates a thermal "inertia" in the simulated ice–ocean system.

Overall, these thermodynamic simulations indicate that, as long as leads are accounted for, many of the characteristics of the advance and retreat of the ice edge correlate well with heat budget estimates based on climatological data. However, it should be remembered that, by driving the simulation with observed air temperatures, which are in turn dependent on the ice conditions, proper results are partially forced. Recognizing this fact, Washington *et al.* (1976) emphasize that their results may be a case of obtaining "good agreement with observations for the wrong reasons." The following examination of a dynamic-thermodynamic simulation of the arctic ice cover suggests that, in the case of the arctic ice thicknesses, such a specious agreement has probably occurred. Specifically, it appears that the single thickness thermodynamic calculations underestimate the growth rate. However, this error is offset by underestimating transport effects, which would tend to reduce the ice thicknesses.

2.   A Dynamic–Thermodynamic Seasonal Simulation

A more complete treatment of the effects of ice dynamics on the seasonal growth and decay cycle of sea ice has been presented by Hibler (1979) in a dynamic thermodynamic simulation of the arctic ice cover. In this study Hibler (1979) made use of a sea ice model in which the strength of the ice interaction is related to a two-level ice thickness distribution by means of a plastic rheology. To allow the ice interaction to become stronger in regions of ice inflow and weaker in regions of outflow, the plastic strength P was taken to be proportional to the ice mass and exponentially related to the fraction of thin ice $1 - A$,

$$P = P^*m \exp[C(A - 1)], \tag{21}$$

where P^* and C are fixed empirical constants. The ice concentration A and mass per unit area m were allowed to evolve according to the continuity Eqs. (2) and (3). Note that ice convergence is allowed in these continuity equations. This convergence thickens the ice, making it stronger. At some point this strength imbalance will counteract the external forcing. For the momentum balance Hibler used Eq. (8), together with quadratic planetary boundary layers [see Eqs. (9)–(12)]. Inertial and momentum advection terms were also included.

The thermodynamic treatment used by Hibler for thin and thick ice was similar to that of Washington *et al.* (1976). In particular, the ice growth is taken to be the sum of growth occurring over open water and the remaining thick ice. Similarly, under melting conditions, the heat absorbed in the open water is used to melt the adjacent ice cover. To approximate the effects of growth and decay on the fraction of thin ice, the open water fraction is allowed to decrease rapidly under freezing conditions and increase slowly under melting conditions. In Hibler's case the decrease was exponentially related to the growth rate with a decay constant chosen so that the "open water" approximated the areal fraction of ice having a thickness between 0 and 50 cm. However, in contrast to Washington *et al.* (1976), no lower limit of open water is forced to exist.

For the standard simulation experiment, Hibler applied this model to the Arctic Basin and performed an integration at 1-day time steps and 125 km resolution in order to obtain a cyclic equilibrium. Input fields consisted of observed, time-varying geostrophic wind over a 1-yr period and fixed geostrophic ocean currents from Coachman and Aagard (1974). For the thermodynamic code, geographically invariant growth rates dependent on ice thickness and time (see Fig. 26) were used. These growth rates were taken from Thorndike *et al.* (1975) and are heavily based on thermodynamic calculations by Maykut and Untersteiner (1971), which include a fixed oceanographic heat flux from below the mixed layer. This simple thermodynamic code simplified the analysis since, in the absence of dynamics, no spatial thickness variations would be present.

The simulated April ice thickness contours and average annual ice velocity field from the standard experiment are shown in Fig. 27. Particularly notable are the thickness results which show a substantial buildup of ice along the Canadian Archipelago in conjunction with a thinning of ice along the North Slope and Siberian coast. The geographical shape of

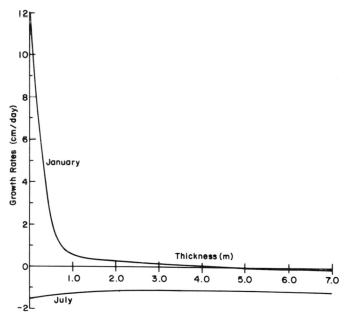

Fig. 26.   Typical winter and summer ice growth rates. [From Thorndike *et al.* (1975).]

these dynamically forced contours agrees better with the available observations (see Fig. 11) than the Washington *et al.* (1976) and Parkinson and Washington (1979) results.

The mechanism of the dynamically forced thickness buildup in the Hibler simulation is apparent from an examination of the mean annual velocity field. While day-to-day velocities can differ substantially, this annual pattern shows that, on the average, advection will tend to remove ice from the North Slope and Siberian coasts and either remove it from the basin or deposit it along the Canadian Archipelago. As the simulation proceeds in time, the buildup strengthens the ice along the Archipelago while thinning weakens the ice off the other coasts. The resulting strength imbalance eventually counteracts the external forcing, thus creating seasonal equilibrium ice thickness and velocity characteristics. By using a plastic rheology, the thickness contours at which this strength imbalance occurs agree well with observed geographical variation in thickness. The simulated contours are, however, dependent on the assumed strength. In

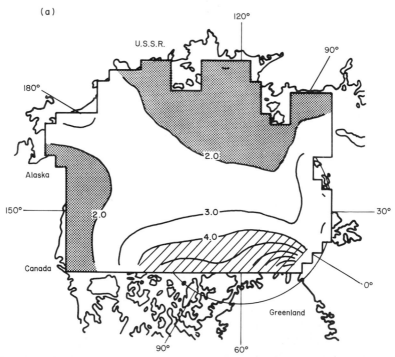

Fig. 27. Simulated arctic sea ice thickness contours in meters (a) and average annual ice velocity field (b) from a dynamic-thermodynamic simulation. In the ice velocity field, a velocity vector one grid space long represents 0.02 m sec^{-1}. [From Hibler (1979).]

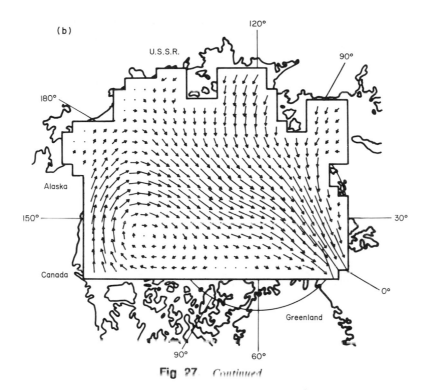

(b)

Fig 27   Continued

the standard case the plastic strength constant was adjusted to give reasonable simulated drift rates (as compared to drifting stations). However, halving P^* will increase the Archipelago thickness by about 50%.

Figure 28 demonstrates the pronounced effect of the dynamics on the geographical variation of ice growth rates (and hence air–sea heat exchange). On the average the offshore advection creates thinner ice and open water off the North Slope and Siberian coast. This effect is particularly pronounced in the summer (when open water is not removed by freezing) and results in a region of low compactness off the North Slope and Siberian coast in good agreement with observations. Largely because of this offshore advection, the annual net ice growth is very large (more than 1.0 m yr^{-1}) off the North Slope and Siberian coast. Off the Canadian Archipelago, on the other hand, the net growth is negative, reflecting the fact that the ice there is so thick that ablation occurs all year round.

An illustration of the complex interaction between the dynamics and the thermodynamics of a variable thickness ice cover is provided by the simulated mass balance results for the Arctic Basin. In Table 2, the mass balance results from the standard experiment [Hibler (1979)] are compared

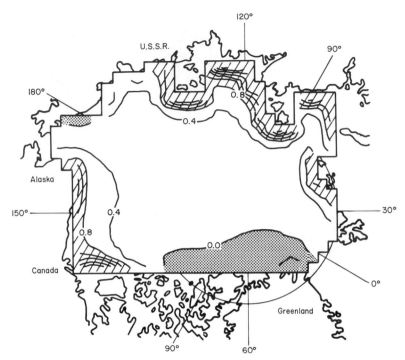

Fig. 28.   Simulated annual net ice growth in meters in the Arctic Basin. [From Hibler (1979).]

with Koerner's (1973) observed estimates. In addition, results of integrating only the thermodynamic portion of the model without dynamics are shown for a zeroth-order comparison. The simulated annual average outflow of about 3200 km³ agrees well with observational estimates. This outflow causes a net ice growth of 0.42 m annually, which has to occur for the ice mass budget to be balanced.

Other changes induced by the dynamics include a 150% increase (from 0.35 to 0.89 m) in the ice production (i.e., the total amount of ice grown) and a 30% reduction in the mean ice thickness. This ice production is in reasonable agreement with observed estimates but the reduced thickness is not. The increase in ice production by an amount greater than the outflow results primarily from increased ablation in the buildup of thick ice along the Canadian Archipelago. Under balance conditions, the total ice growth must balance both the outflow and increased ablation. The reduction in mean ice thickness is due primarily to the fact that thinner ice and/or more open water are required to create the annual growth needed to balance the outflow. In the standard simulation, the thickness required for correct annual growth (about 2.5 m) is rather low compared to observed estimates of 3.7 m. This is in contrast to the dynamics-free simula-

TABLE 2

Comparison of Simulated [Hibler (1979)] and Observed [Koerner (1973)] Mass Balance Statistics for the Arctic Basin Averaged over an Annual Cycle

	Outflow	Ice production (m)	"Open water" ice production	Average percent of "open water"	Net growth (m)	Mean maximum ice thickness (m)	Summer–winter ice thickness change (m)
Standard simulation	3220 km³ (0.451 m)	0.891	0.299 m (34%)	2.84	0.422	2.659	0.524
No dynamics simulation	0.0	0.350		0.54	0.0	3.880	0.350
Koerner's estimates	5580 km³ [a]	~1.1	~40% for ice <1.0 m thick	—	~0.5	~3.7	~0.6

[a] Aagaard and Greisman (1975) estimate the average outflow to be ~0.1 Sverdrup (3154 km³ yr^{-1}).

tion which yields about a 3.7 m thickness, in good agreement with observed estimates.

The most likely source of this thickness discrepancy is insufficient detail in the ice thickness distribution. In particular, due to the nonlinear nature of the ice growth rates and the presence of a variable thickness ice cover, it is probable that the mean growth rate of the "thick" ice is larger than the growth rate of a constant thickness ice cover of equivalent mass.

In general, the results from this dynamic–thermodynamic simulation confirm the importance of ice dynamics to sea ice growth and decay. In particular, by including ice dynamics the sea to air heat losses in the Arctic Basin (as recorded by the ice production) are substantially increased. Also, the geographical character of this heat loss can be greatly modified by the dynamics. A corollary conclusion is that sea–air heat exchange estimates based on a single thickness are probably unrealistically low. In purely "thermodynamic" simulations [e.g., Washington *et al.* (1976)], this inadequacy is masked by underestimating advection which would remove ice and lower the thickness.

3. Coupled Atmosphere–Ice–Ocean Models

In the simulations discussed above, observed atmospheric and oceanic forcing fields are used to drive a sea ice model. As a consequence, the system is not fully coupled since ice-induced modifications in the air–sea heat exchanges are not allowed to modify the atmosphere and ocean circulation. Examples of simulations where a more complete coupling have been employed are numerical studies by Bryan *et al.* (1975) and Manabe *et al.* (1979). In these studies a complete atmospheric circulation model and ocean circulation model are jointly integrated. In this coupled system sea ice is included by a simple parameterization developed by Bryan *et al.* Only one thickness of ice is allowed to exist in this model. The growth and decay of this thickness are calculated using heat budget considerations together with a simple time-independent sea ice thermodynamic model [similar to that of Semtner (1976) except the snow cover is incorporated as part of the ice]. To parametrize the ice motion crudely, the ice is allowed to drift with the upper layer of the ocean until it reaches a fixed cutoff thickness (4 m). At this thickness its motion is totally stopped. To keep track of the ice advection and deformation a single continuity equation [Eq. (3)] is employed.

Using this coupled system, two types of simulations have been done: one with a fixed mean annual insolation [Bryan *et al.* (1975)], and one in which the insolation varies seasonally [Manabe *et al.* (1979)]. In the Antarctic both simulations yield unrealistically small amounts of ice with only about 1 m of ice forming in certain portions of the Weddell and Ross Seas. In the Arctic the seasonal simulations yield reasonable thickness values while the fixed insolation integration yields excessively large values. In particular, the mean annual insolation case [Bryan et al. (1975)], the ice rapidly builds up in thickness until the critical thickness (4 m) is exceeded. At that point all motion stops, making ice export from the Arctic Basin impossible. At the end of the experiment the ice is excessively thick in the Arctic, at some points exceeding 20 m.

With seasonal insolation, however, reasonable Arctic Basin thickness results are obtained. Averaging over about the same region as used by Koerner (1973) and Manabe et al. (1979) find a mean annual thickness of 2.4 m. This thickness has a reasonable seasonal cycle varying from a maximum of about 3.3 m in April to about 2.2 m in August. In explaining the greatly improved results in the seaonal simulation, Manabe et al. (1979) emphasize that this reduction is in large part due to the intense summer insolation which tends to greatly reduce the snow and ice cover. Some of the thickness reduction may also be due to allowing export. However, in an analysis of the ice drift results from this seasonal simulation K. Bryan (private communication) finds the ice export from the Arctic Basin to be substantially less than Koerner's estimates. Consequently, the reduction in ice thickness from the fixed mean annual insolation case appears to result primarily from the nonlinear thermodynamic character of the coupled ice ocean atmosphere system.

Given the complexity of coupled atmosphere–ocean models the reasonable arctic results obtained from this seasonal simulation are encouraging. They provide some hope that, at least in the northern hemisphere, such coupled models may be used as tools to examine the role of sea ice in climate.

V. CONCLUDING REMARKS

Many aspects of the physics of sea ice growth, drift, and decay have been identified within the last decade. It is now becoming possible to investigate in more detail the coupled atmosphere–ice–ocean system. Of particular interest are the extent and the manner in which the ice characteristics are linked to atmospheric and oceanic variability and hence to climatic change.

In this regard, it seems likely that an essential effect of sea ice on climate arises from modifications of the thermal fluxes at the air–sea interface. Perhaps the most obvious effect is the increased sea–air heat flux due to the presence of thin ice or leads, a parameter emphasized by

Fletcher (1965), Badgley (1966), and Maykut (1978). In considering this modification, it is important to note that the annual magnitude of this flux is probably not particularly important in terms of the overall global energy budget, simply because the amount of latent heat in sea ice formed each year is quite small compared to the heat stored and released by the atmosphere and ocean over a seasonal cycle [Oort and Vonder Haar (1976)]. However, sea ice can substantially change the geographical and temporal character of the air–sea heat exchange. These modifications supply a boundary condition whose variation may well affect atmospheric and oceanic circulation [e.g., Herman and Johnson (1978)]. Moreover, this heat exchange is particularly interesting because of its substantial dependence on ice dynamics (see Fig. 28) which, in turn, partially depend on atmospheric and oceanic dynamics.

Other notable modifications of the sea–air interface by sea ice include a change in the momentum transfer into the ocean due to the ice interaction, a change in the surface albedo, and a change in the salinity structure of the ocean due to salt expelled from freezing sea ice. This last effect is felt to be pivotal in antarctic bottom water formation [Foster and Carmack (1976); Gill (1973)]. The albedo change is fundamental to simple Budyko–Sellers climate models [Budyko (1969); Sellers (1969)]. (The characteristic feature of these models is a temperature-dependent albedo that provides a strong positive feedback between ice and snow cover and variations in the insolation.) Finally, modifications of the air–sea momentum transfer may well alter the ocean current and mixed-layer structure.

As these brief comments indicate, sea ice has the potential to possibly modify the climate by modifying the air–sea interface. At present, examinations of the role of sea ice in climatic change are just beginning. From the work reviewed here, however, it is evident that, to understand this role, both the dynamics and thermodynamics of sea ice must be considered.

ACKNOWLEDGMENTS

The author would like to thank Dr. Sam Colbeck for continued advice and encouragement which made completion of this review possible. The author is also grateful for valuable discussions with Dr. Miles McPhee on planetary boundary layers, with Mr. Steve Ackley on antarctic sea ice, with Mr. Terry Tucker on the East Greenland ice, with Dr. W. F. Weeks and Dr. Norbert Untersteiner on the physical properties of sea ice, and with Mr. Austin Kovacs on the morphology of pressure ridges. In addition, comments on the manuscript by Dr. Alan Thondike improved the final paper. Proofreading and editing aid were provided by Mr. Stephen Bowen and Mr. Mark Ray. This work was supported by the Office of Naval Research and by the Army Corps of Engineers.

REFERENCES

Aagaard, K., and Greisman, P. (1975). *J. Geophys. Res.* **80**, 3821–3827.

Ackley, S. F. (1980). *J. Glaciol.* (in press).

Ackley, S. F., and Keliher, T. E. (1976). *AIDJEX Bull.* **33**, 53–76.

Ackley, S. F., Hibler, W. D., III, Kugzruk, F. K., Kovacs A., and Weeks, W. F. (1976). "CCREL Report 76–18," p. 25. U.S. Army Cold Reg. Res. Eng. Lab., Hanover, New Hampshire.

Anderson, D. L. (1961). *J. Glaciol.* **3**, 1170–1172.

Andreas, E. L., Paulson, C. A., Williams, R. M., Lindsey, R. W., and Businger, J. A. (1979). *Boundary-Layer Meteorol.* **17**, 57–91.

Badgley, F. I. (1966). *In* "Arctic Heat Budget and Atmospheric Circulation" (J. O. Fletcher, ed.), pp. 267–278.

Banke, E. G., Smith, S. D., and Anderson, R. J. (1976). *J. Fish. Res. Board Can.* **33**, 2307–2317.

Barnett, D. (1979). *Proc. ICSI/AIDJEX Symp. Sea Ice Process. Models, 1977* (in press).

Bilello, M. A. (1961). *Arcic* **14**, 2–24.

Bilello, M. A. (1979). *Proc. ICSI/AIDJEX Symp. Sea Ice Process. Models, 1977* (in press).

Brown, R. A. (1979). *Proc. ICSI/AIDJEX Symp. Sea Ice Process. Models, 1977* (in press).

Bryan, K., Manabe, S., and Pacanowski, R. L. (1975). *J. Phys. Oceanogr.* **5**, 30–46.

Budd, W. F. (1975). *J. Glaciol.* **15**, 417–428.

Budyko, M. I. (1969). *Tellus* **21**, 611–619.

Bugden, G. L. (1979). *J. Geophys. Res.* **84**, 1793–1796.

Campbell, W. J. (1965). *J. Geophys. Res.* **70**, 3279–3301.

Carsey, F. D. (1979). *Proc. ICSI/AIDJEX Symp. Sea Ice Process. Models, 1977* (in press).

Coachman, L. K., and Aagaard, K. (1974). *In* "Marine Geology and Oceanography of the Arctic Seas" (Y. Herman, ed.), pp. 1–72. Springer-Verlag, Berlin and New York.

Colony, R., and Rothrock, D. A. (1979). *Proc. ICSI/AIDJEX Symp. Sea Ice Process. Models, 1977* (in press).

Coon, M. D. (1974). *J. Pet. Technol.* **26**, 466–470.

Coon, M. D., and Pritchard, R. S. (1980). *J. Glaciol.* (in press).

Coon, M. D., Maykut, G. A., Pritchard, R. S., Rothrock, D. A., and Thorndike, A. S. (1974). *AIDJEX Bull.* **24**, 1–105.

Coon, M. D., Colony, R., Pritchard, R. S., and Rothrock, D. A. (1976). *In* "Numerical Methods in Geomechanics" (C. S. Desai, ed.), Vol. 2, pp. 1210–1277. Am. Soc. Chem. Eng., New York.

Cox, G. F. N., and Weeks, W. F. (1974). *J. Glaciol.* **13**, 109–120.

Cox, G. F. N., and Weeks, W. F. (1975). "CRREL Research Report 345," p. 85. U.S. Army Cold Reg. Res. Eng. Lab., Hanover, New Hampshire.

Croasdale, K. R. (1978). *Proc. Int. Conf. Port Ocean Eng. Under Arct. Cond., 4th, 1977* pp. 1–32.

Crutcher, H. L., and Meserve, J. M. (1970). "Selected Level Heights, Temperatures and Dew Points for the Northern Hemisphere," NAVAIR 50-1C-52 Revised. Naval Weather Service Command, Washington, D.C.

Denner, W. D., and Ashim, N. D. (1979). *Proc. ICSI/AIDJEX Symp. Sea Ice Process. Models, 1977* (in press).

Doronin, Y. P. (1970). *Tr. Arkt. Antartkt. Nanchno-Issled. Inst.* **291**, 5–17; transl. in *AIDJEX Bull.* **3**, 22–39 (1970).

Dunbar, M., and Wittmann, W. (1963). *In* "Proceedings of the Arctic Basin Symposium," pp. 90–103. Arct. Inst. North Am., Washington, D.C.

Egorov, K. L. (1971). *Tr. Arkt. Antarkt. Nachno-Issled. Inst.* **303,** 108–113; transl. in *AIDJEX Bull.* **16,** 119–124 (1972).

Felzenbaum, A. I. (1958). *Probl. Sev.* **2,** 16–46; *Probl. North, (Engl. Transl.)* **2,** 13–44 (1961).

Fletcher, J. O. (1965). "The Heat Budget of the Arctic Basin and Its Relation to Climate," p. 1790 R-444-PR. Rand Corporation, Santa Monica, California.

Foster, T. D., and Carmack, E. C. (1976). *Deep-Sea Res.* **23,** 301–337.

Gill, A. E. (1973). *Deep-Sea Res.* **20,** 111–140.

Glen, J. W. (1970). *AIDJEX Bull.* **2,** 18–27.

Gordienko, P. (1958). *N.A.S.–N.R.C., Publ.* **598,** 210–220.

Gordon, A. L. (1978). *J. Phys. Oceanogr.* **8,** 600–612.

Gordon, A. L., and Taylor, N. W. (1975). *Science* **187,** 346–347.

Hastings, A. D., Jr. (1971). "Report ETL-TR-71-5," p. 98. U.S. Army Eng. Top. Lab., Fort Belvoir, Virginia.

Herman, G. F., and Johnson, W. T. (1978). *Mon. Weather Rev.* **106,** 1649–1664.

Hibler, W. D., III (1974). *J. Glaciol.* **13,** 457–471.

Hibler, W. D., III (1977). *J. Geophys. Res.* **82,** 3932–3938.

Hibler, W. D., III (1978). *Proc. Int. Conf. Port Ocean Eng. Under Arct. Cond., 4th, 1977,* pp. 33–45.

Hibler, W. D., III (1979). *J. Phys. Oceanogr* **9,** 815–846.

Hibler, W. D., III, and Tucker, W. B. (1979). *J. Glaciol.* **22,** 293–304.

Hibler, W. D., III, Ackley, S. F., Weeks, W. F., and Kovacs, A. (1972a). *AIDJEX Bull.* **13,** 77–91.

Hibler, W. D., III, Weeks, W. F., and Mock, S. J. (1972b). *J. Geophys. Res.* **77,** 5954–5970.

Hibler, W. D., III, Weeks, W. F., Kovacs, A., and Ackley, S. F. (1974a). *J. Glaciol.* **13,** 437–455.

Hibler, W. D., III, Ackley, S. F., Crowder, W. K., McKim, H. L., and Anderson, D. M. (1974b). *In* "The Coast and Shelf of the Beaufort Sea" (J. C. Reed and J. E. Sater, eds.), pp. 285–296. Arct. Inst. North Am., Arlington, Virginia.

Hibler, W. D., III, Mock, S. J., and Tucker, W. B. (1974c). *J. Geophys. Res.* **79,** 2735–2743.

Hunkins, K. (1967). *J. Geophys. Res.* **72,** 1165–1174.

Hunkins, K. (1975). *J. Geophys. Res.* **80,** 3425–3433.

Jenne, R. L., Crutcher, H. L., van Loon, H., and Taljaard, J. J. (1971). "NCAR TN/STR-58," p. 68. Natl. Cent. Atmos. Res., Boulder, Colorado.

Kelly, P. M. (1978). *Clim. Monit.* **7,** 95–98.

Koerner, R. M. (1973). *J. Glaciol.* **12,** 173–185.

Kohnen, H. (1976). *Ocean Eng.* **3,** 343–360.

Kovacs, A. (1972). *In* "Sea Ice, Proceedings of an International Conference" (T. Karlsson, ed.), pp. 276–298. Natl. Res. Council, Reykjavik, Iceland.

Kovacs, A. (1976). "CRREL Report 76-32," p. 31. U.S. Army Cold Reg. Res. Eng. Lab., Hanover, New Hampshire.

Kovacs, A., and Mellor, M. (1974). *In* "The Coast and Shelf of the Beaufort Sea" (J. C. Reed and J. E. Sater, eds.), pp. 113–162. Arct. Inst. North America, Arlington, Virginia.

Kozo, T. L., and Tucker, W. B. (1974). *J. Geophys. Res.* **79,** 4505–4511.

Kukla, G. J. (1978). *In* "Climatic Change" (J. Gribbin, ed.), pp. 114–129. Cambridge Univ. Press, London and New York.

Kulakov, I. Yu., Maslovsky, M. I., and Timokhov, L. A. (1979). *Proc. ICSI/AIDJEX Symp. Sea Ice Process. Models, 1977* (in press).

Laikhtman, D. L. (1964). "Physics of the Boundary Layer of the Atmosphere (transl.)" p. 200. U.S. Dept. of Commerce, Washington, D.C.

Lake, R. A., and Lewis, E. L. (1970). *J. Geophys. Res.* **75**, 5243–5246.
Langleben, M. P. (1968). *J. Glaciol.* **7**, 289–297.
Langleben, M. P. (1971). *J. Glaciol.* **10**, 101–104.
Langleben, M. P. (1972). *J. Glaciol.* **11**, 337–344.
Leavitt, E. (1979). *Proc. ICSI/AIDJEX Symp. Sea Ice Process. Models, 1977* (in press).
Lemke, P., Trinke, E. W., and Hasselmann, K. (1980). Submitted for publication.
Lepparante, M. (1977). *In* "100 Years of Winter Navigation in Finland." Oulu, Finland.
LeSchack, L. A., Hibler, W. D., and Morse, F. H. (1971). *AGARD Conf. Proc.* **90**, 5-1 to 5-19.
Ling, C. H., Rasmussen, L. A., and Campbell, W. J. (1979). *Proc. ICSI/AIDJEX Symp. Sea Ice Process. Models, 1977* (in press).
Lowry, R. T., and Wadhams, P. (1979) *J. Geophys. Res.* **84**, 2487–2494.
McPhee, M. G. (1978). *Dyn. Atmos. Oceans* **2**, 107–122.
McPhee, M. G. (1979a). *J. Phys. Oceanogr.* **9**, 389–400.
McPhee, M. G. (1979b). *Proc. ICSI/AIDJEX Symp. Sea Ice Process. Models, 1977* (in press).
McPhee, M. G., and Smith, J. D. (1976). *J. Phys. Oceanogr.* **6**, 696–711.
Malvern, L. E. (1969). "Introduction to the Mechanics of a Continuous Media," p. 713. Prentice-Hall, Englewood Cliffs, New Jersey.
Manabe, S., Bryan, K., and Spelman, M. J. (1979). *Dyn. Atmos. Oceans* **3**, 393–426.
Marko, J. R., and Thomson, R. E. (1977). *J. Geophys. Res.* **82**, 979–987.
Marshunova, M. S. (1961). *Proc. Arct. Antarct. Res. Inst.* p. 229.
Maykut, G. A. (1978). *J. Geophys. Res.* **83**, 3646–3658.
Maykut, G. A., and Untersteiner, N. (1971). *J. Geophys. Res.* **76**, 1550–1575.
Nansen, F. (1902). "The Norwegian Polar Expedition, 1893–1896, Scientific Results." Christiana, Oslo
Nazarov, V. S. (1963). *Okeanologiya* **3**, (2), 243–249.
Neralla, V. R., and Liu, W. S. (1980). *J. Glaciol.* (in press).
Neumann, G., and Pierson, W. J. (1966). "Principles of Physical Oceanography," p. 545. Prentice-Hall, Englewood Cliffs, New Jersey.
Nikiforov, Y. G. (1957). *Probl. Arktiki* **2**, 59–71 (transl. by Am. Met. Soc.).
Nye, J. F. (1973a). *AIDJEX Bull.* **21**, 9–17.
Nye, J. F. (1973b). *AIDJEX Bull.* **21**, 18–19.
Nye, J. F. (1975). *J. Glaciol.* **15**, 429–436.
Oort, A. H., and Vonder Haar, T. H. (1976). *J. Phys. Oceanogr.* **6**, 781–800.
Parkinson, C. L., and Washington, W. M. (1979). *J. Geophys. Res.* **84**, 311–337.
Parmerter, R. R., and Coon, M. D. (1972). *J. Geophys. Res.* **77**, 6565–6575.
Pounder, E. R. (1965). "The Physics of Ice," p. 151. Pergamon, Oxford.
Pritchard, R. S. (1975). *Trans. ASME* **42**, 379–384.
Pritchard, R. S. (1978). *Proc. Int. Conf. Port Ocean Eng. Under Arct. Cond., 4th, 1977* pp. 494–505.
Pritchard, R. S., and Schwaegler, R. T. (1976). *AIDJEX Bull.* **31**, 137–150.
Pritchard, R. S., Coon, M. D., McPhee, M. G., and Leavitt, E. (1977a). *AIDJEX Bull.* **37**, 37–93.
Pritchard, R. S., Coon, M. D., and McPhee, M. G. (1977b). *J. Pressure Vessel Technol.* **995**, 491–497.
Reed, R. J., and Campbell, W. J. (1962). *J. Geophys. Res.* **67**, 281–297.
Reimnitz, E., Toimil, L., and Barnes, P. (1978). *Mar. Geol.* **28**, 179–210.
Rigby, F. A., and Hanson, A. (1976). *AIDJEX Bull.* **34**, 43–71.
Rogers, J. C. (1978). *Mon. Weather Rev.* **106**, 890–897.
Rothrock, D. A. (1975a). *Annu. Rev. Earth Planet. Sci.* **3**, 317–342.
Rothrock, D. A. (1975b). *J. Geophys. Res.* **80**, 387–397.

Rothrock, D. A. (1975c). *J. Geophys. Res.* **80,** 4514–4519.
Rothrock, D. A. (1980). *J. Glaciol.* (in press).
Sanderson, R. M. (1975). *Meteorol. Mag.* **104,** 313–323.
Schwarz, J., and Weeks, W. F. (1977). *J. Glaciol.* **19,** 499–532.
Schwarzacher, W. (1959). *J. Geophys. Res.* **64,** 2357–2367.
Sellers, W. D. (1969). *J. Appl. Meteorol.* **8,** 392–400.
Semtner, A. J., Jr. (1976). *J. Phys. Oceanogr.* **6,** 379–389.
Shirawawa, K., and Langleben, M. P. (1976). *J. Geophys. Res.* **36,** 6451–6454.
Shuleikin, V. V. (1938). *C. R. (Dokl.) Acad. Sci. SSSR* **19,** 589–594.
Skov, N. A. (1970). *Medd Groenl.* **188.**
Sobczak, L. W. (1977). *J. Geophys. Res.* **82,** 1413–1418.
Sodhi, D., and Hibler, W. D., III (1979). *Proc. ICSI/AIDJEX Symp. Sea Ice Process. Models, 1977* (in press).
Sverdrup, H. U. (1928). *Sci. Results Norw. North Polar Exped. Maud* **4,** 1–46.
Swithinbank, C. W. M. (1972). *In* "Sea Ice, Proceedings of an International Conference" (T. Karlsson, ed.), pp. 246–254. Natl. Res. Counc., Reykjavik, Iceland.
Thorndike, A. S. (1973). *Proc. Ocean 73, IEEE Int. Conf. Eng. Ocean Environ. (IEEE Publ. No. 73),* pp. 490–499.
Thorndike, A. S. (1979). *Proc. ICSI/AIDJEX Symp. Sea Ice Process. Models, 1977* (in press).
Thorndike, A. S., and Colony, R. (1978). *Proc. Int. Conf. Port Ocean Eng. Under Arct. Cond., 4th, 1977* pp. 506–517.
Thorndike, A. S., and Colony, R. (1979). *Proc. ICSI/AIDJEX Symp. Sea Ice Process. Models, 1977* (in press).
Thorndike, A. S., Rothrock, D. A., Maykut, G. A., and Colony, R. (1975). *J. Geophys. Res.* **80,** 4501–4513.
Treshnikov, A. F. (1967). *Proc. Pac. Sci. Congr., 11th, 1966* pp. 113–123.
Tryde, P. (1977). *J. Glaciol.* **19,** 257–264.
Tucker, W. B., Weeks, W. F., Kovacs, A., and Gow, A. J. (1979a). *Proc. ICSI/AIDJEX Symp. Sea Ice Process. Models, 1977* (in press).
Tucker, W. B., Weeks, W. F., and Frank, M. (1979b). *J. Geophys. Res.* **84,** 4885–4897.
Udin, I., and Ullerstig, A. (1976). "Research Report No. 18," p. 40. Swedish Administration of Shipping and Naviation, Norrkoping, Sweden.
Untersteiner, N. (1961). *Arch. Meteorol., Geophys. Bioklimatol., Ser. A,* **12,** 151–182.
Untersteiner, N. (1964). *J. Geophys. Res.* **69,** 4755–4766.
Untersteiner, N. (1968). *J. Geophys. Res.* **73,** 1251–1257.
Vijne, T. E. (1976). *Arbok, Nor. Polarinst., 1975* pp. 163–174.
Vowinckel, E., and Orvig, S. (1970). *In* "World Survey of Climatology" (S. Orvig, ed.), Vol. 14, pp. 129–227. Am. Elsevier, New York.
Wadhams, P. (1976). *AIDJEX Bull.* **33,** 1–52.
Wadhams, P. (1977). *Polar Rec.* **18,** 487–491.
Wadhams, P. (1978). *Proc. Int. Conf. Port Ocean Eng. Under Arct. Cond., 4th, 1977* pp. 544–555.
Wadhams, P. (1980a). *J. Glaciol.* (in press).
Wadhams, P. (1980b). *Proc. Seasonal Sea Ice Zone Workshop, 1979* (in press).
Wadhams, P., and Horne, R. J. (1978). Scott Polar Research Institute Tech. Rep. 78–1. Cambridge, England.
Wadhams, P., Gill, A. E., and Linden, P. F. (1979). *Deep-Sea Res.* (in press).
Walsh, J. E. (1977). *Mon. Weather Rev.* **105,** 1527–1535.
Walsh, J. E. (1979). *Proc. ICSI/AIDJEX Symp. Sea Ice Process. Models, 1977* (in press).
Walsh, J. E., and Johnson, C. M. (1979). *J. Phys. Oceanogr.* **9,** 580–591.

Washington, W. M., Semtner, A. J., Parkinson, C., and Morrison, L. (1976). *J. Phys. Oceanogr.* **6,** 679–685.

Weeks, W. F. (1968). *In* "SCAR/SCOR/IAPO/IUPS Symposium on Antarctic Oceanography, Santiago, Chile, 1966, pp. 173–190. Scott Polar Res. Inst., Cambridge.

Weeks, W. F., and Gow, A. J. (1978). *J. Geophys. Res.* **83,** 5105–5122.

Weeks, W. F., Kovacs, A., and Hibler, W. D., III (1971). *Proc. Intl. Conf. Port Ocean Eng. Arctic Cond., 1st, 1971,* Vol. 1, pp. 152–183.

Weeks, W. F., Kovacs, A., Mock, S. J., Hibler, W. D., III, and Gow, A. J. (1977). *J. Glaciol.* **19,** 533–546.

Welander, P. (1977). *Dyn. Atmos. Oceans* **1,** 215–223.

Weller, G. (1972). *AIDJEX Bull.* **14,** 28–30.

Wendler, G. (1973). *J. Geophs. Res.* **78,** 1427–1448.

Williams, E., Swithinbank, C. W. M., and Robin, G. DeQ. (1975). *J. Glaciol.* **15,** 349–362.

Zubov, N. N. (1943). "Arctic Ice," p. 491 (Engl. Transl. NTIS No. AD 426972).

Zwally, H. J., and Gloersen, P. (1977). *Polar Rec.* **18,** 431–450.

4 ICEBERG DRIFT AND DETERIORATION

R. Q. Robe

U.S. Coast Guard Research and Development Center
Avery Point, Groton, Connecticut

I. INTRODUCTION

Historically, man's interest in icebergs has been as much a result of their effect on human imagination as of the danger they present to ocean commerce. Icebergs are immense blocks of freshwater ice that float seven-eighths submerged in the polar and subpolar seas. They have been

DYNAMICS OF SNOW AND ICE MASSES

the cause of many maritime disasters, the most celebrated of which was the sinking of the Titanic in 1912. Yet, despite the hazard they presented to navigation on the open seas, they were frequently beneficial to sailors who ventured into the polar regions. Whalers in Baffin Bay often would tie up in the lee of an iceberg, using it as a massive floating breakwater. The ship not only rode more easily in the lee, but was protected from storm-driven pack ice that could easily crush a ship.

The hazard icebergs presented, however, was far greater than their benefit. As a direct result of the Titanic disaster, several United States and British ships patrolled the shipping lanes south and east of Newfoundland in 1912 and 1913. In 1914 the International Ice Patrol service was established by the First International Conference for Safety of Life at Sea (1913) and assigned to the United States Coast Guard (then the Revenue Cutter Service). Since that time the Coast Guard has conducted iceberg surveillance and oceanographic research in the northwestern Atlantic as a service to mariners, broadcasting twice daily ice messages during the spring and early summer months.

In 1973, with the publication of a paper by Weeks and Campbell concerning icebergs as a possible fresh water source, icebergs began to be considered in a new light. Even modest-sized icebergs contain huge amounts of highly pure water. Several other researchers have pursued the idea of towing icebergs and have dealt in a general way with the difficulties of towing masses weighing many millions of metric tons across thousands of kilometers of water. In 1977, The First International Conference on Iceberg Utilization was held in Ames, Iowa, to review the state of knowledge on icebergs and propose techniques for towing Antarctic icebergs to the Middle East, where the need for fresh water is critical and the cost very high.

Some experience has already been gained in iceberg towing off the east coast of Canada and the west coast of Greenland. In these areas, towboats stand ready to alter slightly the trajectory of any icebergs that present a danger to drill ships and offshore platforms.

Icebergs also present a potential threat to the undersea pipelines and storage facilities that will follow the development of offshore oil fields on the margins of Baffin Bay and the Labrador Sea. The huge mass of an iceberg enables it to scour troughs in the sea floor that could rupture shallowly buried pipelines.

An overview of what is currently known about the production, drift, and final deterioration of icebergs in both the Arctic and the Antarctic is presented here. The author is indebted to researchers of many countries for the material presented.

II. ICEBERGS AND THEIR SOURCES

A. General Iceberg Characteristics

Icebergs are formed from the land ice and shelf ice of the polar regions, particularly from those ice masses associated with Greenland and the Antarctic continent. The composition of icebergs is quite simple, while their mechanical properties and structures are very complex.

Icebergs are composed of snow ice, which has much the same properties as ordinary ice, except it contains bubbles of compressed atmospheric gas trapped when the snow became ice. These bubbles can be either round or tubular. The tubular configuration is thought to be the result of flow stress [Scholander and Nutt (1960)]. Scholander reported that for Greenland icebergs the tubular bubbles have diameters ranging from 0.02 to 0.18 mm and lengths up to 4 mm. The round bubbles ranged from microscopic to 2 mm in diameter. The gas pressures within these bubbles were quite variable even within the same iceberg, ranging from a low of 2.3 atm to as high as 20 atm [Scholander and Nutt (1960); Scholander *et al.* (1956); Urick (1971)]. The bubbles constituted 3–8% of the Greenland icebergs by volume [Scholander *et al.* (1961)]. Similar air-bubble characteristics were reported by Matsuo and Miyake (1966) for the Antarctic. They measured bubbles ranging in size from microscopic to 3 mm in diameter, also noting the presence of tubular bubbles. The percent of bubbles by volume for the Antarctic icebergs was similar to that of the Greenland icebergs, ranging from 2 to 7%.

Entrapped air affects the density of the ice slightly. Pure ice at 0°C has a density of 916.7 kg/m³, and glacial ice can have densities that approach that of pure ice [Hobbs (1974)]. Arctic and Antarctic icebergs, with the exception of those from the ice shelves, have densities that are in the range of 880–910 kg/m³ [Matsuo and Miyake (1966); Smith (1931)]. Very little snow or firn remains on these icebergs since the firn limit is between 600 and 900 m elevation in the Antarctic [Mellor (1959b)] and roughly 1390 m in Greenland [Bauer (1955)]. Consequently, the ice arriving at the sea from outlet glaciers and ice streams has a fairly uniform density structure. In contrast, the firn limit on the ice shelves is virtually at sea level [Mellor (1959b)], which makes the ice shelves areas of net accumulation. As a result, the density of an Antarctic shelf iceberg will increase in a fairly linear manner from a surface density of approximately 450 kg/m³ to a value between 860 and 890 kg/m³ at 60 m depth [Crary *et al.* (1962); Weeks and Mellor (1978)]. Deeper than 60 m the increases in density are rather slight.

Taking an approximate average ice density ρ_i of 850 kg/m³ [Weeks and

Campbell (1973)] for shelf icebergs and $\rho_i = 900$ kg/m³ for all other icebergs, the ratio of the submerged volume V_s to total volume of the iceberg V is

$$V_s/V = d_i = \rho_i/\rho_w,$$

where ρ_w is the density of sea water, ranging from 1020 to 1030 kg/m³. For $\rho_w = 1025$ kg/m³, $d_i = 0.83$ for shelf icebergs and 0.88 for all others.

The portion of the iceberg above water gives very limited information concerning the shape, draft, and relative dimensions of the total iceberg. Icebergs have long been classified by their visible form which can range widely depending on the strength of the ice and the extent of the wave action to which it has been subjected. Customarily, icebergs are classified as follows:

(a) *Tabular* Horizontal or flat-topped icebergs with a length-to-height ratio greater than 5:1 (Fig. 1).

Fig. 1. Tabular iceberg near greenland (USCG photograph).

Fig. 2. Blocky iceberg.

(b) *Blocky* Steep-sided, flat-topped icebergs with a length-to-height ratio of approximately 2.5:1 (Fig. ?)

(c) *Drydock* Icebergs with a wave-eroded U-shaped slot and twin columns or pinnacles (Fig. 3).

Fig. 3. Drydock iceberg.

Fig. 4. Domed iceberg (on left).

Fig. 5. Pinnacle iceberg.

(d)   *Dome*   Smooth, rounded icebergs with low sides (Fig. 4).

(e)   *Pinnacle*   Icebergs with a central spire or a pyramid of one or more spires dominating the shape (Fig. 5).

The largest icebergs are tabular. The largest Antarctic iceberg on record was 170 km long [Weeks and Campbell (1973)]. Icebergs in the other shape categories are often the remnants of tabular iceberg deterioration. As they diminish in size, tabular, blocky, and pinnacle icebergs become domed icebergs if they are unstable and roll, or drydock icebergs if they remain upright.

B.   Glaciers, Ice Streams, and Ice Shelves

1.   Strength of Glacial Ice

The character and form of an iceberg at its source is dependent upon the morphology of the ice mass that produced it as well as the mechanism that fractured it from that ice mass. A grounded glacier will produce icebergs quite different from those produced by a glacier that is floating at its terminus. A rapidly moving floating ice stream will produce icebergs by a mechanism different from that of a slowly moving glacier. Basically, there are three types of iceberg-producing formations: glaciers, ice sheets, and ice shelves (which are described in more detail in Chapters 1 and 2).

(a)   Glaciers which are not afloat at their termini calve very small icebergs whose longest dimension is generally less than the ice thickness.

(b)   Ice streams of glaciers that have a floating length of at least several times their width calve icebergs that have a width on the order of the ice thickness and a length that may be several times the ice thickness.

(c)   Ice shelves, which are very large masses of floating ice, are nourished both by glaciers and by direct precipitation. Ice shelves produce icebergs that have a long dimension from hundreds of meters to more than a hundred kilometers.

2.   Grounded Glaciers

Glaciers that are grounded at their termini typically produce small icebergs that are only of local importance. Since the glacier is not free to flex with changing water levels, fracture is produced by stress generated by movement within the glacier.

The Columbia Glacier in southeastern Alaska has recently been studied extensively due to its proximity to the oil pipeline port of Valdez. It represents a typical active iceberg-producing grounded glacier [Kollmeyer *et al.* (1977); Post (1978)]. The Columbia Glacier has a surface flow

Fig. 6. Face of the Columbia Glacier Terminus [Kollmeyer *et al.* (1977), used with permission].

of approximately 2–6 m/day. Its surface is heavily crevassed (Fig. 6). The face of the glacier is up to 100 m above water. In August 1977 the total ice face (both above and below water within a calving bay) was measured as nearly 300 m high. Calving is very seasonal, occurring in late summer, and appears to be associated with subglacial runoff, which undermines the glacial front and causes a calving bay to form.

Calving above the waterline has been observed to result from shear fracturing. Since calving probably is initiated by undercutting of the glacial face by subglacial runoff, fractures are believed to occur from the crevasse layer to the bottom of the ice. In minor calving areas, many small pieces fall from the glacier face, producing much brash and small icebergs. As the runoff decreases in the fall, calving decreases and much of the forward glacier flow is consumed in filling the space behind the moraine left by the calving bay.

As seen in Fig. 6, crevasses on the order of 20–30 m deep are present and have a spacing that is a small fraction of the depth. These fractures lie between the primary crevasse pattern (Fig. 7). Holdsworth (1969a) gave a range of crevasse depths of 20–40 m for Alaska glaciers and 45–60 m for polar glaciers and a primary crevasse spacing of 2.5 times the depth. A decrease in spacing was noted with increasing age as secondary fractures

Fig. 7.   Surface of Columbia Glacier Terminus [Kollmeyer *et al.* (1977), used with permission].

appeared on intercrevasse blocks. The stress on these blocks doubtless increases as an unsupported face is approached. Colbeck and Evans (1971) in a study of small-scale strain on a temperate glacier also noted the presence of small-scale randomly oriented fractures with a typical spacing of 3 m.

Kollmeyer *et al.* (1977) reported iceberg size distributions. While the number of icebergs varied widely from year to year, it is felt that the relative distribution of sizes remained the same because factors controlling iceberg sizes have not changed in recent years. The largest iceberg measured was 67 m long and had a mass of 6.8×10^7 kg; however, only about 3% of the icebergs exceeded 30 m in length. For any grounded glacier, these sizes would appear typical, with approximately 100 m being the maximum long dimension. Icebergs of this size distintegrate quickly in open water and thus are only of local interest.

3.   Floating Glaciers and Ice Shelves

There are two calving mechanisms for a glacier with a floating terminus or for an ice shelf. The first is a result of stresses within the ice produced by ice movement and lack of hydrostatic pressure on unsupported faces.

The second is caused by buoyancy forces produced by waves, tides, storm surges, and other rapid or periodic changes in water level.

Reeh (1968) has proposed a mechanism to account for iceberg production from a floating glacier or ice shelf. Such icebergs generally have a width roughly equal to the ice thickness.

Consider an infinitely wide ice sheet, of density ρ_i and thickness h, resting on a motionless sea (density, ρ_w) with the ice face a distance L from the grounding point such that $h/L \ll 1$. If all the forces are initially hydrostatic, the force balance at the face of the ice is as shown in Fig. 8.

In Fig. 8 g is the acceleration due to gravity, $d_i = \rho_i/\rho_w$, and $d_i h$ is the depth of water at the bottom of the ice face. Clearly this is not an equilibrium condition, since the hydrostatic forces are balanced at the bottom of the ice and increasingly unbalanced as one approaches the ice surface. The unbalanced forces give rise to a normal tensile force N that acts eccentrically on the ice face, curving the top of the face toward the water. This results in a downward deflection D. The necessary result of this downward movement is a buoyancy force that acts on an adjacent section of the ice just landward of the face. As the buoyancy force, bending movement B, and normal force N are balanced for each section in turn, a surface waveform deformation results which has a decreasing amplitude with distance from the face.

At the point when the magnitude of the deflection $D(L)$ exceeds $-\frac{1}{2}h(1 - d_i)$, the axial force at the face becomes compressive since

$$N(L) \simeq \tfrac{1}{2}\rho_i g h^2(1 - d_i) + \rho_i g h D(L). \tag{1}$$

The bending moment at the face is approximated by

$$B(L) = \tfrac{1}{4}\rho_w g h (d_i h + D(L))^2 - \tfrac{1}{6}\rho_w g (d_i h + D(L))^3 - \tfrac{1}{12}\rho_i g h \tag{2}$$

and its slightly increasing value maintains the downward movement.

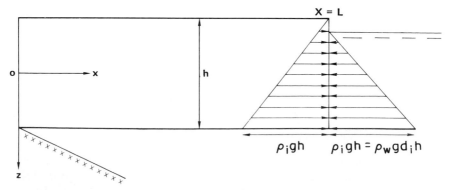

Fig. 8. Floating glacier terminus showing unbalanced hydrostatic forces at the seaward face.

Reeh treated the ice as a purely viscous material where the constant viscosity (μ_a) was assumed to equal

$$\mu_a \simeq \exp(-\theta/4)/\tau^2, \tag{3}$$

where θ is the temperature in degrees centigrade and τ is the effective shear stress, and where θ and τ represent mean values for a particular glacier. In linear viscous theory with the buoyancy force resulting from the deflection D taken into account, the curvature and bending moment are related by

$$(\partial/\partial t) \; \partial^4 D(x)/\partial x^4 = -3\rho_w g D(x)/\mu_a h^3. \tag{4}$$

The boundary conditions at $x = 0$ are $D = 0$ and $\partial D/\partial x = 0$. At $x = L$ the normal force and bending moment have been approximated in Eqs. (1) and (2) and in addition $\partial B/\partial x = 0$. The percent deflection of the glacier with time is shown in Fig. 9. The time factor f for this bending to occur is given by the relation

$$f = 16 \; \exp(-\theta/4)/(\rho_w g h)^3(d_i - d_i^2)^2. \tag{5}$$

Representative values of f are given in Table 1 for $\rho_i = 0.83$ and 0.88, and for a range of temperatures and ice thicknesses. This table can be used in conjunction with Fig. 9 to predict the time period for a given percent deflection.

The effective shear stress produced in the ice reaches its greatest value at a cross section located back from the face a distance equal to the thickness h of the glacier. The maximum effective shear stress to be ex-

TABLE 1

Time Factor f for Various Thicknesses and Temperatures of Ice for Density Ratios (Ice : Water) of 0.83 and 0.88

Temperature (°C)	Thickness (m)		
	200	400	600
Density ratio = 0.83			
0	0.10 yr	0.012 yr	0.0036 yr
−4	0.27	0.033	0.0099
−8	0.73	0.091	0.0270
−12	1.98	0.250	0.0730
Density ratio = 0.88			
0	0.18 yr	0.022 yr	0.0065 yr
−4	0.50	0.060	0.0180
−8	1.30	0.160	0.0480
−12	3.54	0.440	0.1300

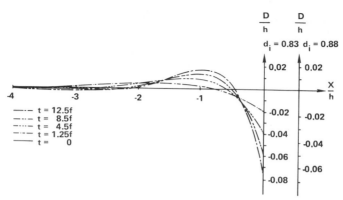

Fig. 9. Relative deflection of a floating glacier as a function of time, $d_i = 0.83$ and 0.88 (used with Table 2). [Redrawn from Reeh (1968) by permission of the International Glaciological Society.]

pected is of the order of 10^5 to 3×10^5 Pa for a wide range of ice thickness and temperature conditions. The maximum surface stress occurs at a distance of $h/2$ from the ice face and is of the same order of magnitude as the effective shear stresses, being about 3×10^5 Pa for a 7–8% deflection of the end of a 300-m thick ice shelf [Holdsworth (1978)].

The fracture, when it occurs, cannot be predicted by stress alone. Ice fracture will depend as much on the location of inhomogenities in the ice as on the stress pattern in a perfectly viscous ice slab. Temperature is also important, as normal stress and effective shear stress increase with temperature, leading to more rapid fracturing.

Figure 10 shows the active front of Humboldt Glacier in Greenland. The widths of the icebergs are quite uniform. Other photographs of this nature show icebergs on their sides appearing to have a width very nearly equal to their thickness. A tabular iceberg with a width very near its total thickness could be expected to have very low stability. If the width-to-ice thickness ratio falls below 0.8:1, the iceberg is quite likely to roll [Allaire (1972)].

To produce icebergs of a width-to-thickness ratio much greater than 1:1, a mechanism that involves the dynamic interaction of ice shelf and ocean is required. Holdsworth (1969b, 1971, 1978) examined the stress patterns produced in a floating ice mass by periodic changes in water level such as tides and wind waves, and by aperiodic impulses such as those produced by storm surges, tsunamis, and earthquakes. These mechanisms are capable of producing fractures far back from the ice edge where hydrostatic pressure is of importance. Two mechanisms that are worth mentioning are (a) long-wavelength variations in water level that bend the

HUMBOLDT 1970 SECTION 6

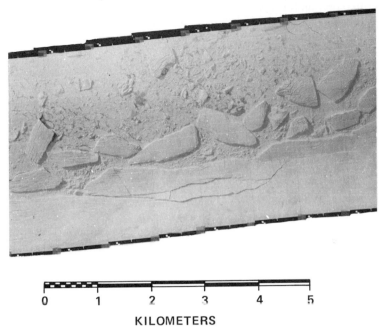

0   1   2   3   4   5

KILOMETERS

Fig. 10. The active front of Humboldt Glacier showing icebergs of consistent width [U.S. Coast Guard (1971)].

shelf at the grounding point and produce the maximum stresses at that point, and (b) vibration within the shelf by wave-induced resonance.

The potential for the very largest icebergs occurs when an ice shelf or floating glacier is fractured at the point where the ice moves from a grounded to a floating situation. This is perhaps the source of some of the icebergs that are on the order of 100 km in length. The breakup of the Ward Hunt Ice Shelf in 1961 and 1962 probably was caused by an unusually large tidal range occurring about 6 February 1962 [Holdsworth (1971)]. This breakup caused the entire outer shelf, an area of 9×75 km, to drift into the Arctic Ocean as five large icebergs (ice islands).

If an ice shelf is treated as an elastic beam of uniform thickness h and length L on an elastic foundation where $L \gg h$ and the beam is fixed at the origin ($x = 0$), the equation for the beam deflection $D(x)$ is

$$\partial^4 D(x)/\partial x^4 = -[12(1 - p^2)/E_b h^3]\rho_w g(D(L) - D(x)), \qquad (6)$$

where p is Poisson's ratio and E_b is the bending modulus of ice.

The boundary conditions are (a) the deflection and slope are zero at the origin $[D(x) = 0$ and $\partial D(x)/\partial x = 0]$, and (b) the deflection $D(L)$ at the free end is the same as the change in sea level and the slope of the upper ice surface is again zero. Thus,

$$D(x) = D(L)[1 - e^{-\lambda x}(\cos \lambda x + \sin \lambda x)], \tag{7}$$

where λ is the damping factor given by

$$\lambda^4 = 3\rho_w g(1 - p^2)/E_b h^3.$$

A tensile stress produced at the upper surface of the ice by a downward movement of the ice shelf is given by

$$\sigma_{xx} = [3\rho_w g D(L)/h^2 \lambda^2] \, e^{-\lambda x}(\cos \lambda x - \sin \lambda x). \tag{8}$$

Thus, at the origin,

$$\sigma_{xx}|_0 = 3\rho_w g D(L)/h^2 \lambda^2. \tag{9}$$

For a 0.5-m deflection of a 200-m thick shelf, $\sigma_{xx} = 3.3 \times 10^5$ Pa.

The calving of icebergs that are larger than those produced by marginal bending but smaller than an entire shelf width requires a disturbing force with a wavelength considerably shorter than the tidal wavelength or the shelf width. If the simple elastic theory is applied again, an ice shelf may be analyzed as a vibrating elastic plate excited by the oceanic wave field [Holdsworth (1968)]. If the dominant period of the wave field coincides with a natural period of a floating ice mass, a resonance condition may occur. A resonance combined with zones of weakness within the ice may cause fatigue failure of the ice. Swithinbank (1978) and Kovacs (1978) detected both top and bottom crevasses on Antarctic icebergs. Bottom crevasses were detected at irregular intervals of 3–10 km.

Holdsworth (1978) modeled the Erebus Glacier Ice Tongue, which is 1.5×14 km, tapering from a 335-m thickness at the glacier to a free-end thickness of 71 m. For an oscillation period of approximately 20 sec and a 0.11-m deflection, a surface bending stress of the order of 10^5 Pa can be developed. Higher vibration modes would require less deflection to produce a given stress. The cyclic nature of the stress and the existing crevasses in the ice indicate that a complete fracture could occur at a stress of 10^5 Pa or less. Kovacs (1978) presented a photographic sequence by V. Lebedev that showed the disintegration of a large 20-m-high tabular iceberg. The fracture pattern was remarkably uniform, with the iceberg fracturing into equally sized pieces. Whether the fracture was a function of a pre-existing weakness in the iceberg or caused by wave-induced stress is not known.

C. Arctic Iceberg Sources

The ice fields, ice caps, and ice sheets that feed the tidewater glaciers on the borders of the Arctic and sub-Arctic Sea are for the most part small and widely scattered. Except for the Greenland Ice Sheet, they exist as remnants of a former ice age. Minor production of small icebergs occurs from glaciers in southern Alaska and from some of the islands of the Arctic Ocean and the Canadian Archipelago. Icebergs such as those in Columbia Bay or Glacier Bay [Ovenshine (1970)] in southern Alaska (Section II.B.2) are generally small and are rarely encountered far from their place of origin. Also, the quantity of icebergs produced by these glaciers is largely unknown, since the production of a small ice cap may be highly seasonal and vary widely from year to year.

The largest producer of icebergs in the northern hemisphere is the Greenland Ice Sheet, which has a surface area of 1.73×10^6 km^2 and covers fully 80% of the island [Bauer (1955)]. The Ice Sheet contains 2.4×10^6 km^3 of water and has a maximum thickness of 3300 m. In comparison, all other of the permanent ice fields in the Canadian Arctic, Soviet Arctic, and Spitsbergen together contain only about 13% of the quantity of ice held in Greenland [Orvig (1972)]. The Greenland Ice Sheet annually calves around 240 km^3 of ice into the neighboring seas. Reeh (1968) gives an estimate of 2.25×10^{17} g/yr for the total calving loss. Bauer (1955) estimates the loss to be 2.15×10^{17} g/yr broken down by region as follows:

North Greenland	0.09×10^{17} g/yr
West Coast (south of Melville Bay)	0.81×10^{17} g/yr
Melville Bay	0.18×10^{17} g/yr
East Coast	1.08×10^{17} g/yr

These regional estimates are gross at best, since in most cases they are based on estimated glacier speed and thickness. The estimates for the east coast are probably high and those for Melville Bay are probably low [Koch (1945); Kollmeyer (1979)]. Greenland glaciers with a speed of 3–10 m/day were considered active, those with a speed of less than 3 m/day were considered stationary, and glaciers moving only 30 m/yr were considered dead. Kollmeyer (1979) measured the speed of the Jacobshavn Glacier on Disko Bay in western Greenland as being up to 22 m/day during July and August.

The icebergs calved from the Greenland glaciers typically have no snow cover as the firn line is at an elevation of 1390 m. This results in icebergs with an average density of around 900 kg/m^3.

Assuming that 20% of the ice calved by the Greenland glaciers is brash ice and the average iceberg calved from a Greenland glacier has a mass of 5×10^6 metric tons (a medium-sized iceberg by Arctic standards), then Greenland would produce over 34,000 of these "average" icebergs each year. Chari and Allen (1974) quote a U.S. Navy estimate of 20,000 icebergs annually from 20 major glaciers on the west Greenland coast alone. Dyson (1972) gives an estimate of 10,000–15,000 large icebergs for all of Greenland.

The east coast of Greenland does not generally produce as many icebergs as the west coast. Along nearly the entire coastal zone from Nordostrundingen in the north to Angmagssalik in the south lies a mountain range that is higher than the inland ice. The ice can drain toward the east only through occasional high passes. Most of the productive glaciers on this coast reach the sea by moving either north or south parallel to the mountains. Once the icebergs are formed on the eastern coast, they are in most cases at the landward end of long fjords or blocked in by numerous small islands. A further impediment to their movement seaward is the mass of sea ice that is carried far southward along the East Greenland coast by the East Greenland current. This sea ice can effectively block the fjords in this area for most of the year.

Koch (1945) identified six glaciers in three basins on the east coast as being very productive: (1) Storstrommen Glacier, L. Bistrops Glacier, and Soraner Glacier; (2) DeGeers Glacier and Jaette Glacier; and (3) the most productive of the east Greenland glaciers, Daugaard–Jensens Glacier. These three glacier systems account for 80% of the east-coast production of icebergs. There are also a number of moderate iceberg-producing glaciers (Fig. 11).

On the west Greenland coast, almost no icebergs are produced south of Disko Bay. The coastal region is free of ice for more than 100 km in places. The glaciers from Jacobshavn to Petermann were visited by Kollmeyer (1979) from 1969 to 1971 as part of his research for the International Ice Patrol (Fig. 11). He identified nine major iceberg-producing glaciers on the west coast. Jacobshavn Glacier, probably the fastest-moving ice stream in the world (up to 8 km/yr in the center), produces about 7.1×10^{15} g/yr of ice, a little more than 7% of the west Greenland production. To the north the next major producer is Rinks Glacier, which calves icebergs over 600 m thick. Both Rinks and Jacobshavn Glaciers are in deep long fjords with sills at their seaward end. These sills may inhibit movement of the icebergs seaward. Farther north, on the southeastern shore of Melville Bay, where the ice sheet comes in contact with the coast, is a series of major iceberg-producing glaciers that calve directly into Melville Bay. These glaciers (Steenstrup, Dietrichson, Nansen, Kong Oscar, and Gade) are spearated by nunataks, or isolated mountains,

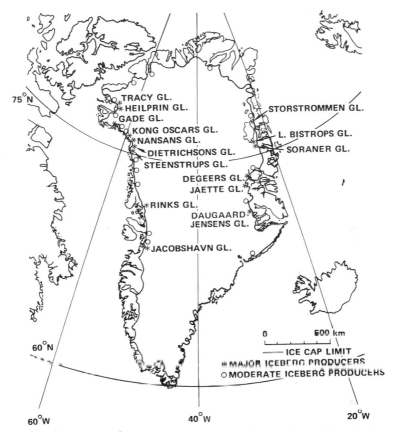

Fig. 11. Major and moderate Iceberg-producing glaciers of Greenland.

which divide the ice sheet. It is thought that most of the Arctic icebergs reaching the open sea originate in this area. Farther to the north in Kane Basin is Humboldt Glacier, 100 km across the face (Fig. 10), which appears to produce many large icebergs. The large Humboldt icebergs apparently are aground in front of the calving face and have been observed to remain stationary for several years. During some years, unusually high tides may permit these icebergs to drift free.

Although the Petermann Glacier in Hall Basin is not believed to produce many icebergs, its front broke up during 1975 [M. Dunbar (personal communication, 1977)]. A small fraction of the parent iceberg was observed in the Labrador Sea in May–June 1976 [Robe et al. (1977)]. This piece was spotted on five occasions and was over 700 m long when first observed.

Other than Greenland, the ice shelves of northern Ellesmere Island are the only source of sizeable Arctic icebergs. Five ice shelves exist there, the largest by far being the Ward Hunt Ice Shelf. All of these shelves experience periodic calving. In 1962 the entire outer shelf of the Ward Hunt Ice Shelf calved, producing five large icebergs. The M'Clintock and Ayles Ice Shelves both broke up sometime between June 1962 and 1966 [Holdsworth (1971)].

D. Antarctic Iceberg Sources

The production of icebergs from the Antarctic Ice Sheet has been estimated to range from 0.4×10^{17} g/yr to a high value of 16.5×10^{17} g/yr [Mellor (1967)]. Zotikov et al. (1974), using precipitation data without taking into account ablation rates, give an estimate of 19.8×10^{17} g/yr. Most estimates [Mellor (1967); Wexler (1961)] appear to be in the range of $5.0 \times 10^{17} - 12 \times 10^{17}$ g/yr, with the high end of the range from more recent estimates. This can be compared with estimates for Greenland of 2.15×10^{17} g/yr.

The Antarctic Ice Sheet covers 14×10^6 km² (including the Antarctic Peninsula area) and has a volume of approximately 21×10^6 km³ of water, almost ten times that of Greenland.

Ice reaches the sea along approximately 20,000 km of coastline, of which 11,000 km are ice shelves [Mellor (1959a)]. These ice shelves account for major differences in size and character of the Antarctic icebergs compared to the Greenland icebergs. Mellor (1967) gives his 1963 estimate of combined iceberg production of all land-based ice discharge as 2.4×10^{17} g/yr. This figure is surprisingly close to the 2.15×10^{17} g/yr given for Greenland. The ice shelves (Fig. 12) cover a combined area of 1.5×10^6 km². According to Swithinbank (1955), the stability of ice shelves is dependent to some extent upon their being protected by land on either side and to the ice being grounded locally. An ice shelf that extends seaward of this protection will suffer periodic retreat. Ice shelves are partly formed in place by the accumulation of firn and ice. The firn line for ice shelves is effectively sea level. Consequently, the average density of the shelf ice is lower than that of glacier ice.

Icebergs produced by the Antarctic ice shelves can be enormous. Swithinbank (1978) reports an iceberg 75×110 km and containing 2000 km³ (17×10^{17} g) of ice. This is a year's worth of Antarctic iceberg production using the higher estimates given. Icebergs of more typical size, 1–2 km across [Korotkevich (1964)], were observed near the Shackleton Ice Shelf. Weeks and Mellor (1978) present size distributions of Antarctic icebergs based on Landsat imagery. A modal size for the 648 icebergs

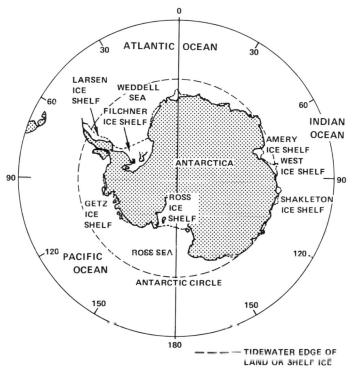

Fig. 12.   Major ice shelves of Antarctica.

within the sea ice was 0.65–0.75 km. For the 74 at the edge of the pack it was 0.35–0.45 km.

The thickness of the major ice shelves is about 200 m at the seaward edge [Robin (1953); Mellor (1967); Korotkevich (1964)], and increases with distance from the edge. The Ross Ice Shelf, lying between McMurdo Sound and Little America, and the Filchner Ice Shelf, on nearly the opposite side of the continent, are ice shelves on a grand scale and certainly calve the majority of Antarctic icebergs. The Ross Ice Shelf has a surface area of roughly 0.5×10^6 km^2 and a thickness that varies from 200 m at the edge to over 700 m at the land edge (Fig. 13). Most of the shelf appears to be afloat. The front moves seaward at a rate estimated between 1240 m/yr and a high value of 1870 m/yr [Zumberge, quoted by Dolgushin (1966)]. The Filchner Ice Shelf has an area of 0.40×10^6 km^2, which is slightly smaller than the Ross Ice Shelf. Its edge moves seaward at approximately 280 m/yr [Dolgushin (1966)].

The Amery Ice Shelf, although quite small in comparison, is formed by the Lambert Glacier, which drains nearly one-eighth of the Antarctic Ice

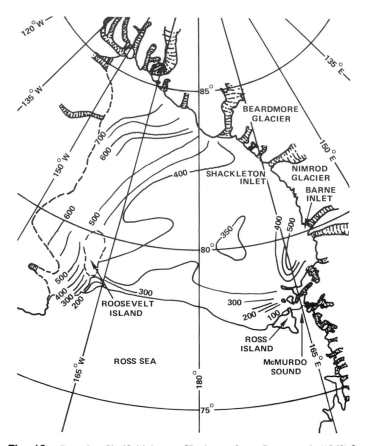

Fig. 13. Ross Ice Shelf thickness. [Redrawn from Crary *et al.* (1962).]

Sheet. Budd [quoted by Weeks and Mellor (1978)] estimated that the output of the Amery Ice Shelf was 0.27×10^{17} g/yr, approximately 2–5% of the iceberg production of the continent. In 1963, one-fifth of the Amery Ice Shelf broke off. This one calving incident produced approximately 8.2×10^{17} g of ice. Depending on the estimates that are accepted, this constitutes nearly one year's production for the entire continent.

Along the rest of the Antarctic coast the movement of ice toward the sea is in the form of sheet flow and ice stream flow. This flow is somewhat slower than that seen in Greenland. Mellor (1959b) reported on the movement of several glaciers on Holme Bay. Their speeds were in the range of 1–2 m/day. Sheet flow in the area was approximately 0.04 m/day. Dolgushin (1966) gave movement rates for outlet glaciers of 100–1200 m/yr with a mean rate for 49 outlet glaciers of 588 m/yr.

The floating seaward edges of the Greenland glaciers and the Antarctic glaciers and ice shelves seem to be quite similar, having a common ice thickness of 200–300 m. This can probably be explained by the distribution of stresses at steep ice cliffs.

III. GLOBAL DRIFT PATTERNS

A. General Discussion

In comparing the drifts of icebergs in the northern and southern hemispheres it quickly becomes obvious that in the north much is known about number and track and relatively little is known about size distribution, while in the south much is known about size and very little about number and drift track. This situation is a natural outgrowth of the reasons people study or seek out the icebergs. The northern icebergs, which drift so far south that they cross major shipping lanes south of the latitude of New York, are at best a bother and at worst a deadly hazard. They have been tracked and counted since the early part of this century. Now they pose a threat to development of resources in the eastern North American Arctic. They are a working problem. The Antarctic icebergs have been, until quite recently, a minor scientific curiosity and a big tourist attraction for ships going to and from Antarctica. They do, of course, pose a danger for the limited number of scientific and fishing ships that ply these waters. The one thing that is always remembered about the Antarctic icebergs is that the big ones are truly huge. The normal population is sometimes forgotten in comparison. Antarctic icebergs at lat 65° S are often compared with Greenland icebergs that are at lat 50° N. Since the ice thickness of the floating glaciers of Greenland and the Antarctic is on the order of 200–300 m thick at the seaward edge, the bulk of the population distribution in both areas (the larger shelf icebergs aside) can be expected to have a long dimension of under 1000 m [Romanov (1973); Weeks and Mellor (1978)]. The similarity of the iceberg sizes should be considered in dealing with the problems of iceberg drift for both polar regions.

B. Arctic Drift

In the polar basin, icebergs are a rare occurrence. The few sighted are thought to originate from Zemlya Frantsa-Isifa and Severnaya Zemlya [Armstrong (1957)]. Drift speeds west of Spitsbergen at 72° N to 79° N have been measured at 7–22 km/day toward the south and southwest, as the ice follows the general circulation pattern [Vinje (1975)]. The majority of the icebergs (or ice islands as they are often called) in the Arctic Ocean are from the ice shelves of Ellesmere Island. These icebergs generally remain in the Beaufort Sea Gyre [Weeks and Campbell (1973)], where 85–100 have been sighted. They follow a zigzag course, averaging perhaps 2 km/day. These icebergs start at Ellesmere Island, move westward along the Canadian Archipelago, then turn toward the pole north of the Bering Strait, and, after 10–12 yr, drift back to their approximate starting point north of Ellesmere island. Kovacs (1978) reported nine of these tabular icebergs in the fast ice near Flaxman Island, Alaska, in May 1977 with a tenth iceberg reported northeast of Cross Island. A few of these tabular icebergs get caught in the East Greenland Current and are carried south with the sea ice.

In February 1964, 20 large tabular icebergs were sighted between Hamilton Inlet and Cape Chidley on the Labrador coast. These were thought to be pieces of WH-5, an iceberg from the 1962 breakup of the Ward Hunt Ice Shelf. Some of the icebergs were as long as 600 m. By May 1964 some of them had drifted as far south as 44° N [Franceschetti (1964)].

The major outlet for surface water from the Arctic Ocean is the East Greenland Current flowing toward the south between Spitsbergen and Greenland. This current carries vast amounts of the Arctic pack ice as far south as Cape Farewell, on the southern tip of Greenland. The pack ice rarely retreats north of 70° N along this coast. Since all the major iceberg-producing glaciers are north of Scoresby Sound, at 70° N, the presence of the pack ice tends to confine icebergs in the fjords and sounds where they were calved. In years when the pack ice is unusually light, a greater number of icebergs reach the open sea [Koch (1945)]. The majority of the icebergs reaching the sea drift southward with the pack ice on the cold (surface temperature less than 0°C) East Greenland Current (Fig. 14). Some, however, get caught in the northern branch of the warm (surface temperature 4–6°C) Irminger Current and are transported along the northern coast of Iceland [George (1975)]. An iceberg cannot be expected to move very far eastward because of the higher water temperature and the exposure to waves [Robe (1978)].

South of 70° N, the icebergs continue a southward drift until they round Cape Farewell. In this region icebergs are increasingly exposed to the deteriorating effects of warmer water and wind waves. An unknown number, thought to be small, begin the drift north in the warmer West Greenland Current. No icebergs are added until the current is north of Disko Bay. Near Davis Strait a branch of the West Greenland Current moves west and joins the southward-flowing, cold Baffin Island Current. This was observed in 1976 by "tagging" an iceberg that drifted in a generally westerly direction at between 0.1 and 0.75 m/sec [Brooks (1977)]. A portion of the West Greenland Current continues north until it crosses the

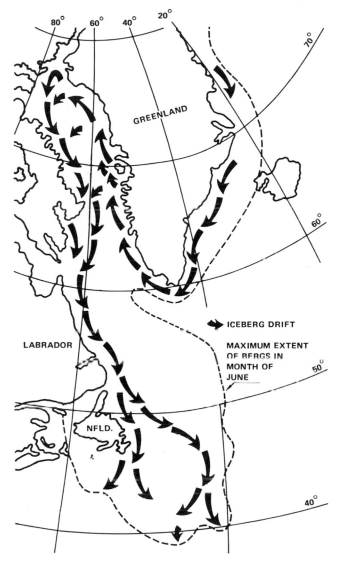

Fig. 14. Drift path of Greenland icebergs. [Redrawn from Soulis (1975).]

north end of Baffin Bay, after picking up the icebergs calved into Melville Bay. The flow out of Kane Basin at this point may contain icebergs from the Humbolt Glacier and those farther north. The concentration of icebergs is at a maximum just north of Disko Island and around Cape York (Fig. 15). From the north of Baffin Bay, icebergs follow a very consistent

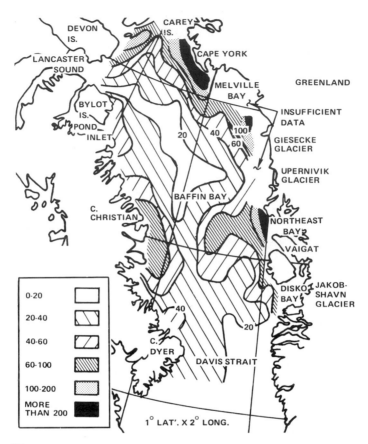

Fig. 15. Average density of icebergs in Baffin Bay for July, August, September, and October 1948, 1949, and 1964–1969 [U.S. Coast Guard (1971)].

path along the continental margin of North America and Baffin Island. First they travel in the Baffin Island Current and then the Labrador Current south of Hudson Straits. Dempster (1974) reported the daily flux of icebergs past a radar site near Saglek, Labrador, as 5 per day within a 77-km range. Finally the icebergs reach the Grand Banks of Newfoundland where they drift either eastward north of Flemish Cap or southward between Flemish Cap and the Grand Banks of Newfoundland. The southern limit of drift is generally defined by the northern edge of the warm (higher than 12°C) North Atlantic Current. It is possible for icebergs to be transported across the warm current in cold-water eddies [Robe (1975)] and on extremely rare occasions they can survive as far east as the

Azores and as far south as Bermuda [Dyson (1972)]. The average drift speed varies from 0.25 to 0.5 m/sec [Soulis (1975)].

During this journey south, no icebergs are added and many are lost by (a) being grounded in the many fjords along the Baffin Island and Labrador coasts or (b) moving outside the pack ice, where they will gradually decay from wave action even in cold water. The pack ice serves several functions in protecting the iceberg: (1) waves are quickly damped at the edge of the pack; (2) the water temperature is maintained near the freezing point; and (3) icebergs are prevented from grounding in fjords and embayments.

The effect of sea ice and seasonal warming can be seen in Fig. 16, which shows the average seasonal iceberg density distributions for the Labrador coast [Gustajtis and Buckley (1978)]. Since there is no source of icebergs in this region, the flux of icebergs must decrease toward the south. Figure 16 does not reflect the flux, only the observed distribution. During the fall (Fig. 16a), the surface temperatures are highest, and there is little or no sea ice coverage. Icebergs deteriorate rapidly under these conditions. The lack of fast ice along the shore leads to a great deal of grounding. The winter distribution (Fig. 16b) is more uniform as surface temperatures drop and fast ice forms along the shore. The disappearance of the southernmost icebergs continues into early winter. The spring distribution (Fig. 16c) reflects the return of the sea ice and lower water temperatures. The main distribution has moved farther seaward as fast ice prevents grounding of the icebergs on the coast. In the summer distribution (Fig. 16d), the seaward edge of the sea ice retreats and is broken up, thus increasing the melting rate.

South of 48° N the International Ice Patrol has been maintaining a count of icebergs since 1913. The number drifting south of 48° N is highly variable and displays no cyclic pattern. The yearly count has been as low as zero in 1966 and as high as 1572 in 1972 (Fig. 17). The limits of the southern drift for recent decades is shown in Fig. 18. While the limits of the yearly drift vary widely, the decade extremes show remarkable consistency. Icebergs as large as 2×10^{13} g have been seen off the Grand Banks. This is equivalent to an iceberg $600 \times 150 \times 200$ m [Chari and Allen (1974)].

Efforts to predict the number of icebergs that will pass 48° N, based on representative air temperatures and barometric pressure differences, have not been very successful [Corkum (1971); Murty and Bolduc (1975)]. The number of variables contributing to iceberg movement and survival are too great and their quantification too uncertain to permit a simple and accurate prediction at this time.

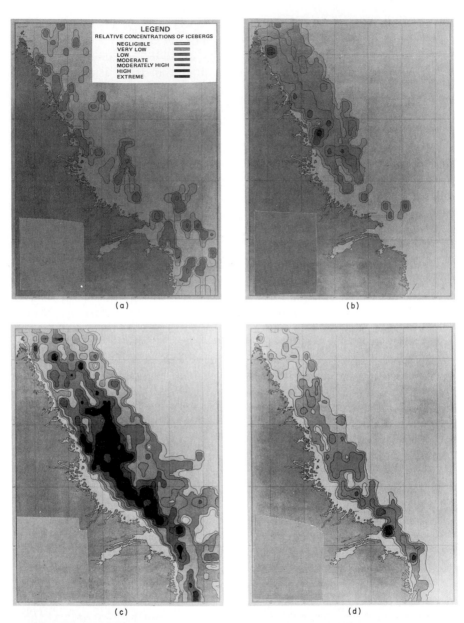

Fig. 16. Seasonal iceberg densities along the Labrador Coast [Gustajtis (1977), Centre for Cold Ocean Resource Engineering]: (a) average fall; (b) average winter; (c) average spring; (d) average summer.

236

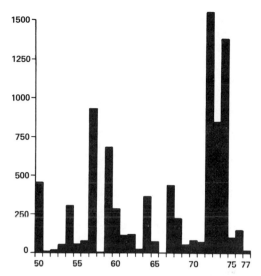

Fig. 17. Number of icebergs south of 48° N for 1950–1977. [From U.S. Coast Guard data.]

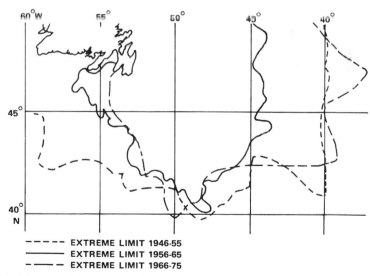

EXTREME LIMIT 1946-55
EXTREME LIMIT 1956-65
EXTREME LIMIT 1966-75

Fig. 18. Decade extreme limits of iceberg drift (from U.S. Coast Guard data). [Redrawn from Jelly and Marshall (1967).]

C. Antarctic Drift

In the Antarctic, drift tends to be more zonal than meridional as is the case in the Labrador Sea and Baffin Bay. The general pattern of ocean circulation near the continent tends to be from east to west because of prevailing geostrophic winds. This zonal drift increases the length of time icebergs remain in the 100-km coastal zone. Zotikov et al. (1974) estimates this period to be 13 yr. On the eastern side of the Antarctic Peninsula this zonal flow becomes meridional and provides the only definite location where the drift is northward. This flow is the western side of the cyclonic Weddell Gyre. The westward drift in the coastal zone could be as high as 0.5/sec [Deacon (1977)]. Ledenev (1962) noted that icebergs are unevenly distributed along the shores of the continent. This is partly due to the location of sources, but greater concentrations are found near shoals and capes. Korotkevich (1964) observed that the eastern edge of West Ice Shelf did not produce icebergs due to the mass of icebergs driven into it by the easterly winds. In fact the West Ice Shelf actually grows on its eastern side as icebergs become incorporated into the shelf proper.

Average speeds for the westward drift is not available except for a few large and, therefore, identifiable icebergs. Korotkevich (1964) measured average speeds of about 0.05 m/sec near shore and twice that offshore. This is in agreement with the speed computed on a fragment of the Amery Ice Shelf measuring 160 × 72 km [Koblents and Budretskiy (1966)], which broke loose during the middle of 1964. A fragment measuring 50 × 70 km drifted 1600 km at a speed of 0.05 m/sec [Shpaykher (1968)].

Counts of icebergs in the coastal zone have been spotty and estimates made are usually based on gross extrapolations. Shil'hikov (1965) estimated that over 30,000 icebergs were between 44° E and 168° E. Dmitrash (1973) reported figures for the average size of icebergs observed in four regions: White Island, Prydz Bay, Banzure Coast, and Enderby Land. A total of 747 icebergs were measured in these four areas. The regional length averages ranged from 223 to 1740 m, while the width averages ranged from 142 to 1170 m, the largest figures being for icebergs off the Banzure Coast. The average length values are doubtless affected by a distribution heavily skewed toward the more numerous smaller icebergs. The length-to-width ratio of the averaged dimensions varied only between 0.63 and 0.67.

A limited number of icebergs have been tracked for extended periods of time using satellite photographs or satellite-tracked instrument packages [Tchernia (1977); Swithinbank et al. (1977); Swithinbank (1969)]. These drift tracks show an east to west drift within approximately 200 km of the coast. A strong northward movement was apparent in an iceberg tracked on the eastern side of the Antarctic Peninsula as the drift continued to parallel the coast in this region. Tchernia (1977) observed a north or northeasterly drift of several icebergs in the region 80° E to 100° E. Drift speeds of 11–14 km/day were observed. The meridional component of drift around the Antarctic Continent generally is toward the north [Deacon (1977)] due to surface divergence caused by wind stress. Average conditions on an oceanic scale do not exist on any meridian [Gordon *et al.* (1977)] and the northward flow is calculated mainly on the necessity to balance the fluxes of heat and salt. A sufficient number of iceberg tracks to define all locations where a northward drift occurs and to locate areas of concentrated northward drift have not been produced for the Antarctic except as noted in the Weddell Gyre.

Since drift toward the north in the surface layers is difficult to quantify or to pinpoint in most areas, a reasonable approach is to calculate climatic means that would influence the greatest number of icebergs for the greatest possible time. Baronov *et al.* (1976) calculated the drift of sea ice from a model using average winds for April (a time when the pack ice is just beginning to grow). The location of the offshore drift zones identified by his model are presented in Fig. 19.

The icebergs will move offshore under the influence of winds and current until they reach the circumpolar current in the region of the prevailing westerlies. Morgan and Budd (1978) give a northerly drift of 6° of latitude per year near 65° S, decreasing to about 3° of latitude per year at 58° S and near zero drift northward by 45° S. As the icebergs move north, they decrease rapidly in length. Romanov (1973) presented data on average iceberg length and height above water level. He noted little longitudinal dependence of the data, but a significant latitudinal dependence. The data averages are presented for two zones: (1) south of 65° S, which approximates the coastal zone, and (2) north of 65° S to the limit of icebergs. The average length decreases considerably in the northern zone (Fig. 20), particularly for the "superbergs," those of 1000 m in length or greater. Twenty percent of all icebergs have a length greater than 1000 m south of 65° S, while north of 65° S the largest 20% have a length greater than 500 m. In the upper 10% bracket, the values are 3000 and 700 m. Weeks and Mellor (1978) presented similar data derived from Landsat imagery of the Bellingshausen Sea. They observed that approximately 10% of the icebergs protected by the sea ice had lengths in excess of 1000 m while none outside the sea ice exceeded that length. The obvious conclusion is that the "superbergs" break up rapidly when they are removed from the protection provided by cold water and sea ice cover. In contrast, the height of the icebergs (Fig. 20) north or south of 65° S differs by only a few meters. This should not be interpreted as indicating that the tops and bottoms of icebergs do not melt. As the tabular icebergs decay, they prob-

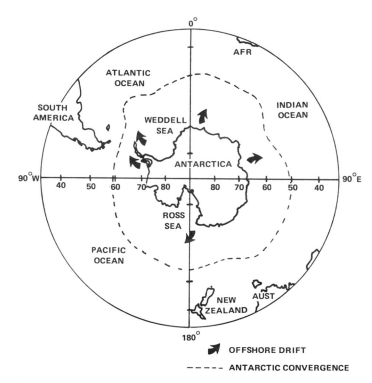

Fig. 19.   Location of offshore drift of sea ice modeled from average April winds.

ably become pinnacle or drydock icebergs which may have little mass but towering spires. The typical height-to-draft ratio for a tabular iceberg is 1:6 to 1:7 for a perfectly rectangular shape. However, the ratio may be as low as 1:1 for a pinnacle iceberg. Thus, a tabular iceberg with a 50-m height and 300-m thickness could conceivably become an iceberg with a 150-m height and the same ice thickness.

The effective northern limit of the drift is the boundary between Antarctic Surface Waters ($-1-4°C$ and approximately $34^0/oo$ of dissolved salts) and Sub-Antarctic Surface Water ($6-14°C$ and $34.5-35.0^0/oo$ of dissolved salts). This region is often referred to as the Antarctic Convergence, but more recently has been referred to as the Antarctic Front [Emery (1977)] or Polar Front [Gordon *et al.* (1977)]. This front generally falls on the southern boundary of the Polar Front zone. No single definition will serve to define the Polar Front zone in all areas, but it appears to be circumpolar. Since strong pressure gradients exist at this frontal zone, a strong geostropic current flowing eastward can be expected.

Figure 21 gives the approximate location of the Antarctic Convergence

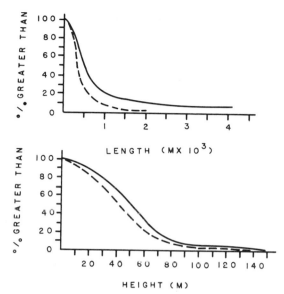

Fig. 20. Length and height distributions for Antarctic icebergs: (solid lines) south of 65° S, (dashed lines) north of 65° S. [Redrawn from Romanov (1973).]

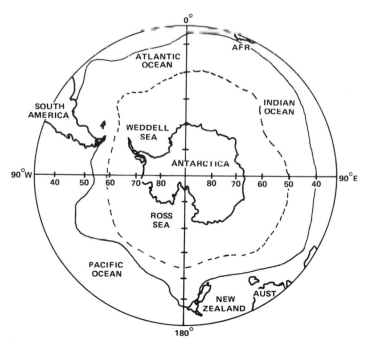

Fig. 21. Location of antarctic convergence (dashed line) and the northern extent (solid line) of Antarctic iceberg drift.

(Antarctic Front) and the maximum northern extent of all iceberg sightings (data combined from numerous sources). Gaskin (1972) presented iceberg sightings southeast of New Zealand between 22 January and 22 March 1967. Note that most of the icebergs were sighted at the convergence zone (as defined by the maximum temperature gradient). North of the convergence zone, no icebergs were sighted and very few were sighted south of it. Only eight of the 85 icebergs sighted were tabular. Brodie and Dawson (1971) reported iceberg sightings east of New Zealand from historical records. In September and October 1892, a large number of icebergs were sighted south and east of the South Island. The presence of these icebergs so far to the north appeared to be in response to unusually strong southerly and south–southwesterly winds blowing for several days. The likelihood that icebergs would survive for more than a week north of the Antarctic Convergence is very small.

IV. LOCAL ICEBERG DRIFT

A. General Discussion of Forces

The prediction of small-scale iceberg drift is of major importance to builders and operators of fixed offshore structures. The problem has been attacked both empirically and analytically with limited but increasing success.

A floating body, such as an iceberg, is acted upon by forces arising from the relative motions of the body to air and water, plus inertial forces that arise from its mass and state of motion. These forces, which are not linear, combine and interact in a most complex way. All of the forces are time dependent and space dependent with their own scales of motion. As an iceberg moves, it undergoes constant acceleration. Sodhi and Dempster (1975) examined the response of icebergs to a change in currents. They reported that when an iceberg enters a new current field, 50% of the velocity change takes place within a characteristic time T that is given by

$$T = 2M/\rho_w C_w A_w \, \Delta U, \tag{10}$$

where M is the iceberg mass, ρ_w the water density, C_w the steady-state drag coefficient, A_w the underwater cross-sectional area, and ΔU the theoretical velocity change.

Using data from Russell et al. (1978), where M is 2×10^9 kg, ρ_w is 1025 kg/m^3, C_w is 1, A_w is 1.2×10^4 m^2, and ΔU is 0.12 m/sec, this relationship gives a value of T equal to 45 min. This value is for a relatively small iceberg. For larger icebergs, the time lags would increase in direct proportion to the mass. Soulis (1975) suggested an inertial time of response for a small iceberg of 3.5 h. In response to subsurface currents, icebergs have been observed moving against the surface current, either with or against the wind. This phenomenon was the result of the high degree of variability in the speed of the ocean currents at various depths. The high-density stratification reduces the turbulent viscosity between water layers, allowing regions of large current shear.

Two major difficulties are inherent in modeling the effects of currents on icebergs. First, the input environmental data is not well known and very expensive to obtain. Second, verification of the model through actual drift observations is difficult and time consuming. Data on winds are somewhat more available than those on ocean currents. Geostrophic winds can be calculated for most oceanic regions from charts of barometric pressure, given certain assumptions about their decay in the surface boundary layer. Ocean current data, on the other hand, are not directly available in the open ocean on a real-time basis. In order to obtain direct, absolute currents, moored current meters are required. Not only would the collection of sufficient data be expensive, but the data would not be available until long after a particular iceberg had melted. Indirect current determinations, calculated from the field of pressure and computed relative to some arbitrary zero value, are available. They involve several very restrictive assumptions, however, such as frictionless flow and steady-state conditions. The indirect current calculations can be made for a large area over a period of a few days, depending upon the speed of the ship collecting subsurface density data. These data are not, however, available on a real-time basis, and a single determination of a current field must be used for several weeks.

In addition to these uncertainties in the environmental factors, there are the uncertainties of the iceberg itself. While the density of the ice and the surrounding water is known to a close approximation, the shape and surface roughness of an iceberg can be determined only by a close surface examination.

The difficulty of verification also plagues the iceberg drift modeler. Verification requires that a comprehensive data set be collected for a number of icebergs such as those reported by Russell et al. (1978). These data must include wind, current, iceberg shape and mass, and accurate measurement of the motion of the icebergs. Satellite navigation systems have only recently solved the problem of accurate position location on the open ocean. Prior to their use, moored reference markers were required [Kollmeyer et al. (1965)]. Currents are best measured by integrating current drogues [Russell et al. (1978)], which can be tuned to the depth of the iceberg and repositioned rapidly when the drogue-to-iceberg distance exceeds a few kilometers.

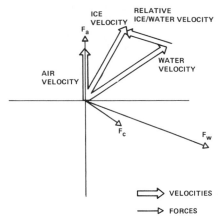

Fig. 22.   Forces on an iceberg for stationary drift due to a uniform wind and current. [Redrawn from Wolford (1972).]

The forces acting on an iceberg are, in general, unbalanced. Therefore, the iceberg is constantly accelerating and decelerating. The force balance can be written

$$M \, dU/dt = F_w + F_a + F_c, \tag{11}$$

where M is the iceberg mass, dU/dt the acceleration, F_w the water drag, F_a the air drag, and F_c the Coriolis force. A balance of forces appears in Fig. 22. Each of the component forces must be evaluated as a function of time.

B.   Water Drag

There are three kinds of drag forces that act on the submerged portion of an iceberg. They are: (1) form drag, a pressure force; (2) frictional drag, which is due to the viscous stress acting at the surface of the iceberg; and (3) inertial drag, which is due to the acceleration of the water relative to the iceberg.

In most applications, all of these drag forces have been combined into one drag coefficient C_w, which relates the relative velocity of the iceberg and the water to the drag force. That is,

$$F_w = \rho_w A_w C_w |U| U/2. \tag{12}$$

A physically more correct form was given by Russell *et al.* (1977):

$$F_w = C_d \rho_w A_w |U| U/2 + C_m M_p \, dU/dt + \text{history expression}, \tag{13}$$

after Hamilton and Lindell (1971), where C_d is the steady-state drag coefficient, C_m the inertial drag coefficient, A_w the area of a cross section of the wetted portion of the body, ρ_w the water density, U the relative iceberg velocity, t time, and M_p the mass of fluid displaced by the iceberg.

The history expression is such that its value is zero if the acceleration begins from rest or from a constant velocity.

The steady-state form drag coefficient is a function of the Reynolds number Re associated with the iceberg; Re is equal to LU/γ, where L is a length characteristic of the iceberg and γ is the kinematic viscosity. At higher Reynolds numbers where the flow is more turbulent, the steady-state drag coefficient generally decreases.

For the iceberg whose underwater profile is shown in Fig. 23, the relative velocity with respect to the water column was measured by Russell *et al.* (1978) and is shown in Figs. 24 and 25. The mean speeds of these drifts were 0.053 and 0.119 m/sec, respectively. These speeds gave Reynolds numbers that ranged from 1.8×10^6 to 1.4×10^7 given a length scale of 100 m.

Direct measurements of C_d were conducted by Banke and Smith (1974)

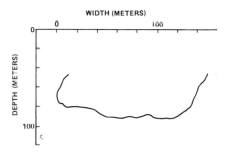

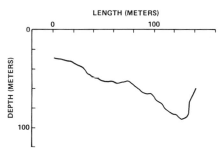

Fig. 23.   Underwater profiles of an iceberg showing shape on nearly perpendicular planes. [Redrawn from Russell *et al.* (1977).]

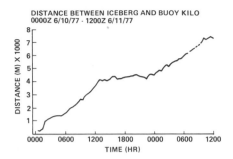

Fig. 24. Relative drift of an integrating current drogue with respect to an iceberg–drogue depth 10–95. [Redrawn from Russell *et al.* (1977).]

under uncertain conditions and for a very small iceberg mass of 1.5×10^8 g and underwater area of 46 m². The Reynolds numbers for these experiments were in the range 7×10^3 to 8×10^4. The water drag coefficient C_w was found to equal 1.2. For higher Reynolds numbers, the drag coefficient can be expected to decrease.

Steady-state motion will not be possible unless there is no relative motion between the water and the iceberg. In that case, C_w is equal to C_d and F_w is zero. Otherwise the iceberg is responding to unbalanced forces and interacting with the adjoining water mass.

Measurements of iceberg motion are not normally made with a sampling interval short enough for the acceleration to be calculated effectively. The calculations reflect only the average motion over some time in-

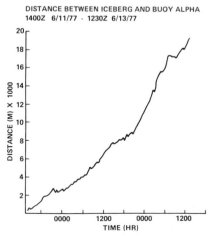

Fig. 25. Relative drift of an integration current drogue with respect to an iceberg–drogue depth 10–96 m. [Redrawn from Russell *et al.* (1977).]

terval. Since the water drag force is nonlinear with respect to relative velocity, this causes a deviation from true force balance. Consequently, the drag coefficient that must be used in a steady-state model in order to produce reasonably realistic paths is large compared to the actual steady-state drag coefficient. It must incorporate the inertial drag term and the history factor. The value of C_w is greater than C_d in general.

Tatinclaux and Kennedy (1978) examined the effect of surface ripple formation on the turbulent boundary layer of an iceberg. Due to the pressure difference caused by flow separation at the crest of a ripple, the drag increased almost linearly with ripple steepness (amplitude/length). At a steepness of 0.1, a common value, the frictional resistance can be 50–100% higher than for a ripple-free surface.

It is quite possible for an iceberg to experience currents that differ in both speed and direction at various depths. A wind drift layer is commonly superimposed on a geostrophic or tidal current. One difficulty arises because of the assumption that the drag force is related to the square of the relative velocity. The square of the average relative velocity is not equivalent to the average of the squared relative velocities. Figure 26 shows an example of this effect. In this example, the direction of the

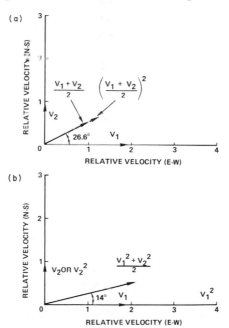

Fig. 26.   Effect of the velocity squared term on the magnitude and direction of the drag force: (a) square of the averaged relative velocities; (b) average of the squared relative velocities.

velocity square term differs by 12.6° in direction and 40% in magnitude. When writing the component equations, the component of the squared term must be used rather than the square of the components of the relative velocity.

C. Wind Drag

The effect of wind drag on an iceberg is far less than that of water drag, because most of the surface of the iceberg is submerged and because the density of air is much less than that of water. The wind effect can be great when much higher wind velocities occur, due to the effect of the velocity-squared term in the drag force equation,

$$F_a = \rho_a A_a C_a |U|U/2, \tag{14}$$

where ρ_a is the density of air, A_a the area of a cross section exposed to air, and C_a the drag coefficient for air.

Winds of less than 8 m/sec are thought to have very little effect on the drift of icebergs; the effect of stronger winds is due mainly to the creation of wind drift currents in the surface layer of the ocean.

The "sail effect" of the wind on the portion of the iceberg above the water would be quite difficult to separate from other wind effects since the creation of an airfoil would depend on a wind that may be highly variable and an iceberg that is free to rotate.

D. Coriolis Force

The Coriolis force is an artificial term used to descrbe the apparent curvature of straight-line motion in a rotating coordinate system. This force is usually represented by

$$F_c = 2\omega \sin(\phi) \, UM, \tag{15}$$

where ω is the angular velocity of the earth's rotation, ϕ the latitude, U the velocity with respect to the earth's surface, and M iceberg mass.

This force acts 90° to the right of the velocity in the northern hemisphere and 90° to the left in the southern hemisphere. Because of the time over which it acts, the Coriolis force can have a significant impact on the drift track. It amounts to a small but continuous force. The water drag acts only to the extent that there is a difference in the water and iceberg motion. The Coriolis force can act even in a steady drift situation, creating a small acceleration that must then be balanced by frictional drag.

The effects that the Coriolis force has on a drifting iceberg and on the surrounding water are quite different. The Coriolis force in the water column is balanced by a pressure gradient, friction, or wind stress, while the Coriolis force on the iceberg is balanced by an increase in the water drag force.

E. Variability of Drift

Soulis (1975) created an empirical model of iceberg drift based on a 1972 study at Saglek, Labrador, where a total of 110 icebergs were tracked by radar. Selecting the 33 best drifts together with surface current and wind data, a parametric study of the iceberg drifts was conducted. The results are shown in Table 2.

Items of interest in this table are (a) the mean speed of the icebergs was 2.5 times that of the measured surface current, and (b) the high variability in all parameters relating to the drift. The drift relative to the wind and current, for example, ranged more than 90° in each direction, with the mean drift directly down current and slightly to the right of the wind. These data amply demonstrate the difficulty of relating measurements of air and water factors to the observed movement of an iceberg.

TABLE 2

Estimated Parameters for 33 Icebergs Drift Tracks[a]

Parameters	Units	Weighted mean	Standard error of the mean	Minimum	Maximum
Magnitude of speed relative to mean current	—	2.46	0.36	0.32	9.0
Direction relative to mean current	deg[b]	−0.83	8.62	−152	104
Amplitude relative to rotary current	—	0.499	0.09	0.109	3.14
Phase angle relative to rotary current	deg[b]	−13.7	8.31	−150	97
Magnitude relative to wind speed	%	3.99	0.49	1.06	14.8
Direction relative to wind	deg[b]	−25.0	7.40	−116	94
Best correlation time lag—current	min	204	18	0.0	>360
Best correlation time lag—wind	min	214	15	0.0	>360

[a] Soulis (1975) with changes.
[b] Degrees counterclockwise.

V. DETERIORATION

A. General Discussion

Iceberg deterioration has not been extensively studied. Field measurements are virtually nonexistent. On the simplest level, icebergs melt. They receive energy by radiation, convection, and conduction from the air, water, and sun. The action of these physical processes on the iceberg is complicated by the necessary phase change from solid to liquid and the presence of dissolved salts in the surrounding fluid. Kollmeyer (1966) conducted the first thorough theoretical study of iceberg deterioration and identified turbulent heat transfer as being of prime importance.

If icebergs melted as an entirety, they would require an extremely long time period to melt. The surface-to-volume ratio is quite small, on the order of 0.013 m²/m³ for an iceberg $1 \times 0.5 \times 0.25$ km, and approximately two-thirds of this surface area is below water. For a much smaller iceberg, with dimensions of $0.5 \times 0.25 \times 0.1$ km, the ratio would be 0.4 m²/m³, again with two-thirds of the surface area below water. The effect of iceberg breakup on deterioration should be obvious, since the source of the melting energy is external.

Icebergs are fairly stable as long as they are protected by sea ice from the stresses of the open sea, but even in a protected environment breakup can occur due to stresses created by changes in buoyancy and temperature. R. C. Kollmeyer (personal communication, 1976) observed a 1.8-km-long iceberg in the Jacobshavn fjord, well protected by other icebergs and brash from the open sea, roll to one side to the point where the crevassed surface layer fell off like a set of dominoes. On a return roll, the iceberg fractured into nearly equal portions. No readily apparent cause for the breakup was observed.

Studies with impulse radars [Kovacs (1978); Swithinbank (1978)] have shown that icebergs with no external flaws often have both firn-covered crevasses on the surface as well as bottom crevasses. The crevasses are related to the flow history of the ice and can occur at intervals of several hundred meters.

In order to study iceberg deterioration, an estimate of the rate of mass loss is required. Because of the irregular shapes of icebergs and the unobservable underwater portion, any estimate would be very uncertain. Attempts have been made to provide a rule of thumb for volume estimates. Methods depend on the three linear measurements that would box-in the above water portion of the iceberg, the length l, width w, and height h. Shil'hikov (1965) in a study of Antarctic icebergs estimated the volume as 4.9lwh for tabular, 4.1lwh for domed, and 2.5lwh for pinnacle. Farmer and

Robe (1977) estimated the volume for Greenland icebergs as 3.35lwh. For icebergs of unusual shape, these approximations may be rough at best.

B.   Wave Effects

Any mechanism that acts to increase the surface-to-volume ratio of an iceberg will greatly inhance melting. Chief among the agents that can lead to breakup is the wave field. The wave field may produce a resonance in an iceberg that is longer than the dominant wavelength, leading to fatigue and breakup. This process, discussed in Section II.B.3 as a mechanism for the calving of shelf ice, is supported by observations of very large tabular icebergs breaking in more or less uniformly sized fragments [Koblents and Budretskiy (1966); Kovacs (1978)] and by the near absence of icebergs longer than 1500 m north of 65° S [Romanov (1973)].

Once an iceberg has reached a size small enough that wave flexure and fatigue are no longer as important (less than 1000 m), other wave deterioration mechanisms begin to dominate. Wave turbulence prevents a static cold-water boundary layer from forming around the iceberg at the surface and acts to transport the warmer surface water toward the iceberg. While icebergs are protected by sea ice, they undergo little deterioration at their margins. When the move out of the sea ice, rapid deterioration begins. Robe *et al.* (1977) followed the deterioration of a tabular Arctic iceberg for 26 days (Fig. 27). The wave-induced deterioration followed planes of weakness in the iceberg and concentrated at indentations and embayments in the iceberg's perimeter. The melting clearly did not extend far below the surface since the iceberg remained in one piece even after a channel cut completely through the above-water portion. This concentration of wave turbulence and heat transfer is often observed in icebergs that remain stable and do not roll during deterioration. These become drydock icebergs that have their center cut through, leaving towering walls on each side. Drydock icebergs are frequently observed to have roughly parallel waterlines on the sides that indicate the absence of rolling during melting. Icebergs that roll become increasingly rounded and domed.

Martin *et al.* (1978) examined the effect of waves on a sample of ice in the laboratory. They observed that if nearly perfect reflection existed, the wave cut notch had a vertical extent of approximately twice the wave amplitude a above the mean free surface and $1/K$ below the mean surface, giving a total wave notch extent Z_n equal to $2a + K^{-1}$, where K is the wave number. If the waves transport water toward the iceberg with a velocity $U_S(Z)$, the Stokes drift is given by

$$U_s(Z) = fa^2K \exp(-2KZ), \tag{16}$$

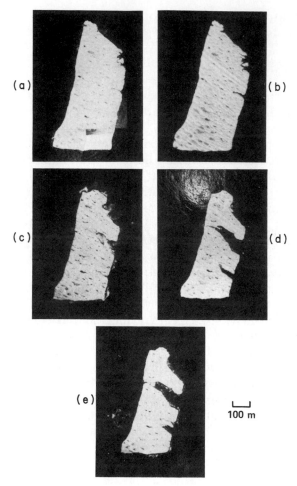

Fig. 27. The deterioration of an iceberg showing the progressive englargement of embayments: (a) 1708Z, 12 May 1976, 49°19.9'N, 49°33.3'W, area 190,485 m²; (b) 1804Z, 13 May 1976, 49°28.9'N, 49°26.6'W, area 189,126 m²; (c) 1404Z, 31 May 1976, 48°09'N, 49°34'W, area 137,361 m²; (d) 1343Z, 4 June 1976, 47°39.7'N, 49°15.8'W, area 121,113 m²; (e) 1812Z, 6 June 1976, 47°31'N, 49°07'W, area 109,067 m². [Robe *et al.* (1977), reproduced with permission from *Nature.*]

where f is the wave frequency, K the wave number, a the wave amplitude, and Z the depth below the mean-free surface. Then the depth integrated Stokes drift toward the iceberg is

$$\overline{U}_s = \tfrac{1}{2}fa^2. \tag{17}$$

The heat transport is

$$Q_{\mathrm{SD}} = \rho_w C_p \, \Delta\theta \, \overline{U}_s, \tag{18}$$

where C_p is the heat capacity of sea water, and $\Delta\theta$ is the temperature difference between the iceberg melt temperature and the far field temperature. It was experimentally determined that $0.01Q_{\mathrm{SD}}$ is an approximate value for the efficiency with which the Stokes drift contributed to melting the ice.

When the undercutting of the iceberg face is sufficient, the extended ice will calve. The approximate conditions when this will take place can be determined by examining the failure of an elastic beam. If the protruding ice beam is assumed to be rectangular, the shear stress at the end of the beam can be calculated by simple beam theory. For a beam of unit width, the shear stress at the beam end is

$$\sigma_{xy} = \rho_i g h d, \tag{19}$$

where ρ_i is the ice density, h the thickness of the beam, d the length of the beam, and g the acceleration of gravity.

Holdsworth (1973) gives a value for failure in pure shear of about 10^5 Pa. If ρ_i is taken as 0.85, then failure will take place when $d = 12$ m.

A beam length of 12 m is the maximum for a beam of sufficient thickness so that the bending moment is not important. If the beam is thin enough, the moment becomes important and failure occurs through bending. Simplifying the development by Holdsworth (1973) of the stress of a floating ice beam by the assumption that the beam is in air and the compression modulus is equal to the tension modulus, an elastic rectangular ice beam will have a surface stress in tension

$$\sigma_{xx} = 6B(x)/h^2, \tag{20}$$

where x is the coordinate along the beam and $B(x)$ the bending moment of the beam.

If 2×10^5 Pa are sufficient to initiate fracture, then the transition from bending failure to shear failure occurs at a beam thickness of approximately 18 m. For an ice face thicknesses of less than 18 m, the depth of undercutting is related to face height by

$$h = 3\rho_i g d^2 / \sigma_{xx}^c, \tag{21}$$

where σ_{xx}^c is the critical surface stress required to initiate fracture.

Since 30–40 m is a fairly typical figure for the freeboard of a large tabular iceberg, failure due to flexure or simple shear is possible when wave undercutting is present. If the geometry of the ice beam is well known, a more exact analysis is possible using the exact shape and different values for the tension and compression moduli. Using a similar analysis for a below-water beam (called a "ram") failure in pure shear would occur at a length of 81 m.

C. Heat Transfer (Water)

The transfer of heat from the air or water to the iceberg occurs in a boundary layer that is generally turbulent and is made more complex by the phase change from ice to water and by the dissolved salts that are in the water and not in the ice.

Tien and Yen (1965) studied the effect of melting by forced convection on the effective heat transfer. They examined several theories for the interface and related the ratio of heat transfer with melting to that of heat transfer without melting. They also related it to a dimensionless parameter, the Stefan number given by

$$S = C_p(\theta_\infty - \theta_0)/\Delta H, \tag{22}$$

where θ_∞ is the temperature at infinity, θ_0 the temperature of the surface, ΔH the enthalpy change due to melting, and C_p the heat capacity of water. For a situation where θ_∞ equals 5°C and θ_0 equals -0.45°C, the change in heat transfer due to the presence of melting is less than 3%.

The transfer of heat from the water to the ice involves the interdependence of the momentum, thermal, and salinity boundary layers. The rate of melting is primarily a function of the Stefan number, which represents a thermal driving parameter for the mass generation at the interface. The Stefan number is dependent on the concentration of salt at the interface. A higher concentration reduces the melting temperature of the ice, resulting in a higher Stefan number. This in turn increases melting and dilution of the salt at the interface causing the melting temperature of the ice to increase. At far field temperatures near 0°C, the freezing-point depression causes a great increase in the Stefan number due to the presence of salt, but very little melting takes place. At higher far field temperatures, rapid melting at the boundary increases the melting temperature by salt dilution until it approaches the melting temperature of fresh water (0°C).

Griffin (1973, 1975, 1978) examined the effect of saline water and freezing point depression on the melt rate of ice. He used a laminar boundary layer model for uniform flow. Marschall (1977) used a laminar boundary layer model for free convective melting of ice. The flow around an iceberg, however, has been observed to have a Reynolds number in the range 10^6–10^7, well within the turbulent region. Josberger (1978), in experiments on the melting of ice in saline water, determined that the laminar region of convective flow probably would not be longer than 0.5 m in length in an oceanic environment. His experiments revealed three zones of melting (Fig. 28): (a) a laminar region near the bottom where the difference in thermal and saline diffusion created a cold low saline flow upward near the ice wall with a cold high saline flow downward outside the upward flow; (b) a fully turbulent region due to the buoyancy of the diluted melt water; and (c) a transition zone where far field water flows in a jet to supply the divergence formed by the upward and downward flows. The melt rate is highest in the transition zone where the warmer, more saline far field water is transported directly to the ice face. The melt rate in the turbulent region was 30% higher than in the laminar region.

The melt rate R is given to within $\pm 5\%$ by

$$R = 3.7 \times 10^{-2}(\theta_\infty - \theta_0)^{1.5} \quad \text{m/day} \tag{23}$$

and the average melt rate $\overline{R}$ for a wall of length L is

$$\overline{R} = \frac{1}{L}\int_0^L R(x/l)^{1/4}\, dx, \tag{24}$$

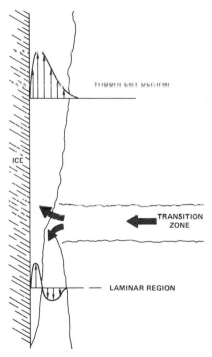

Fig. 28.   The observed flow field adjacent to an ice face in saline water. [Reprinted with permission from *The Proceedings of the First International Conference on Iceberg Utilization,* Josberger (1978), Pergamon Press Limited.]

where x is the coordinate in the flow direction and l an arbitrary scale factor.

For a far field temperature of 15°C, the average melt rate of a 1-m face was estimated to be three times as great as that of a 100-m face. Thus, iceberg deterioration is at least partly an accelerating process since, as an iceberg calves or deteriorates, its rate of calving and deterioration increases.

D. Radiation

Kollmeyer (1966) studied the radiation balance of an iceberg on theoretical grounds. He concluded that direct and indirect radiation cannot account for even 1% of the energy needed to melt an iceberg. Radiation may have a considerable effect on calving, however, through the creation of thermal stresses in the surface layer of the iceberg. Holdsworth (1969a) approximated the thermal stress in ice as

$$\sigma_x = -E \, \Delta\theta / \alpha(1 - p), \tag{25}$$

where E is Young's modulus (bars), α the coefficient of thermal expansion, p Poisson's ratio, and $\Delta\theta$ the temperature change. Stresses in ice as high as 7×10^5 Pa/°C could be generated in the surface layers.

The maximum intensity of solar radiation occurs very near the minimum absorption coefficient of ice, giving solar radiation a fairly great penetration in ice. White ice with a scattering coefficient of $0.7-1.2$ cm^{-1} has an albedo of 60% for ice masses of thicknesses greater than $10-20$ cm [Gilpin *et al.* (1977)]. The relatively high scattering caused by the presence of bubbles has the effect of reducing the depth of penetration of the radiation while increasing the heating near the surface because of the increased path length of the scattered radiation in that region. The internal heating drops an order of magnitude in the first 10 cm, with over 10% of the heating occurring in the first centimeter. At 45° lat, the incoming solar radiation during the spring is between 0.1 and 0.4 cal/cm²/min on a 24-h basis. If the ice is colder than the melting point, as could be expected at night with the subcooling of glacial ice and cold nighttime temperatures, the top centimeter of ice could experience a warming of 1°C in a time given by

$$t = C_p \rho_i d / rIQ, \tag{26}$$

where t is time, C_p the heat capacity of ice, r the albedo of white ice, I the intensity of radiation, d the layer depth, and Q internal heating distribution. For $C_p = 0.5$ cal/gm, $r = 60\%$, $I = 0.2$ cal/cm²/min, $d = 1$ cm, and $Q = 10\%$, t would be approximately 35 min. It is quite likely that considerable thermal stress could be produced in an iceberg under conditions of rapid radiation change.

Icebergs have been observed to show a pronounced increase in calving in the early morning, shortly after being struck by direct solar radiation or after the iceberg has emerged from the fog. In perfectly calm weather icebergs have been observed to break up completely with a loud report leaving only growlers and brash behind. Barnes (1927) and others have felt that this type of calving and destruction was the result of internal thermal stresses and perhaps surface stresses caused by sudden heating or cooling.

VI. FUTURE TRENDS IN RESEARCH

The future of research on icebergs will be directed largely at reducing their impact on transportation, resource development, and in using them as source of fresh water. Emphasis will be placed on all weather detection and tracking and on deterioration (either how to accelerate it or how to slow it). Long-range projections of future iceberg populations will require consideration of the mass balance of the Greenland and Antarctic Ice Sheets.

ACKNOWLEDGMENTS

The author would like to thank Dr. R. C. Kollmeyer, Dr. Willy Weeks, and Mr. Stephen Ackley for their reviews and helpful comments on the manuscript and Charlotte Robe for her editorial comments and proofreading.

REFERENCES

Allaire, P. E. (1972). *J. Can. Petl. Technol.* **11,** 21–25.
Armstrong, T. (1957). *J. Glaciol.* **3,** 106.
Banke, E. G., and Smith, S. D. (1974). *IEEE Int. Conf. Eng. Ocean Environ., 1974* Vol. 1, pp. 130–132.
Baranov, G. I., Ivchenko, V. O., Maslovskiy, M. I., Yreshnikov, A. F., and Kheysin, D. E. (1976). *Probl. Arktiki Antarkt.* **47,** 118–139, CRREL Bibliogr. No. 31-256.
Barnes, H. T. (1927). *Proc. R. Soc. London* **114,** 161–168.
Bauer, A. (1955). *J. Glaciol.* **12,** 456–462.
Brodie, J. W., and Dawson, E. W. (1971). *N. Z. J. Mar. Freshwater Res.* **5,** 80–85.
Brooks, L. D. (1977). *Proc. —Annu. Offshore Technol. Conf.* **00,** 279–286.
Chari, T. R., and Allen, J. H. (1974). *Proc. Int. Conf. Port Ocean Eng. Under Arct. Cond., 2nd, 1973* pp. 608–616.
Colbeck, S. C., and Evans, R. J. (1971). *J. Glaciol.* **10,** 237–243.
Corkum, D. A. (1971). *J. Appl. Meteorol.* **10,** 605–607.
Crary, A. P., Robinson, E. S., Bennett, H. F., and Boyd, W. W., Jr. (1962). "IGY Glaciological Report," No. 6. Am. Geogr. Soc., New York.

Deacon, G. E. R. (1977). *Proc. Polar Oceans Conf., 1974* pp. 11–16.

Dempster, R. T. (1974). *IEEE Int. Conf. Eng. Ocean Environ.* Vol. 1, pp. 125–129.

Dmitrash, Zh.A. (1973). *Sov. Antarct. Exped. Inf. Bull (Engl. Transl.)* **8**, 441–442.

Dolqushin, L. D. (1966). *Sov. Antarct. Exped. Inf. Bull (Engl. Transl.)* **6**, 41–43.

Dyson, J. L. (1972). "The World of Ice." Alfred A. Knopf, New York.

Emery, W. J. (1977). *J. Phys. Oceanogr.* **7**, 811–822.

Farmer, L. D., and Robe, R. Q. (1977). *Photogramm. Eng. Remote Sens.* **43**, 183–189.

Franceschetti, A. P. (1964). *U.S. Coast Guard Oceanogr. Rep.* No. 5, pp. 1–36.

Gaskin, D. E. (1972). *N.Z. J. Mar. Freshwater Res.* **6**, 387–389.

George, D. J. (1975). *Polar Rec.* **17**, 399–401.

Gilpin, R. R., Robertson, R. B., and Singh, B. (1977). *J. Heat Transfer* **99**, 227–232.

Gordon, A. L., Taylor, H. W., and Georgi, D. T. (1977). *Proc. Polar Oceans Conf., 1974* pp. 45–76.

Griffin, O. M. (1973). *J. Heat Transfer* **95**, 317–323.

Griffin, O. M. (1975). *J. Heat Transfer* **97**, 624–626.

Griffin, O. M. (1978). *Proc. Int. Conf. Iceberg Util., 1977* pp. 229–244.

Gustajtis, K. A., and Buckley, T. J. (1978). *Proc. Int. Conf. Port Ocean Eng. Under Arct. Cond., 4th, 1977* pp. 972–983.

Hamilton, W. S., and Lindell, J. E. (1971). *ASCE J Hydraul. Div.* **97**, 805–817.

Hobbs, P. V. (1974). "Ice Physics." Oxford Univ. Press (Clarendon), London and New York.

Holdsworth, G. (1969a). *J. Glaciol.* **8**, 107–129.

Holdsworth, G. (1969b). *J. Glaciol.* **8**, 385–397.

Holdsworth, G. (1971). *Can J. Earth Sci.* **8**, 299–305.

Holdsworth, G. (1973). *J. Glaciol.* **12**, 235–250.

Holdsworth, G. (1978). *Proc. Int. Conf. Iceberg Util., 1st, 1977* pp. 160–175.

Jelly, K. E. P., and Marshall, N. B. (1967). *Mar. Obs.* **37**, 86–93.

Josberger, E. G. (1978). *Proc. Int. Conf. Iceberg Util., 1st, 1977* pp. 245–264.

Koblents, Ya. P., and Budretskiy, 00 (1966). *Sov. Antarct. Exped. Inf. Bull. (Engl. Transl.)* **6**, 272–273.

Koch, L. (1945). *Medd. Groenl.* **130**, No. 3, 109–116.

Kollmeyer, R. C. (1966). *U.S. Coast Guard Oceanogr. Rep.* No. 11, CG 373-11, pp. 41–52.

Kollmeyer, R. C. (1979). *Proc., Int. Workshop World Glacier Inventory, 1978* (in press).

Kollmeyer, R. C., O'Hagan, R. M., and Morse, R. M. (1965). *U.S. Coast Guard Oceanogr. Rep.* No. 10, CG 373-10, pp. 1–24.

Kollmeyer, R. C., Motherway, D. L., Robe, R. Q., Platz, B. W., and Shah, A. M. (1977). "A Design Feasibility Study for the Containment of Icebergs Within the Waters of Columbia Bay, Alaska." Coast Guard Res. Dev. Cent., Groton, Connecticut.

Korotkevich, E. S. (1964). *Sov. Antarct. Exped. Inf. Bull. (Engl. Transl.)* **1**, 40–45.

Kovacs, A. (1978). *Proc. Int. Conf. Iceberg Util., 1st, 1977* pp. 131–145.

Ledenev, V. G. (1962). *Sov. Antarct. Exped. Inf. Bull. (Engl. Transl.)* **35**, 146–151.

Marschall, E. (1977). *Lett. Heat Mass Transfer* **4**, 381–384.

Martin, S., Josberger, E., and Kauffman, P. (1978). *Proc. Int. Conf. Iceberg Util., 1st, 1977* pp. 260–264.

Matsuo, D., and Miyake, Y. (1966). *J. Geophys. Res.* **71**, 5235–5241.

Mellor, M. (1959a). *J. Glaciol.* **3**, 377–384.

Mellor, M. (1959b). *J. Glaciol.* **3**, 522–533.

Mellor, M. (1967). *In* "The Encyclopedia of Atmospheric Sciences and Astrogeology" (R. W. Fairbridge, ed.), Vol. II, pp. 16–19. Van Nostrand-Reinhold, Princeton, New Jersey.

Morgan, V. I., and Budd, W. F. (1978). *Proc. Int. Conf. Iceberg Util., 1st, 1977* pp. 220–228.

Murty, T. S., and Bolduc, P. A. (1975). *Proc.—Annu. Offshore Technol. Conf.* **0**, 785–789.

Orvig, S. (1972). *Proc. Symp. World Water Balance, 1972* pp. 41–49.

Ovenshine, A. T. (1970). *Geol. Soc. Am. Bull.* **81**, 891–894.

Post, A. (1978). *Geol. Surv. Rep.* pp. 78–264.

Reeh, N. (1968). *J. Glaciol.* **7**, 215–232.

Robe, R. Q. (1975). *Int. Comm. Northwest Atlantic Fish., Spec. Publ.* No. 10, pp. 121–127.

Robe, R. Q., (1978). *Nature (London)* **271**, 687.

Robe, R. Q., Maier, D. C., and Kollmeyer, R. C. (1977). *Nature (London)* **267**, 505–506.

Robin, G. de Q. (1953). *J. Glaciol.* **2**, 208.

Romanov, A. A. (1973). *Sov. Antarct. Exped. Inf. Bull. (Engl. Transl.)* **8**, 499–500.

Russell, W. E., Riggs, N. P., and Robe, N. P., and Robe, R. Q. (1978). *Int. Conf. Port Ocean Eng. Under Arct. Cond., 4th, 1977* pp. 784–798.

Scholander, P. F., and Nutt, D. C. (1960). *J. Glaciol.* **3**, 671–678.

Scholander, P. F., Kanwisher, J. W., and Nutt, D. C. (1956). *Science* **123**, 104–105.

Scholander, P. F., Hemmingsen, E. A., Couchman, L. K., and Nutt, D. C. (1961). *J. Glaciol.* **3**, 813–822.

Shil'hikov, V. I. (1965). *Sov. Antarct. Exped. Inf. Bull. (Engl. Transl.)* **3**, 23–26.

Shpaykher, A. O. (1968). *Sov. Antarct. Exped. Inf. Bull. (Engl. Transl.)* **7**, 182.

Smith, E. H. (1931). *U.S. Coast Guard Rep. Int. Ice Patrol Serv. North Atlantic Ocean Bull.* No. 19.

Sodhi, D. S., and Dempster, R. T. (1975). *IEEE Conf. Eng. Ocean Environ.* pp. 348–350.

Soulis, E. D. (1975). *Can. Soc. Pet. Geol., Mem.* **4**, 879–890.

Swithinbank, C. (1955). *Geogr. J.* **121**, 64–76.

Swithinbank, C. (1969). *Polar Rec.* **14**, 477–78.

Swithinbank, C. (1978). *Proc. Int. Conf. Iceberg Util., 1st, 1977* pp. 100–107.

Swithinbank, C., McClain, P., and Little, P. (1977). *Polar Rec.* **18**, 495–501.

Tatinclaux, J., and Kennedy, J. F. (1978). *Proc. Int. Conf. Iceberg Util., 1st, 1977* pp. 276–282.

Tchernia, P. (1977). *Proc. Polar Oceans Conf., 1974* pp. 107–120.

Tien, C., and Yen, Y. (1965). *J. Appl. Meteorol.* **4**, 523–527.

Urick, R. J. (1971). *J. Acoust. Soc. Am.* **50**, 337–341.

U.S. Coast Guard (1971). *Rep. Int. Ice Patrol Serv. North Atlantic Ocean, Bull.* No. 56.

Vinje, T. E. (1975). *Arbok, Nor. Polarinst. 1973* pp. 197–202.

Weeks, W. F., and Campbell, W. J. (1973). "CRREL Research Report 200." U.S. Army Cold Reg. Res. Eng. Lab., Hanover, New Hampshire.

Weeks, W. F., and Mellor, M. (1978). *Proc. Int. Conf. Iceberg Util., 1st, 1977* pp. 45–98.

Wexler, H. (1961). *J. Glaciol.* **3**, 867–872.

Wolford, T. C. (1972). Dissertation, Catholic University of America, Washington, D.C.

Zotikov, I. A., Ivanov, Yu.A., and Barbash, V. R. (1974). *Oceanology* **14**, 485–490.

5 FRESHWATER ICE GROWTH, MOTION, AND DECAY

George D. Ashton

Snow and Ice Branch
U.S. Army Cold Regions Research and Engineering Laboratory
Hanover, New Hampshire

I. INTRODUCTION

Freshwater ice has always been of interest to people in cold climates. At least as early as A.D. 1000, the Viking saga Havámál offered advice still good today in the caution, "Praise not . . . the ice until it has been crossed" [Bronstad (1965)]. In winter the people of the north traveled using the relatively flat and smooth ice-covered lakes and rivers. Ice-covered lakes and rivers are still used in northern regions as routes of transport. Increasingly, however, the concern with lake and river ice is to

minimize interference with man's activities, most of which are fashioned
for the warmer, ice-free time of year but which have become increasingly
prevalent during the ice season.

In summarizing our present knowledge of freshwater ice growth, motion, and decay, we shall emphasize two main categories of occurrence,
river ice and lake ice. Attention will center on the processes involved,
quantitatively discussed where possible, descriptively where the processes are less well understood or so complex as to defy analysis as yet.
This review will concentrate on the more common types of ice occurrence, with little discussion of the more rare, but no less intriguing, phenomena.

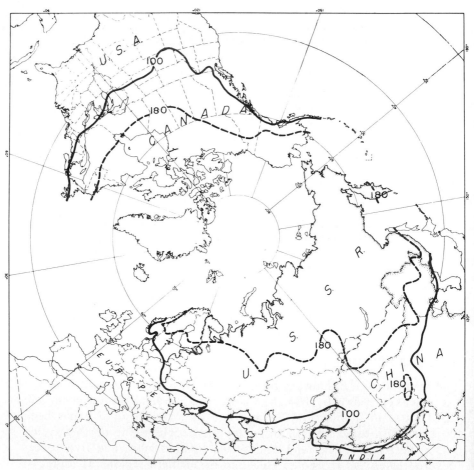

Fig. 1. The average number of days per year during which water is unnavigable due to
ice. Period generally described as freeze-over to breakup [Bates and Bilello (1966)].

The feature of rivers that causes river ice to be different from lake ice is the dominating effect of the flow velocity. River types range from those rivers that are large and deep with relatively simple planforms (e.g., canals) to streams so small that the ice can span unsupported from bank to bank or, in the extreme case, completely block the water flow beneath, forcing it to flow over the ice, resulting in extensive formation of icings. The nature of lake ice, on the other hand, depends largely on the lake size. The wind has a dominating effect on formation, growth, motion, and decay. In both cases, we shall also be concerned with the water temperatures, since the energy exchanges between the water, ice, and air determine the quantities of ice present as wind and water velocities determine the ice distribution.

Literature on river and lake ice is widely dispersed through the technical literature. There are a few important summaries such as the early works of Hoyt (1913), Barnes (1928), and Altberg (1936). More recent general summaries include those of Michel and Triquet (1967), Michel (1971), Starosolsky (1969), Pivovarov (1973), and Ashton (1978a). Bibliographies are also available, most notably the literature summaries of the *Journal of Glaciology* and the extensive annual bibliography of the U.S. Army Cold Regions Research and Engineering Laboratory. This latter bibliography also allows access to the extensive Russian literature.

There is considerable yearly variation in the spatial and temporal occurrence of ice covers on rivers and lakes. There have been a number of studies dealing with the extent, duration, and maximum thickness. In Fig. 1, the extent of unnavigable water due to ice in the northern hemisphere is mapped [Bates and Bilello (1966)]. Figure 2 describes the isopleths of greatest maximum thickness of ice for periods of record on lakes in northern North America [Bilello and Bates (1966)]. Similar regional interpretations of historical records have been made and include average dates of first ice, freeze-over, and ice clearing. Such data summaries are useful for general planning purposes but are severely limited by both the large annual variability and the anomalies associated with specific sites, particularly in the more temperate regions.

II. RIVER ICE

A. Thermal Regimes and Energy Budgets of Rivers

1. Energy Budget of Open Surface Flow

Most rivers are turbulent, a consequence of the Reynolds number (Re, a measure of inertial forces relative to viscous forces defined as $UD\rho/\mu$), generally being considerably greater than the laminar to turbulent thresh-

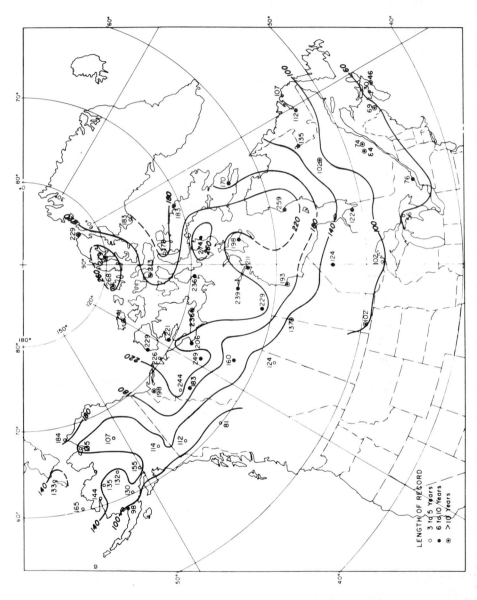

Fig. 2. Isopleths of greatest ice thickness (cm) observed for period of record on lakes in northern North America [Bilello and Bates (1966)].

old value of about 600. If the turbulence is great enough, the resultant mixing is sufficient to overcome any initial stratification associated with the density variations between 0 and 4°C (the temperature of maximum density) to 8°C (where the density is the same as at 0°C). The rate of vertical mixing of an initially stratified flow is determined by the magnitude of the "densimetric" Froude number (Fr_d, a measure of the effects of gravitational stabilizing effects relative to inertial effects, defined as $U/(gD \Delta\rho/\rho)^{1/2}$. For $Fr_d < 1$, the turbulence is insufficient to overcome the stabilizing effects of gravity acting on the density difference and the flow will remain stratified. This is usually the case for lakes and reservoirs. For Fr_d greater than one, mixing occurs over some distance, generally of the order of several hundred times the depth. For the moment we shall assume complete vertical mixing. As a result heat loss from the surface acts to cool the flow over its depth fairly uniformly, and it is possible to predict changes in the water temperature caused by meteorological variables using straightforward energy budget methods applied to the water surface; that is,

$$\partial T_w/\partial t + U \, \partial T_w/\partial x = -\phi_{wa}/\rho C_p D,$$ (1)

where ϕ_{wa} is the surface heat flux from the water to the air. The surface heat flux is composed of the major terms.

$$\phi_{wa} = \phi_s + \phi_l + \phi_e + \phi_c + \phi_{sn},$$ (2)

where ϕ_s is the short-wave radiation, ϕ_l the long-wave radiation, ϕ_e the evaporative heat flux, ϕ_c the sensible heat flux, and ϕ_{sn} the effective heat flux due to snow falling on the surface. Detailed mathematical expressions for each of the components are given [e.g., Dingman *et al.* (1968)]. The only terms different in winter (as opposed to summer) for open water surfaces are the evaporative flux ϕ_e and the radiation fluxes ϕ_s and ϕ_l. Dingman *et al.* (1968) adopted the empirical "Russian winter equation" of Rimsha and Donchenko (1957) for ϕ_e, which predicts somewhat greater evaporative fluxes (see Fig. 3) than other more extensively used formulas such as that of Ryan *et al.* (1974). Both formulas apply a wind function $f(V_a)$ to the vapor pressure difference at the water surface and at an elevation z above the surface such that

$$\phi_e = f(V_a)(e_s - e_a).$$ (3)

Rimsha and Donchenko's (1957) expression takes the form

$$f(V_a) = 60.1 + 2.61(T_w - T_a) + 29.3V_a$$ (4)

while the expression of Ryan *et al.* (1974) takes the form

$$f(V_a) = 26.9(T_w - T_a)^{1/3} + 30.9V_a,$$ (5)

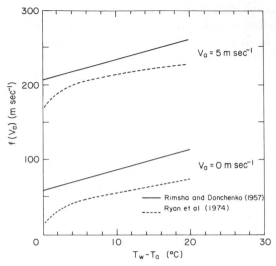

Fig. 3. Comparison of the Rimsha and Donchenko (1957) (solid line) and Ryan *et al.* (1974) (dashed line) wind function in the expressions for evaporative heat flux.

where $T_w - T_a$ is the temperature difference (°C), V_a the wind velocity (m sec⁻¹), and $f(V_a)$ has dimensions of meters per second. In Fig. 3 a comparison of the two relationships is made. The Rimsha–Donchenko equation is an empirical result of studies on rivers in winter while the equation of Ryan *et al.* is an empirical result of energy-budget studies of water bodies during nonwinter conditions. Considering the accuracy usually available for the input variables, the two equations are not greatly different.

The short-wave and long-wave radiation fluxes ϕ_s and ϕ_l are believed to be different in winter because of the interference by vapor clouds above the water surface when the air–water temperature difference is very large. The writer knows of no way to estimate the magnitude of this effect.

Using Eq. (1), the cooling of the river water temperature until it reaches 0°C may be predicted. As long as the surface is free of ice the flow temperature responds to the changing meteorological variables with shallow streams responding more rapidly than deeper streams because of the difference in depth. As the water cools down to 4°C, mixing occurs over the entire depth since the cooled surface water is more dense than the warm water below. From 4°C to 0°C mixing depends on the turbulence, although as noted earlier the turbulence in rivers generally is sufficient for a well-mixed condition. When the water temperature is 0°C, further cooling causes an ice sheet or (if the water supercools) frazil ice to form. The diurnal variations in ϕ_s and ϕ_I cause cooling to be more rapid at night. Hence, early in the winter, frazil ice formed at night will melt the following day. This process continues until sufficient heat is lost from the entire river to protect the frazil, or until a complete ice cover is formed. This initial formation period will be discussed in more detail later.

2. Energy Budget of Ice-Covered Flow

Once an ice cover has formed, the thermal regime of the flow may be characterized as extremely close to the freezing point, on the order of 0.1°C or less and usually 0.02°C or less. Detailed measurements of the flow temperature [e.g., Ashton and Kennedy (1970)] show only extremely small temperature differences in any vertical line, and only slight variations across the width, with slightly warmer temperatures generally found near the shore areas. These warmer temperatures result primarily from tributary inflow that has not dispersed fully in the lateral direction, and secondarily from groundwater influx. The ice cover acts as a seal on the river, preventing heat gain from the atmosphere, while cooling any thermal effluents introduced into the flow by heat transfer to the ice undersurface.

In the absence of tributary sources of thermal effluent and neglecting generation of heat by viscous dissipation (considered later), Eq. (1) takes the form

$$\partial T_w/\partial t + U \, \partial T_w/\partial x = -\phi_{wi}/\rho C_p D, \qquad (6)$$

where ϕ_{wi} is the heat flux from the water to the ice. No field measurements of ϕ_{wi} are known, but laboratory experiments [Ashton and Kennedy (1972)] show $\phi_{wi} = h_{wi}(T_w - T_m)$ to be consistent with the correlations for closed conduit turbulent heat transfer of the form

$$\mathrm{Nu} = C_n \, \mathrm{Re}^{0.8} \, \mathrm{Pr}^{0.4}, \qquad (7)$$

where the *Nusselt number* Nu, which is a nondimensional measure of heat transfer, is defined as $\phi_{wi}D/(2k_w(T_w - T_m))$, the Reynolds number Re is defined as $UD\rho/(2\mu)$, and the Prandtl number Pr is defined as $\mu C_p/k_w$. The Prandtl number, which is a measure of the ratio of thermal diffusivity to momentum diffusivity, is 13.6 for water at 0°C. The value of C_n is uncertain but seems to be above the value (0.023) commonly used in tube flow.

It is convenient to adopt the Lagrangean viewpoint of following a parcel of water in space (as contrasted with the Eulerian viewpoint of a coordinate system fixed in space). Introducing the operator

$$D/Dt = \partial/\partial t + U \, \partial/\partial x \qquad (8)$$

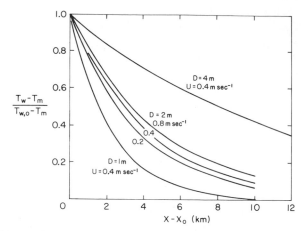

Fig. 4.   Decay of water temperature downstream beneath an ice cover (for C_n of 0.023) for various values of depth D and flow speed U.

allows Eq. (6) to be written in the Lagrangean form

$$DT_w/Dx = -\phi_{wi}/\rho UDC_p \qquad (9)$$

and integrated to yield

$$(T_w - T_m)/(T_{w,0} - T_m) = \exp(-h_{wi}(x - x_0)/\rho UDC_p), \qquad (10)$$

where $T_{w,0}$ is the temperature at x_0 and h_{wi} is a heat transfer coefficient obtained from Eq. (7). In Fig. 4 solutions for Eq. (10) are presented for various flow depths and velocities. The effect of velocity is not great since, while higher velocities increase the rate of convective heat transfer to the ice cover, the water moves proportionately faster and farther downstream.

3.   Energy Budget of the Ice Cover

The thickening or melting of the ice cover is governed by the heat fluxes to the upper and lower surfaces. At the lower surface the heat balance is

$$\phi_i - \phi_{wi} = \rho_i \lambda \frac{d\eta}{dt}, \qquad (11)$$

where η is the ice thickness and ϕ_i the heat flux by conduction through the ice. The upper surface is at a temperature T_s less than or equal to 0°C and is exposed to the ambient air temperature T_a (see Fig. 5). Because the heat

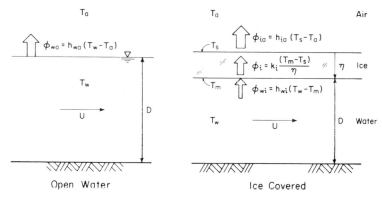

Fig. 5. Definition sketch for energy budget of rivers with an ice cover.

of fusion is so much greater than the specific heat capacity of the ice and thicknesses are ordinarily not very large, a linear temperature profile may be assumed through the ice cover. Although this neglects the penetration of solar radiation, it greatly simplifies the solution of Eq. (11). Physically this means a change in upper surface temperature T_s has an instantaneous effect on the heat flux through the ice ϕ_i, or

$$\phi_i = -k_i(T_s - T_m)/\eta. \tag{12}$$

In turn, the heat transfer from the top surface of the ice to the air ϕ_{ia} is assumed in the simplest case equal to ϕ_i. If ϕ_{ia} is expressed in terms of a heat transfer coefficient equal to $h_{ia}(T_s - T_a)$, then T_s may be eliminated and

$$\phi_i = \frac{T_m - T_a}{\eta_i/k_i + 1/h_{ia}} \tag{13}$$

Most analyses of the thickening of an ice cover assume that T_0 is equal to T_a and ϕ_{wi} is equal to zero, then integrate Eq. (11) to yield

$$\eta = \left(\frac{2k_i}{\rho_i \lambda}\right)^{1/2} \left(\int_0^t (T_m - T_a)\, dt\right)^{1/2}. \tag{14}$$

It has long been the practice to relate thickening to the accumulated degree-days of freezing in the form

$$\eta = \alpha \left(\sum S_d\right)^{1/2}, \tag{15}$$

where α is a correction factor and $\sum S_d$ is the accumulated degree-days of air temperatures below 0°C. This is the same form as Eq. (14). Substitution of appropriate values for k_i, ρ_i, and λ yields α equal to 0.00012 m sec$^{-1/2}$ °C$^{-1/2}$ (conveniently $\simeq$ 1.0 in. day$^{-1/2}$ °F$^{-1/2}$). Actual observed values of α range from 50 to 90% of this value with the larger values associated with thicker ice sheets or windy conditions. Both effects are consistent with the neglect of the difference between the air temperature and the surface temperature. A practical compromise is to predict thickening by stepped integration of

$$\frac{d\eta_i}{dt} = \frac{1}{\rho_i \lambda} \left[\frac{-(T_a - T_m)}{\eta_i/k_i + \eta_s/k_s + 1/h_{ia}} \right], \tag{16}$$

where η_s and k_s are the thickness and conductivity of any snow layer on top of the ice. While more elaborate variations of the above analyses may be done, there is seldom sufficiently detailed information on the heat transfer from the ice to the air, the air temperature, the thickness, or the conductivity of the snow to justify the extra effort. In the case of river ice, however, there are times when the analysis should include the effects of the water–ice heat flux ϕ_{wi} since, even for a value of T_w just above 0°C, the ϕ_{wi} term can have a value comparable to ϕ_i or greater. If ϕ_{wi} is greater than ϕ_i, net melting occurs on the undersurface. Generally the difficulty in implementing such analysis is associated with inadequate knowledge of T_w. To illustrate the relative effect of T_w, the equilibrium thickness of ice ($\phi_{wi} = \phi_i$) without a snow cover, that is, the condition for which the tendency to thicken is exactly balanced by the tendency to melt, has been calculated as a function of air temperature T_a and water temperatures T_w for a typical case. The results in Fig. 6 show even small water temperature differences $T_w - T_m$ are important to inhibiting the ice growth. Finally we

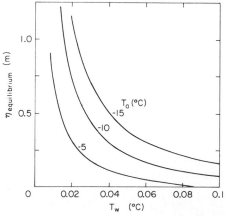

Fig. 6. Effect of T_w and T_a on the equilibrium thickness of ice for which the heat fluxes are in balance [for $h_{ia} = 25$ W m^{-2} °C^{-1}, $U = 1$ m sec^{-1}, $D = 1$ m, and ϕ_{wi} determined from Eq. (7) with $C_n = 0.023$].

note that very thin ice covers [$\eta \to 0$ in Eq. (16)] grow linearly in time because of the air boundary layer effect which is represented in Eq. (16) by the ice–air heat transfer term h_{ia}.

4. Initial Ice-Cover Formation

The initial ice cover on a river is almost always the result of small crystals forming at the surface of the water and freezing together. When the initial crystals are entrained away from the surface, the formation is the result of frazil evolution, which is treated in more detail below. At very low velocities, however, the nucleating crystals rapidly grow on the surface and form a thin ice cover that subsequently thickens because of heat loss to the cold atmosphere above. At usual river velocities, however, the process of initial ice cover formation is one of accumulation, either of small floes of plate ice or of frazil accumulations.

B. Frazil Ice

The formation of frazil ice, small crystals suspended in the flow, has been summarized by Osterkamp (1975). The initial crystals are generally in the form of thin disks with diameters of about 5 mm which grow into irregular platelets which are about 10–15 mm in their maximum dimension. These platelets agglomerate to form flocs; the flocs float to the surface and accumulate in more or less hemispherically shaped pans that subsequently freeze together to form the initial ice cover. Figure 7 shows four stages of the evolution of frazil, ranging from the initial crystals to the large floes that make up the ice cover.

The change in temperature versus time of water undergoing supercooling and frazil formation is generally as shown in Fig. 8. The initial cooling from A to B is governed by the loss of heat from the surface. At point B initial nucleation begins and the cooling deviates from its previous path $A-B$. When the water temperature is below 0°C, the ice–water mixture is out of thermal equilibrium and, as the crystals grow, the temperature reaches its maximum depression C before asymptotically approaching 0°C from below. At point C the energy used by freezing equals the rate of heat loss from the volume of water containing the frazil. Similarly, as the curve approaches 0°C asymptotically, further production of frazil occurs more or less in balance with the heat loss from the surface of the flow, perhaps lessened somewhat by interference with turbulent mixing at the surface caused by the frazil itself. The energy conservation principle allows reasonable estimates of frazil production in an open reach of river by integration of the surface heat loss over time. For example, for

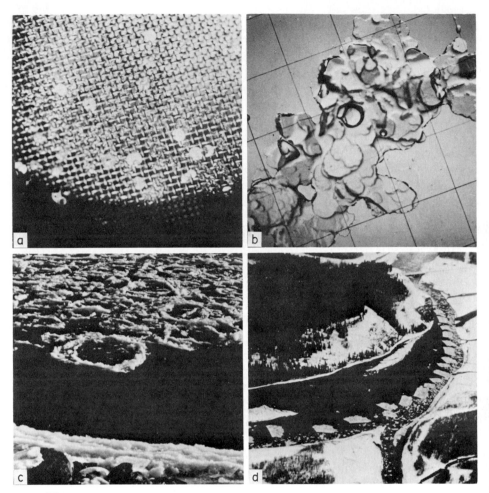

Fig. 7. Frazil ice: (a) Initial individual platelets, (b) frazil flocs, (c) frazil pans or "pancakes," (d) an initial ice cover resulting from frazil accumulation. [From Osterkamp, T. E. (1978). *J. Hydraul. Div. ASCE* **104**, 1239–1256.]

a reach of open area A_o the volumetric production V_f of frazil is approximately

$$\rho_i \lambda V_f = A_o \int_{t_0}^{t} -h_a(T_a - T_w) \, dt. \tag{17}$$

Massive amounts of frazil can be produced in relatively short reaches of open water over a winter season. For example, for an area of 1000 m², h_{ia} equal to 25 W m⁻² sec⁻¹, and T_a of $-15°C$, the rate of production is 106

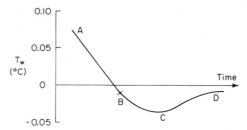

Fig. 8. Temperature evolution during frazil formation. [After Carstens (1966)].

m³ day⁻¹, all of which is eventually deposited somewhere downstream. The most effective way to suppress frazil production is to induce an ice cover to form, thus blocking heat loss from the flow to the subfreezing air. In practice this is accomplished by maintaining a flow velocity less than about 0.6 m sec⁻¹, often with the additional installation of a floating barrier (ice boom) to initially arrest the floating ice.

While we have described the general nature of frazil formation, there is considerable controversy over the initial mechanism of frazil nucleation. The controversy arises because observations of frazil formation in rivers show a maximum supercooling (point C in Fig. 8) much less than the observed supercooling of several degrees below freezing required for nucleation in water. Among the various mechanisms put forth are spontaneous heterogeneous nucleation [Altberg (1936); Devik (1942)], heterogeneous nucleation in a thin supercooled surface layer of the flow adjacent to air layers of below-freezing temperatures [Michel (1967)], crystal multiplication processes [Chalmers and Williamson (1965)], and a mass exchange mechanism between the surface water and the cold air above [Osterkamp (1975)]. The spontaneous heterogeneous nucleation concept seems unlikely because closed flow systems (where the water is not exposed to the air) undergo considerably greater supercooling before nucleation. The thin supercooled surface layer hypothesized by Michel (1967) no doubt occurs but not to levels of supercooling necessary for heterogeneous nucleation. Crystal multiplication is possible only if there is an initial crystal present. The mass exchange mechanism put forth by Osterkamp (1975) seems most reasonable.

The basic concept of the mass exchange mechanism is that small crystals of ice nucleate in the cold air above the surface and fall to the water; nucleation then occurs by some form of crystal multiplication such as that put forth by Chalmers and Williamson (1965). This is consistent with the general observation that initial frazil formation does not occur until air temperatures are several degrees below freezing, although the actual threshold temperature is ill defined. In experiments involving spraying above freezing river water through cold air, frazil formation was not observed at air temperatures above about $-4°C$ [Billfalk and Desmond (1979)].

When the matrix water surrounding the frazil crystals is supercooled, the individual crystals are not in thermal equilibrium but are actively growing in size. This "active" frazil is characterized by a tendency to adhere to other objects. As a result, frazil in the active state attaches to underwater objects and may cause severe blockage of water intake structures, particularly trash racks. Such blockage can be avoided either by heating the intake structure elements or by inducing an ice cover upstream. The covered flow allows time for the crystals to approach a state of quasi-equilibrium (point D in Fig. 8), partly by the warming of the water by viscous dissipation. The viscous heating increases temperature at a rate given by

$$dT_w/dx = (g/C_p) \, dH/dx, \tag{18}$$

where dH/dx is the hydraulic gradient, related to the flow velocity U and depth D by the Darcy–Weisbach friction factor f,

$$dH/dx = fU^2/2gD. \tag{19}$$

Combining Eqs. (18) and (19),

$$dT_w/dx = fU^2/2C_pD. \tag{20}$$

The values for f of ice-covered flow are somewhat uncertain. Nezhikovskiy (1964) has summarized a large amount of data for Russian rivers and found large values of f at the beginning of the period of ice cover with decreasing values over the winter season. It is not clear if the decrease was due to reduced blockage of the flow by melting of frazil deposits, to a smoothing of the ice undersurface, or to a combination of the two effects.

Active frazil also readily adheres to river bottom materials, resulting in underwater formations called *anchor ice*. As anchor ice accumulates, its buoyancy eventually may exceed the adhesive force of attachment and it rises, often with fair-sized rocks and bottom sediment that are subsequently incorporated into the ice cover.

Once the frazil has formed it is carried downstream and either forms an ice cover by agglomeration or is swept beneath a downstream ice cover, eventually coming to rest beneath the ice cover. These deposits can be massive and often extend to the bottom of the flow. When the resulting blockage causes significant rises in water level the accumulation is termed a *hanging dam*. Typical porosities of deposited frazil are approximately 0.4–0.6 with permeabilities on the order of 15×10^{-10} m^2. The water flowing through a frazil-filled cross section tends to concentrate in small

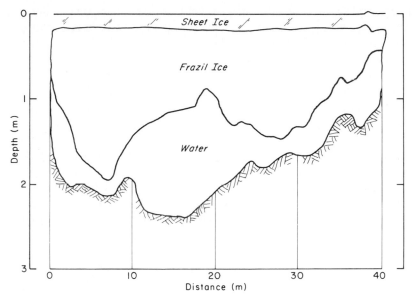

Fig. 9. Cross section of a river with large frazil ice accumulation.

areas of the total river cross section. A typical example is shown in Fig. 9. The axes of these flow sections often meander within the total cross section. There is very little flow of water through the frazil mass itself. As the ice cover above thickens it incorporates the deposited frazil, and exhibits a fine-grained crystal texture, in contrast to the larger columnarlike crystals that consitute an ice cover thickening into water without frazil.

C. Floe Accumulation

The actual formation of the initial ice cover, as noted earlier, usually results from accumulation. It is not at all clear what conditions are necessary to initiate a cover in a particular reach of river unless there are obvious points of initiation such as a complete surface barrier. The usual concept of the location of initiation is a narrow reach at a river constriction. The opposite is usually the case, however, as can be concluded by the following simplified analysis. For convenience we assume rectangular cross sections of flow and constant volume flux of water Q_0 equal to UBD, where U is mean velocity, B width, and D depth. Associated with this flow is a flux of water surface S_w equal to UB. Since Q_0 is constant, $S_w = Q_0/D$. The surface flux of ice S_i is equal to $C_i S_w$, where C_i is the ice concentration. Hence

$$C_i = S_i/S_w = S_i/UB. \tag{21}$$

Since S_i and Q_0 are constant, the concentration C_i increases or decreases as the depth increases or decreases, respectively. The effect of changes in velocity U and width B on the surface concentration may be examined in a similar manner by substituting UBD for Q_0 in Eq. (21). At a constriction with similar cross sections, as B decreases the concentration increases. An analysis much like this was made by Schoklitsch (1937).

The location of initiation is thus where the transport capacity is exceeded by the surface flux of ice. Frankenstein and Assur (1972) have characterized three stages in this process: simple floating of ice fragments with no significant interaction, initial clogging with significant fragment interaction but not enough to resist the fluid shear beneath, and final stoppage when the interfragment forces are sufficient to bridge the flow and resist the fluid shear beneath. As yet, however, there is no adequate characterization of the constitutive relation between moving ice fragments to evaluate the last, critical condition. Partly as a result of this uncertainty about where the ice first stabilizes to form a cover, most analyses of accumulation postulate an initial surface barrier and proceed from that point.

The accumulation of the ice cover upstream has been examined from two viewpoints, that of the behavior of the accumulation thickness itself, and that of the behavior of individual ice floes. Initial work on the behavior of individual ice floes was performed by Pariset and Hausser (1961) who obtained a criterion for submergence at the leading edge of an ice cover (see Fig. 10) by performing a vertical force balance between the hydrodynamic forces tending to submerge a floe vertically and the buoyant force. This analysis led to a relationship of the form

$$U_c \bigg/ \left[2g \left(\frac{\rho_w - \rho_i}{\rho_w} \eta \right) \right]^{1/2} = K, \qquad (22)$$

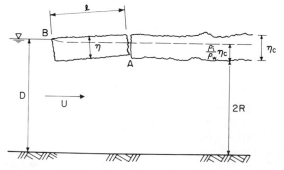

Fig. 10. Definition sketch for floe submergence and ice accumulation analysis.

where U_c is the critical velocity for submergence and K is a "form coefficient" dependent upon the floe shape; K varied from about 0.6 to 1.1 as a weakly increasing function of the length–thickness ratio of l/η. Michel (1971) introduced porosity e as a separate factor to account for cases where an ice floe is porous (frazil floes) but still has a measure of structural integrity, with the result

$$U_c \Big/ \left[2g \frac{\rho_w - \rho_i}{\rho_w} (1 - e)\eta \right]^{1/2} = K. \tag{23}$$

Uzuner and Kennedy (1972) performed a large number of experiments on simulated ice floes of various densities and analyzed the critical condition for rotational underturning (the most usual behavior of a floe at the edge) by an analysis of the moments acting about the floe's downstream point of contact with the stationary ice cover (point A in Fig. 10). The critical condition for entrainment was the "no-spill" condition, i.e., when the stagnation water level (at point B in Fig. 10) exceeds the top edge of the floe. Ashton (1974a,b) used a simplified but similar moment analysis and found the experimental results of Uzuner and Kennedy (1972) to be well described (see Fig. 11) by

$$\frac{U_c}{[g\eta(1 - \rho_i/\rho_w)]^{1/2}} = \frac{2(1 - \eta/D)}{[5 - 3(1 - \eta/D)^2]^{1/2}}. \tag{24}$$

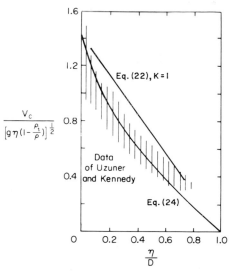

Fig. 11. Threshold stability of ice floes expressed in terms of a densimetric Froude number using floe thickness as the length scale.

Equation (24) may also be expressed in the form of the more traditional Froude number $U_c/(gD)^{1/2}$ by taking $\rho_i/\rho_w = 0.916$ (see Fig. 12). Figures 11 and 12 are equivalent plots of the same data. The implication of Figs. 11 and 12 is that individual floes will accumulate by simple juxtaposition until the condition of Eq. (24) is exceeded. They then will accumulate in a jumbled mass to an equilibrium thickness that we denote by η_c. This thickness was analyzed in a manner similar to that of vertical submergence of a single ice floe by Pariset and Hausser (1961) and Pariset *et al.* (1966). They found an ice cover of accumulated floes progresses upstream with a thickness η_c described by

$$\frac{U_c}{[2g\eta_c(1 - \rho_i/\rho_w)]^{1/2}} = 1 - \frac{\eta_c}{D}. \tag{25}$$

Uncertainties in these results relate to the no-spill condition, which is not strictly valid. At very small values of the ratios thickness to length η/l and thickness to depth η/D, the analysis above does not agree with data, as shown by Larsen (1975) particularly. Similarly, if the river is wide, the internal resistance of the ice accumulation will be unable to resist the increasing force in the streamwise direction as the cover progresses upstream, largely because of the fluid shear stresses on the underside, and the gravitational component acting along the slope of the river. When the ice cover reaches some threshold value of thickness it will further thicken by internal collapse and piling. At some point the shearing capacity of the ice cover, which transmits the forces to the banks, will be exceeded and the ice cover will become unstable. This situation has been analyzed by Pariset *et al.* (1966). A brief outline of their analysis follows.

An accumulating ice cover of thickness η_c floating on a river of width B

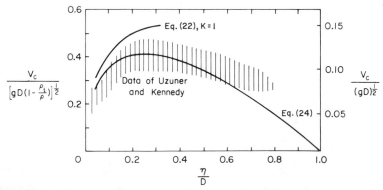

Fig. 12. Threshold stability of ice floes expressed in terms of a densimetric Froude number using depth as the length scale (left-hand ordinate) and in terms of the flow Froude number (for $\rho_i/\rho_w = 0.916$) (right-hand ordinate).

with surface slope S is subjected to a shear stress τ_w on the underside and a gravity force per unit area τ_g acting along the slope. The reaction of the banks per unit length consists of a cohesive contribution $\tau_c\eta_c$ and an "ice-over-ice friction term" μF, where μ is a coefficient and F is the longitudinal force per unit width of a section of ice cover. Equilibrium force balance of a differential length dx (where x is measured downstream from the edge of the cover) over the width of the river then yields

$$(\tau_w + \tau_g)B \ dx = B \ dF + 2(\tau_c\eta_c + \mu F) \ dx, \tag{26}$$

which is integrated to yield

$$F = (B(\tau_w + \tau_g)/2\mu - \tau_c\eta_c/\mu)[1 - \exp(-2\mu x/B)], \tag{27}$$

with the boundary condition that the longitudinal force is zero at the edge of the ice.

Before continuing we note that if the first bracketed term is negative the river is characterized as "narrow," because the ice cover can withstand the shearing forces of the water and the gravitational forces and hence does not thicken by shoving and piling. For this case the thickness will be given by Eq. (25). If the first term is positive, the thrust increases with distance from the edge and approaches the limit of the first term. The cover will then thicken by shoving and piling until its internal resistance equals the first term, and hence

$$\rho_i(1 - \rho_i/\rho_w)g\eta_c^2/2 = (B/2\mu)(\tau_w + \tau_g) - \tau_c\eta_c/\mu. \tag{28}$$

The left-hand term in Eq. (28) is the internal resistance of a jumbled floating mass of ice assuming the mass is "fluidized." If there is also a passive resistance due to cohesion this term may be larger, but, particularly for breakup conditions, the cohesion may be neglected. Using the Chezy formulation for open channel flow [where C is related to the Darcy–Weisbach coefficient f by $C = (8g/f)^{1/2}$],

$$U_u = CR^{1/2}S^{1/2}, \tag{29}$$

where U_u is the velocity beneath the cover, and R is the hydraulic radius, then the shear stress because of water flow becomes

$$\tau_w = \rho_w g U_u^2/C^2 \tag{30}$$

and the force on a unit area due to gravity is

$$\tau_g = \rho_i g \eta_c U_u^2/C^2 R. \tag{31}$$

Pariset *et al.* (1966) found from field measurements that μ has a value of about 1.28 and the product $\tau_c\eta_c$ varied in the range 5.1–6.2 N m^{-1}. Eq. (28) represents an equilibrium condition expressing a balance between the forces on the ice cover and the ability of the ice cover to transmit those forces to the shores. Using Eqs. (30) and (31), Eq. (28) becomes

$$\frac{BU_u^2}{C^2 R^2} = \left[\mu \frac{\rho_i}{\rho_w} \left(1 - \frac{\rho_i}{\rho_w} \right) \left(\frac{\eta_c}{R} \right)^2 + \frac{2\tau_c \eta_c}{\rho_{wg} R} \right] \bigg/ \left(1 + \frac{\rho_i \eta_c}{\rho_w R} \right), \qquad (32)$$

which is now expressed in a form using the variables of U_u and R rather than τ_w and τ_g.

It is convenient to relate U_u and R to conditions upstream through the use of continuity (UD equals $2RU_u$). We also make the very good approximation that D is equal to $2R + \rho_i \eta_c / \rho_w$. Substituting into Eq. (32) we then obtain

$$\frac{BU^2}{C^2 D^2} = \mu \left(1 - \frac{\rho_i}{\rho_w} \right) \left(\frac{\rho_i}{\rho_w} \right) \left(\frac{\eta_c}{D} \right)^2 \left(1 - \frac{\rho_i}{\rho_w} \frac{\eta_c}{D} \right)^3 \bigg/ \left(1 + \frac{\rho_i}{\rho_w} \frac{\eta_c}{D} \right), \qquad (33)$$

the values of which are shown in Fig. 13. Pariset et al. (1966) traced the evolution of an ice cover through several stages of accumulation using Fig. 13 for a variety of cases. We shall consider only the following case of a steady flow discharge. Referring to Fig. 13, point A represents an accumulation thickness unable to resist the flow forces. As it thickens it moves to point A' on the curve, and will stabilize there. If, however, additional ice is deposited, the ice cover will thicken and eventually may reach point

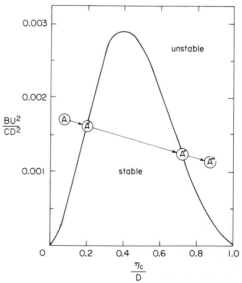

Fig. 13. Stability diagram for the accumulation of ice covers. [Adapted from Pariset, E., Hausser, R., and Gagnon, A. (1966). *J. Hydraul. Div. ASCE* **92**, 1–24.]

A''. Further thickening will cause the ice cover to again become unstable reaching point A'''. Pariset *et al.* (1966) also describe a number of practical applications that have been made of the above analysis, including its use as a framework for conducting model tests.

Before continuing we also note that by this analysis a floating ice cover will be unstable regardless of its thickness if $BU^2/(C^2D^2)$ is greater than about 2.9×10^{-3}. Written in terms of the Froude number $[U/(gD)^{1/2}]$, this condition for instability may also be expressed as Fr greater than $0.152D/(Bf^{1/2})$. From a descriptive point of view this means that the cover will continue to thicken and pile until eventually the head losses in the reach of river will cause an increase in depth sufficient to allow stability to be regained. The above relationships can also be used to examine whether or not changing the geometry of a given river cross section will contribute to the stability of ice accumulations.

D. River Ice Motion

When in motion, river ice is primarily moved by water currents. During the period of frazil formation, before a stationary cover is formed, the frazil particles behave much as a fine sediment that is buoyant. Thus frazil motion is the upside-down counterpart of sediment transport. Just as secondary currents cause sediment to settle on the inside of bends, the same currents cause frazil to collect on the outside of the bends. Similarly, frazil floats upward in regions of low-velocity flow. Measurements of frazil deposits in reservoirs generally show an upside-down deltalike formation beneath the surface ice cover near the inlet. When frazil and ice fragments have glutted a channel section and the flow has decreased, the flow often develops a channel within the ice mass that may meander from bank to bank even though the original channel is straight.

During the period of intact ice cover, the hydraulics remain essentially "open channel flow" since the ice cover is generally sufficiently flexible to rise and fall with changes in the water level, or sufficiently weak to break at the banks and allow vertical movement. The water levels for a given discharge of water, however, are higher than for corresponding discharges during non-ice periods. The major effect is the addition of a second boundary on the flow and the displacement effect of the ice cover. Except for short-term effects of storage and release of water associated with the imposition or removal of the ice cover, the discharge in a given reach is fixed by other processes acting well upstream. In each reach, then, the bottom slope and discharge are fixed and the water must accommodate the increased resistance by adjustment in the depth and surface slope. An additional complication is introduced by the fact that the fric-

tion factor associated with the undersurface of the ice cover is different
from that of the bottom and varies during the winter. Uzuner (1975) compared a number of analyses of the composite friction factor of an ice-
covered flow. The writer prefers the procedure of Larsen (1969), which
consists of deriving the velocity distribution associated with each boundary, finding a common maximum, and equating the sum of the discharges
of the two subareas to the total discharge. Larsen derived the result using
the Manning n coefficient, but a similar derivation may be performed
using the Darcy–Weisbach friction factor f defined implicitly by

$$U = [(8g/f)RS]^{1/2},\qquad(34)$$

where R is the hydraulic radius, S the slope, and U the mean velocity. The
result is

$$f = \frac{R_1 + R_2}{2[(R_1/f_1)^{1/2} + (R_2/f_2)^{1/2}]^2},\qquad(35)$$

where the subscripts 1 and 2 refer to the smoother and rougher boundaries. A value of R_1/R_2 different from unity has little effect for practical
ranges of f_1/f_2 so the result may be well approximated by

$$f = [(1/f_1)^{1/2} + (1/f_2)^{1/2}]^{-2}\qquad(36)$$

Equation (36) is presented in Fig. 14 in terms of the ratios f/f_2 and f_1/f_2 and
also in terms of the Manning n coefficient as presented by Larsen (1969).
Thus, given the values of f_1 and f_2, the composite friction factor may be

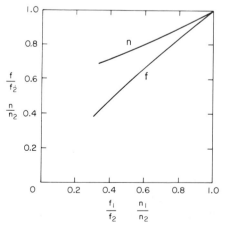

Fig. 14. Composite friction factor for ice-covered channels as a function of the relative
roughnesses of the two boundaries; n is the Manning coefficient and f is the Darcy–
Weisbach friction coefficient.

determined and subsequent calculations made without explicit reference to the differing roughnesses of the ice cover and the bottom.

To examine the effects of the ice cover on the depth of flow, we consider only the following simple case where the discharge is fixed and adopt the Chezy formula for open channel flow

$$U = CR^{1/2}S^{1/2}, \tag{37}$$

where the Chezy coefficient C, a measure of flow resistance, equals $(8g/f)^{1/2}$, f is the composite friction factor, R the hydraulic radius, and S the friction slope. For open-water conditions the volume flux per unit width is

$$Q_o = U_oD_o = C_oR_o^{1/2}S_o^{1/2}D_o \tag{38}$$

and, for ice-covered conditions,

$$Q_i = U_iD_i = C_iR_i^{1/2}S_i^{1/2}D_i, \tag{39}$$

where D is the depth of the flow section and the subscripts o and i denote open-water and ice-covered conditions, respectively. For ice conditions R_i is about equal to $D_i/2$ while for open water conditions R_o equals D_o. For the same discharge, then,

$$S_i/S_o = 2(C_o/C_i)^2(D_o/D_i)^3. \tag{40}$$

In actual cases both the depth of flow and the slope increase compared to open-water conditions, and river geometry generally requires backwater calculations to estimate water levels accurately. Note that Larsen (1973) found S_i/S_o to be 1.99 and 1.86 for two reaches of the Gallejaur headrace channel in northern Sweden, and 1.62 for the Kilforsen channel, also in Sweden. These values are consistent with Eq. (40) since the channels had little change in total depth. In both cases the ice undersurface had some relief features on the underside.

Finally due consideration must be made for the displacement effect of the ice cover which causes the water surface to be $(\rho_i/\rho_w)\eta$ (where ρ_i/ρ_w is the specific gravity of the ice and η the thickness) above the undersurface of the ice cover.

In practice the bottom coefficient is generally assumed to be the same in winter as in summer and the roughness of the ice boundary is estimated. The latter estimate is the most tenuous of the whole procedure since the roughness coefficient changes through the winter. If the initial ice cover is composed of angular floes accumulated on the surface, the initial roughness will be quite high but will decrease as either the sharp edges are smoothed by melting or the ice cover thickens beyond the original bottom surface of accumulation. A thickening ice cover generally has a smooth planar bottom. As the ice cover begins to melt at the lower surface a wavy dunelike relief pattern develops on the underside, causing an increase in the roughness coefficient. The wavy relief pattern, termed "ice ripples," has a wavelength l inversely proportional to the mean velocity with $l = 0.3$ m corresponding approximately to a mean velocity of 0.4 m sec^{-1}. Depth effects of flow appear to have some effect on the wavelength and a correlation of all field and laboratory data has been accomplished [Ashton (1972)]. Typical measured values for the roughness associated with the ice surface are reported by Nezhikovskiy (1964), Carey (1966, 1967), and Larsen (1969, 1973).

E. River Ice Decay

The deterioration of a river ice cover is a complex process. As tributaries add warm water to the flow, heat is transferred from the water to the ice undersurface [the heat transfer can be estimated using Eq. (7)]. Usually the snow cover melts at the same time and the ice becomes isothermal at 0°C. The snow melting changes the albdo significantly and allows radiation to penetrate the ice cover. Accompanying this isothermal state is a general deterioration in the strength of the ice cover due to melting at the ice grain boundaries. In its most extreme form, particularly when the ice cover has thickened by simple freezing and the ice crystals are columnar in nature, the rotting or "candling" that results can cause the ice cover to become very weak. Even relatively small forces can cause breakup. The rotting process described above has been observed to originate at the bottom, at the top, or from both surfaces simultaneously. Warm tributary inflows, originating along the shores, cause the ice cover to melt first near the bank areas. These banks are also the anchor points for the ice cover, holding it against the shearing stresses of the flow beneath, and the melting makes the ice cover more vulnerable to mechanical destruction.

F. Breakup Process and Ice Jams

The breakup process is very complex and varies in nature from site to site and from season to season. As yet, the breakup process has not been adequately described in mathematical terms, but there are many specific site descriptions available in the literature. The breakup process generally results from an increased flow together with a weakening of the ice cover. The breakup can cause major problems, usually because of ice jams.

An ice jam can take many forms, but two types are most common. The first occurs when the wave of discharge displaces and moves the ice cover, which then clogs a downstream cross section of the river. This causes a temporary rise in water level due to storage. The rise in water level may eventually break up the jam, which is then released to move downstream and repeat the process. Figure 15 shows such a jam.

There is little quantitative data on water levels and geometry of ice jams of a form suitable for verifying theoretical predictions. Deryugin (1975) analyzed 20 years' data for two reaches of the Lena River, plotting the maximum water height observed at any gaging station during the passage of an ice jam against the height that would have been associated with the same flow discharge in open water conditions, thus obtaining the "least underestimate of the maximum stage" associated with the breakup period. The result is depicted in Fig. 16 and clearly shows the increased water height associated with ice jams. It is not known whether or not the bounding line with slope of 2:1 is a general characteristic of flood levels associated with ice in larger rivers. Note, however, that Pariset *et al.* (1966) found stages associated with ice jams in the Saint Lawrence River near Montreal to have the same general ratio when compared with summer levels at equivalent water flows.

The other common type of jam occurs when a discharge of ice encounters either an intact ice cover downstream or a change to a lesser

Fig. 15. Photograph of an ice jam on the Connecticut River near Windsor, Vermont.

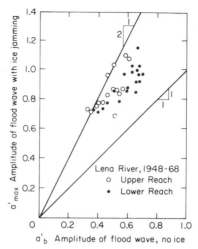

Fig. 16. Relative amplitudes of water level rises in the Lena River with and without ice jamming. ○, Upper reach; ●, lower reach. Lines of slopes 1 and 2 are shown for reference. [After Deryugin (1975)].

river slope. The ice transport capacity decreases (sometimes becoming zero), and the ice piles and blocks a cross section. No verified analytical method is available to predict such ice jams, and the best guide is prior experience when available.

Donchenko (1978) described the formation of ice jams downstream resulting from fluctuations in discharge from the tailwaters of large hydroelectric projects and concluded destruction of the downstream edge occurs when the water level fluctuation is 3–4 times the ice thickness. He also found the water heights associated with such jams to increase with increasing discharges.

Among the more detailed analytical treatments of ice jam formation, besides that of Pariset *et al.* (1966) described earlier, are those of Uzuner and Kennedy (1976) and of Tatinclaux (1977). Kennedy (1975) has also summarized the relationships that theoreticians have at their disposal for the analysis, but a rigorous solution to the problem appears very complex. The theoretical model of ice jams put forth by Uzuner and Kennedy (1976) assumes static force equilibrium within the ice cover and uses the differential equations of unsteady nonuniform momentum and continuity to develop the general features of ice accumulation in a straight rectangular channel. Critical assumptions made include a priori jam initiation and subsequent accumulation by a steady supply of ice from upstream. The internal strength of the ice mass is characterized by a Coulomb-type law for granular materials. The details of the model are too extensive to be recounted here and, as yet, there has not been validation of the model by field data. The largest uncertainties in the model are considered by Uzuner and Kennedy (1976) to be the relationships adopted for the shear and compressive strengths. Practical usage of the model would require knowledge of the initiation location and the ice supply. On the other hand, the model does predict certain general features such as an equilibrium thickness of accumulation and provides estimates of the time required to reach that thickness, which is of the order of a few hours.

The theoretical model of Tatinclaux (1977) is somewhat less ambitious than that of Uzuner and Kennedy (1976) and concentrates on the equilibrium thickness of ice jam accumulations. Again, initiation is postulated and the internal strength uncertainty avoided by limiting the analysis to the hydrodynamic accumulation of floes and not considering internal failure; the case treated is that characterized as a "narrow" river by Pariset and Hausser (1966). The crucial part of the analysis is the hypothesis that a floe arriving at the upstream end of an accumulation has kinetic energy imparted to it by the surrounding flow, which in turn determines the maximum depth of submergence of the floe and hence the thickness of the accumulation. Whether that concept is correct or not, the analysis does lead to a relationship of the form

$$\frac{\rho_i \eta_c}{\rho_w \eta_i} - 1 = \beta \left(1 + \frac{\alpha \rho_i \eta_i}{1 - p \rho_w D} C \right) (F_d - F_c), \tag{41}$$

where

$$F_d = V^2/2g \left(1 - \frac{\rho_i}{\rho_w} \right) \eta_i \left(1 - \frac{\rho_i \eta_c}{\rho_w D} \right)^2, \tag{42}$$

$$F_c = V_c^2/2g \left(1 - \frac{\rho_i}{\rho_w} \right) \eta_i \left(1 - \frac{\eta_i}{D} \right)^2, \tag{43}$$

and η_c and η_i are the accumulation and individual floe thicknesses, β is an experimentally determined proportionality factor related to the kinetic energy concept, C is the concentration of ice floes in the upstream flow, D is the approach flow depth, V the approach flow velocity, V_c the critical velocity for submergence of a single floe, p the porosity of the accumulation, and α is the ratio of the surface velocity to the mean velocity. Tatinclaux (1977) reported a large number of experiments on plastic and ice blocks conducted in flumes and the data were well correlated by

$$\rho_i \eta_c / \rho_w \eta_i - 1 = 1.27 (F_d - F_c), \tag{44}$$

where F_c was found to be 0.35 for ice blocks and 1.18 for plastic blocks. The essence of Eq. (44) is that the thickness η_c of an accumulated ice cover is related to the individual floe thickness η_i and that the thickness increases as a modified Froude-type number F_d exceeds a threshold Froude-type number F_c characteristic of individual floe behavior. While explicitly different in form, this result is not a great deal different from the results found by Pariset *et al.* (1966) as represented by Eq. (25).

As mentioned above, the theoretical models of ice jams postulate initiation and generally include a supply of ice in a regular manner to the upstream front of the accumulation. The initiation problem is critical to practical usage of the theories, and there has been little systematic work on the conditions for initiation. A number of studies have examined initial arching across an opening in a surface obstruction such as an ice boom. These include studies by Calkins and Ashton (1975) and Tatinclaux and Lee (1978), both of which used small blocks in flumes. The more vexing problem, however, is to characterize those channel geometries that cause stoppage of the flowing ice mass. Most jams seem to be initiated at a change in channel slope from steep to mild. The problem is complicated by the movement of ice often occurring en masse, rather than as a regulated supply of discrete ice floes. There is a considerable literature on ice jams, mostly descriptive, among which are works by Chizov (1974), Michel (1971), Korzhavin (1971), Williams (1973), and Bolsenga (1968). The last presents a review of the available literature before 1968.

III. LAKE ICE

The nature of a lake ice cover depends on the type of lake on which it forms and the weather conditions of that particular winter. We shall distinguish the following kinds of lakes:

(1) *Small* lakes are of small area, relatively shallow, and have little inflow or outflow of water during the period of ice cover. They are small enough that an ice cover forms quickly in the fall and is little influenced by wind because of limited fetch.

(2) *Reservoir* lakes have significant throughflow of water but are not so large that wind plays a significant role in the movement of ice except for a short period during the initial ice formation and a short period during the breakup.

(3) *Large* lakes are sufficiently large that wind plays a significant role in the movement of the ice cover. They seldom have a complete, intact ice cover. The Great Lakes of North America are of this type, but smaller lakes with dimensions of a few kilometers often behave similarly.

It is difficult to quantify the parameters that prevail on any given lake.

To some measure it depends on how cold the winter is, the nature of the winds which the lake experiences, and the magnitude of the throughflow.

A.   Energy Budgets and Initial Ice Formation

Extensive treatises on the thermal regimes and energy budgets of lakes prior to the period of ice cover are available. Here it suffices to note that during the fall cooling period loss of the heat from the surface results in a density increase. The cooled water at the surface mixes with the water below to obtain isothermal conditions until the entire lake has cooled to 4°C (the temperature of maximum density). Further cooling results in lower-density water at the surface and a stable stratification tends to develop. If the lake is small and sheltered, there will be little mixing induced by the wind and the initial ice cover will form with only a shallow layer of water below 4°C at the top surface. Generally, however, the wind mixing is sufficient to provide a well mixed region extending to some depth, often to the bottom. The temperature of this well-mixed region depends on the particular sequence of wind and air temperature variations. The first calm night of low air temperatures will result in an ice cover that then insulates the water below from heat loss and further wind mixing. The larger the lake, of course, the longer the period of cold and calm needed for the ice cover to gain sufficient integrity to resist later periods of wind. On very large lakes a complete ice cover is the exception rather than the rule and cooling of the water will continue through the entire period of subfreezing air temperatures.

Once a complete ice cover has formed, the water below will gradually warm through the winter if there are no inflows into the lake. Much of this warming is a result of release of energy from the bottom sediments, energy that was stored in the top few meters over the summer season. If the thermal diffusivity of the bottom sediments is known, standard heat conduction analyses are applicable and this heat flux may be calculated, although some uncertainty results from the necessity of specifying the boundary temperature over the future course of the winter. Pivovarov (1973) has treated this case, as well as other minor components of the under-ice water energy budget, in considerable detail. At the surface of the lake, loss of heat results in a thickening of the ice like that described earlier for river ice. In fact, the methods outlined earlier for river ice covers are applicable, although the heat flux from the water to the ice [ϕ_{wi} in Eq. (11)] is very small, particularly if there are no significant lake currents. It is important, however, to include the effect of snow cover in the expression of the heat flux through the ice [ϕ_i in Eq. (12)] with the most expedient means being a representation of the snow cover by a thickness η_s with an effective thermal conductivity k_s, for which case

$$\phi_i = \frac{T_m - T_a}{\eta_i/k_i + \eta_s/k_s + 1/h_{ia}}. \tag{45}$$

The thermal conductivity of snow is generally related to the density [a summary of observations is provided by Mellor (1977)]. As is the case for river ice, practical application of Eq. (45) generally requires stepped integration of Eq. (10) to account for such effects as increments in snow thickness and the temporal variations in air temperature. The semiempirical (degree-day)$^{1/2}$ method using a coefficient determined from past records has been used frequently (see Chapter 3 for growth of sea ice).

In cases where snowfall is heavy, the above-described thermal thickening analysis is inappropriate since the snow layer may attain sufficient weight to submerge the ice cover. Water then flows on top of the ice layer, permeating the snow, and the snow–water mixture freezes to form a fine-grained ice layer. This may occur several times during a season. Examination of the crystal structure profile often reveals a signature characteristic of adjacent lakes which have experienced the same meteorological events [Gow (1977)].

For lakes with throughflow, the temperature of the water beneath the ice cover depends in part on the temperatures of the inflow and outflow. Natural inflows are generally near 0°C and hence less dense than lake water. After an initial entrainment of some of the warmer lake water due to momentum exchange, the flow passes through the lake in a layer above the warmer water and exists with a temperature remaining at 0°C, at least if the outflow is near the surface [Stigebrandt (1978)]. If the outflow is submerged below the thermocline (as at a dam) the exiting water will gradually deplete the warmer water and the cold entrance water will "fill" the lake from above. Because this top water is near 0°C and because of the slow velocity, the transfer of heat to the undersurface is small except very near the inlet and outlets.

For very large lakes, formation of an intact ice cover strong enough to resist the forces of the wind and wind-induced currents is difficult. As a consequence much of the surface remains open and the lake cools throughout the winter, slowly for deep lakes but relatively rapidly for shallower lakes. Large lakes generally are nearly isothermal with depth and have water temperatures only a few tenths of a degree above 0°C, as was observed in detail by Stewart (1973) on Lake Erie where maximum midwinter water temperatures were of the order of 0.2°C. The wind on such lakes causes waves, and the wave-generated turbulence is sufficient to entrain small crystals at the surface, thus producing a top layer of frazil slush. During periods of relative calm, plate ice also forms, but is broken up by the next period of strong winds.

B. Shoreline Ice Formation

On large lakes the most varied and the most important ice formations occur along and are influenced by the shoreline. In small, nearly enclosed bays the ice cover forms early and behaves similarly to ice covers of small lakes and reservoirs. At other points the ice can take on many forms for which almost every observer seems to have his own terms. A typical classification system [Michel (1971)] involves coverage (ice-free: no ice; open water: areal coverage less than 0.1; scattered: 0.1–0.5; broken: 0.5–0.8; close ice: 0.8–1.0; and consolidated ice: complete coverage), size of typical fragment (brash: less than 2 m in dimension; blocks: 2–10 m; small floes: 10–200 m; medium floes: 200–1000 m; and giant floes: greater than 1000 m), and age of the ice (slush, young or recently formed ice, and winter ice over 0.2 m thickness). Similarly the surface can be characterized as puddled, snow-covered, frozen, or rotten, and the topography of the surface can be described as rafted, ridged, or hummocked. There are also cracks, leads (narrow open water areas), and polynyas (areas of open water among an otherwise ice-covered surface). Each of the various forms depends on the past history of the meteorological variables, particularly the wind, air temperature, and radiation components and, of couse, the water temperature. Attempts have been made to standardize terminology including coordination of the multilingual terms [International Association for Hydraulic Research (1977)].

Faced with such a complexity of ice types the task of prediction is difficult. There has been one study [N. Gruber (personal communication)] in which a logic diagram was constructed and used to predict the nature of ice formation using only the variables of wind speed and direction, air temperature, cloud cover (which has considerable influence on the radiation losses), time of day, and water temperature. The logic diagram incorporates history and leads to the following shore ice types: smooth sheet, frazil ice, slush balls, conglomerate sheet, broken sheet and pack ice, frazil slush, pancake ice, rafted ice, ridged ice, and piled ice.

In a test of the method some 20 available observations were correctly hindcast using the observed variables recorded at a nearby weather station as input to the logic diagram. The criteria used are presented in Table I and, while there is no doubt that the magnitude of the decision parameters could probably be improved with additional testing, its performance

TABLE 1

Shoreline Ice Formation Criteria

	Prior formation required	Air temperature (°C)	Wind speed (m sec⁻¹)	Wind direction	Cloud cover
		Meteorological parameters			
Primary ice formations					
1. Smooth sheet	None	≤ 0	≤ 5	—	—
2. Frazil ice (daytime)	None	≤ -1	≤ 5	—	>0.6
3. Frazil ice (night)	None	≤ 0.5	—	—	<0.5
4. Slush balls	None	≤ 0	≥ 5	—	—
Secondary ice formations					
5. Conglomerate sheet	6 or 7 or 4 or 8	≤ 0	≤ 5	Onshore	—
6. Broken sheet and	1^a or 5^a	≥ 0	—	—	—
pack ice	1^a or 5^a	—	≥ 3.6	Offshore	—
7. Frazil Slush	2 or 3	≤ 0	≤ 5	Onshore	—
8. Pancake ice	(7 or snow) and 1	≤ 0	≤ 5	—	—
9. Rafting	6^a	—	>5	Onshore	—
10. Piling	6^a	≥ 0	≥ 5	Onshore	—
11. Ridging	1^a or 5^a	—	>5	Onshore	—

[a] Within 24-hr period

has been quite good. The criteria were then used to examine 20 years of meteorological data at the site to yield a hazard assessment of times of occurrence for the different ice types.

Foulds (1974) has also provided an excellent description of the extreme shore ice events that can occur on large lakes and noted that the severity of the ice piling and accumulation is often not appreciated simply because it occurs at times when observers prefer to be in more protected places. It has been observed that severe ice piling on shore only occurs upon return of an ice sheet that has previously blown off shore since the piling largely results from dissipation of the momentum of the ice sheet when it collides with the shore [Tsang (1975)].

The seiching of lakes, particularly larger ones, causes a vertical oscillation of the ice cover and commonly causes the ice cover to fracture in a line near to and roughly parallel with the shore (see discussion of ice flexure in Chapter 4). This same vertical oscillation also often results in lightly loaded piles, like those used in small dock construction, to be repeatedly "jacked" upwards, often several meters over a winter season. The upward net motion of the piles results from the upward resistance to load being less than the downward resistance. Various measures are used to counteract this motion, most of which are aimed at isolating the pile from the ice movement.

The expansion and contraction of the ice cover of lakes caused by changes in air temperature often results in the formation of ice "ramparts," particularly on shores with gentle slopes. Much of the net motion is due to cracks forming within the ice sheet then filling with water and freezing, thus providing a net shoreward expansion. The forces exerted by the ice on a vertical-faced object due to this motion can be large, limited only by the failure strength of the ice.

C. Midlake Ice Formation and Motion

There has been little analysis of the behavior of ice in the central areas of large lakes, although numerous descriptive surveys are available. In part such analysis has been severely hampered by the inability to obtain synoptic coverage of ice type and ice thickness. The advent of remote sensing techniques shows considerable promise for the future and include side-looking airborne radar (which allows interpretation of the surface roughness), conventional aerial photography, and satellite sensors. A recent development is a pulsed-radar technique that allows determination of the thickness of the ice cover without recourse to physical penetration and will even profile thicknesses of rafted ice and frazil accumulation [Dean (1977)]. Techniques of analysis that have proven successful in predicting Arctic sea ice movement and extent may eventually be applied to large freshwater lakes with suitable modifications.

D. Lake Ice Decay

The deterioration processes of lake ice are more complex than the thickening processes. Daily air temperatures above freezing cause melting on the top surface, changing the albedo from that of snow to that of water. While the 0°C temperatures on top and bottom result in no conductive heat transfer, there is absorption of solar radiation, which manifests itself in melting at the grain boundaries and causes the ice to become somewhat porous and rotten. At about the same time, runoff from the surrounding land causes melting near the shorelines and may raise the lake water temperature. If enough wind comes up to break the ice cover in this rotted state, it may disappear very quickly because the ice fragments will mix with the warmer water below. Attempts to apply energy budget methods to the melting of lake ice covers have had limited success [Williams (1965)], largely because of the isothermal nature of the ice cover and the complexity of the top melting process. Also, mechanical destruction is often more important than inplace melting processes making energy budget analyses ineffectual. On the other hand some sites do exhibit characteristic time–thickness decay "signatures" from year to year that can offer some guidance for specific sites [Bilello (197)].

IV. ARTIFICIAL EFFECTS ON RIVER ICE

A. Thermal Effluents

The disposal of thermal wastes, usually from power plants, affects river ice in several important ways. The initial formation of an ice cover is delayed while its breakup is hastened. The ice cover will tend to be thinner, perhaps nonexistent close to the thermal source. Open water also will exist to various distances from the source depending on the magnitude of the heat released and the varying meteorological conditions. The interest in these effects stems from concern for environmental change, a desire to extend navigation for longer periods of the year, and the concern that such ice suppression might induce greater ice formation downstream resulting from frazil formed in the larger open-air areas.

The case of a fully mixed thermal discharge such as would exist downstream from a deep reservoir discharging water at an initial temperature $T_{w,0}$ is now briefly examined. Similar results would be obtained if a thermal effluent from a power plant were fully mixed with a river flow. The steady-state version of Eq. (1) is

$$dT_w/dx = -\phi/\rho C_p UD. \tag{46}$$

For the open-water case, the heat flux ϕ is taken as $h_{wa}(T_w - T_a)$, where the water–air heat transfer coefficient h_{wa} is calculated from energy budget analysis. If T_a and h_{wa} are constant, then, with water temperature T_w equal to $T_{w,0}$ at x_0, the temperature at any point x is given by

$$(T_w - T_a)/(T_{w,0} - T_a) = \exp[-h_{wa}(x - x_0)/\rho C_p UD] \tag{47}$$

and, of course, the state relationship requires that $T_w \geq 0°C$. The location at which T_w equals $0°C$ is then given by

$$x - x_0|_{T=0} = -(\rho C_p UD/h_{wa}) \log_e[-T_a/(T_{w,0} - T_a)]. \tag{48}$$

Except for minor variations, this equation represents the analysis of Dingman *et al.* (1968) who used the location of the $0°C$ isotherm as the criterion for location of the ice edge downstreams. In attempting to verify that criterion they found considerable difficulty, both because of the complexity of actual site conditions and because of the highly unsteady nature of T_a [Weeks and Dingman (1972)]. Paily *et al.* (1974) and later Paily and Macagno (1976) analyzed the same open-water case with the additional inclusion of longitudinal dispersion effects and a slightly different formulation of the heat flux ϕ. The effect of longitudinal dispersion was not significant. They also used the criterion of the 0°C isotherm for the location of the ice edge.

A different criterion for the location of the ice edge can be obtained starting with the heat balance of the undersurface of the ice cover given by Eq. (11):

$$\phi_i - \phi_{wi} = \rho_i \lambda \, d\eta/dt, \tag{11}$$

where ϕ_{wi} and ϕ_i may be approximately represented by

$$\phi_{wi} = h_{wi}(T_w - T_m) \tag{49}$$

and

$$\phi_i = (T_m - T_a)/(\eta_i/k_i + 1/h_a), \tag{50}$$

where h_{wi} is a heat transfer coefficient applied to the temperature difference $T_w - T_m$. Assuming $d\eta_i/dt$ is zero (no thickening), the heat flux through the ice ϕ_i equals the heat flux from the water to the ice ϕ_{wi} and the ice edge is the location at which η_i equals 0, then the location of the ice edge corresponds to a water temperature

$$T_w - T_m = (h_a/h_{wi})(T_m - T_a) \tag{51}$$

and the corresponding distance is given by use of Eq. (47),

$$x - x_0 \Big|_{\substack{d\eta/dt=0 \\ \eta=0}} = \frac{\rho C_p UD}{h_{wa}} \left[-\log\!\left(\frac{1 + (h_a/h_{wi})(T_m - T_a)}{T_{w,0} - T_a} \right) \right]. \tag{52}$$

This distance is always less than the distance given by Eq. (48); the ratio is shown in Fig. 17. In some cases the distance given by Eq. (52) is negative. This corresponds to the case when the heat transfer to the undersurface of the ice cover is greater than the heat transfer from the surface to the cold air above or ϕ_{wa} less than ϕ_{wi}. Since the water ice heat transfer coefficient h_{wi} is also more or less proportional to the flow velocity, this criterion helps explain cases where open water exists in regions of high flow velocity downstream of an ice cover that overlies a region of low velocity.

Since this second criterion is obtained by an analysis of the melting of the ice cover, it is applicable when the ice edge is retreating downstream under the influence of warming air temperatures. When the air temperatures are cooling, the 0°C isotherm criterion is applicable since ice will not form until the water has been cooled to 0°C.

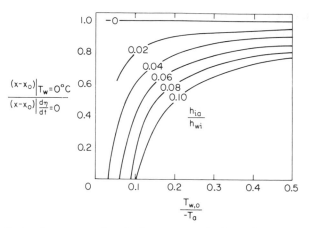

Fig. 17. Ratio of open water lengths downstream of a thermal source predicted by the criteria of 0°C water temperature and a melting criteria. Expressed as a function of water to air temperature ratio.

All the analyses presented above are steady-state analyses. In nature the variations in air temperature result in the ice edge moving downstream in warmer weather and upstream in colder weather. The energy required for melting the ice cover causes the downstream retreat to be slower than the change in air temperature would indicate. Conversely the requirement of an initial accumulation thickness for ice cover stability causes the movement of the ice front in the upstream direction during periods of cooling to be slower than would result from the criteria given above.

The rare cases in which thermal effluents are fully mixed with the river flow include the withdrawal from depth of a stratified reservoir, outflow of some lakes, cases of power plants where the waste heat is deliberately diffused across the entire flow, and a few cases where the entire river flow is used in a once-through cooling system. The more usual cases are side discharges and discharges at the bottom of rivers. In the side discharge case lateral mixing is important and full vertical mixing is assumed. The conditions under which the assumption of full vertical mixing is inappropriate are examined later.

Assuming full vertical mixing and including the lateral dispersion term, the governing equation for dispersion of thermal effluents is

$$\partial T_w/\partial t + U\, \partial T_w/\partial x = (\partial/\partial z)(E_z\, \partial T_w/\partial z) - \phi/\rho C_p D. \qquad (53)$$

As before, the surface heat flux ϕ depends on whether or not an ice cover is present. The dispersion coefficient, which for practical cases of river flow is dominated by turbulent exchange processes rather than molecular diffusion, is generally represented in the form

$$E_z = k_z U_* R, \tag{54}$$

where k_z is a coefficient weakly dependent on the Darcy–Weisbach friction coefficient f and is typically equal to about 0.23 for straight rivers, U_* the shear velocity equal to $U(f/8)^{1/2}$, and R is the hydraulic radius. For an ice-covered flow R approximately equals $D/2$, so E_z for ice-covered conditions is half the value for open water of the same depth [Engmann (1977)]. This effect plus the fact that an ice cover can exist on water at a temperature slightly above 0°C [Eq. (48)] accounts for the narrow areas of open water downstream from point sources of thermal effluent. These open areas often extend many kilometers downstream.

The vertical mixing of flows with different densities is characterized by the "densimetric" Froude number defined by

$$\text{Fr}_d = U/(gR\,\Delta\rho/\rho)^{1/2}, \tag{55}$$

where $\Delta\rho$ is the density difference associated with the temperature differences. For flows with Fr_d less than one, the stabilizing effect of the density difference is greater than the mixing effect of the turbulent exchange processes; such flows are characterized as stratified. Above Fr_d equal to 1, the flows are characterized as partially mixed and above Fr_d equal to 8 the flows are generally considered to be well mixed. By using known variations of water density with T_w temperatures [Dorsey (1940)], the densimetric Froude number Fr_d may be related to the usual Froude number [Fr equals $U/(gR)^{1/2}$] as a function of effluent temperature T_2 greater than the ambient water temperature T_1. The relationship of the Froude numbers is shown in Fig. 18 as a function of T_2, with $T_1 = 0$°C. One may then determine Fr_d as a function of Fr for various values of T_2 as shown in Fig. 19. Since typical values of Fr for most rivers are greater than 0.05, most rivers are well-mixed during the winter.

Some additional insight into the vertical mixing process is obtained by examining the distance over which mixing reduces an original density difference to some specified fraction of its original value. Schiller and Sayre (1975) found the distance $(X_{0.2})$ over which $\Delta\rho/\rho$ is reduced to two-tenths of its original value to be a function of the shear velocity and the initial value of Fr_d. Their results are reasonably represented by a relation

$$X_{0.2} U_*/RU \cong 28\,\text{Fr}_d^{-0.75}. \tag{56}$$

By substituting $(f/8)^{1/2} = U_*/U$ and using the relationships just developed for Fr_d as a function of temperature, we may then determine the normalized mixing distance $X_{0.2}/R$ as a function of temperature. The norma-

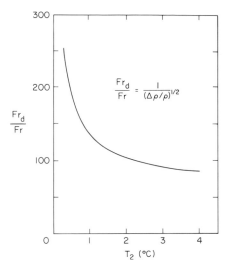

Fig. 18. Ratio of densimetric Froude number Fr_d to flow Froude number Fr as a function of effluent temperature T_2 for an ambient water temperature T_1 of 0°C.

lized $X_{0.2}/R$ is shown in Fig. 20 as a function of the Froude number and effluent temperature and using a value of $f = 0.05$. Except for values of Fr less than 0.01, mixing is substantially complete in distances of the order of a few hundred times the depth, which are fairly short for river situations. The conclusion is that the assumption of full vertical mixing is a good one, even for artificially introduced warm water.

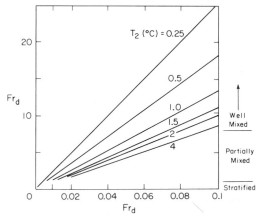

Fig. 19. Densimetric Froude number Fr_d versus the usual Froude number Fr for various values of effluent temperature T_2 at ambient water temperature of 0°C. The degree of mixing is also shown.

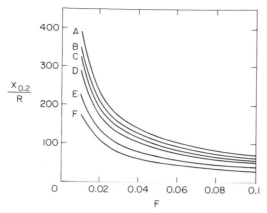

Fig. 20. Normalized mixing distance $(X_{0.2}/R)$ versus Froude number for various values of effluent temperature (for ambient water temperature of 0°C). T_2 (°C): A, 4.0; B, 2.0; C, 1.5; D, 1.0; E, 0.5; F, 0.25.

B. Navigation

Most larger rivers are used for substantial amounts of commodity transport, so it is naturally desirable to use them year round. A wide variety of techniques are used to minimize the ice interference. Icebreakers are used to keep the channel ice broken, vessel hulls are strengthened to resist the ice forces, and a variety of measures are employed to combat the ice problems associated with navigation structures (locks and dams) and navigation aids. These measures include coating of lock walls to reduce the adhesion of ice that can constrict lock widths, a variety of direct heating methods, and use of bubbler systems and water jets to divert the ice from the forebay and gate areas. In some cases ice booms have been used to retain ice and prevent it from moving into restricted channel areas where it may jam and block the channel. At the present time there is considerable interest in extending navigation seasons to include the period of ice cover. Associated with this interest are increasing research efforts on the problems of both river and lake ice.

V. ARTIFICIAL EFFECTS ON LAKE ICE

A. Thermal Effluents

Whereas introduction of a thermal effluent into a river generally results in suppression of the ice cover, the same does not hold true for lakes since the thermal effluent, after some initial mixing, sinks to the bottom and "fills" the lake from below. Thus, even at outlets with significant heat load the ice cover will be melted only in the near-vicinity of the outlet. In one case a thermal effluent was distributed along a diffuser pipe and discharged through local ports such that the exit velocity was directed upwards for the purpose of melting the ice above, but to little effect. The momentum of the discharge jet soon dissipates and the warm effluent spreads over the bottom.

On the other hand, air-bubbler systems have often been used as a means of tapping the warmer water on the bottom of lakes and delivering it to the surface where it locally melts the ice at the undersurface. Unlike a submerged jet of water in which the maximum velocity rapidly decreases with distance from the outlet, the maximum velocity of a plume of water rising with the bubble stream is more or less constant with distance above the outlet. After the bubbler-induced plume contacts the surface it divides and spreads laterally with the characteristics of a submerged water jet; the velocity rapidly decays with distance from the point of contact. Since the heat flux from the water to the ice undersurface is approximately proportional to the product of the velocity and the water temperature above 0°C, the lateral extent of the ice suppression is generally small and the open-water width responds quickly to air temperature changes such that narrow openings are associated with cold air temperatures and wider openings are associated with warmer air temperatures. Ashton (1978b) has presented an analysis that seems to agree reasonably well with observations. He has also pointed out the possibility of extended operation of a bubbler system depleting the thermal reserve that is stored in smaller water bodies at the onset of the period of ice cover, and which may render such a bubbler system ineffective after sustained operation.

B. Navigation

Navigation on ice-covered lakes largely depends on the ability of the vessels to break the ice. With frequent traffic this is often possible through an entire winter season since the broken ice tends to collect in a thickened accumulation at the sides of the channel while leaving only thin ice in the channel itself. These side accumulations sometimes themselves freeze and act to armor the channel sides. Bubbler systems are sometimes used to aid the channel clearing, particularly when the ice cover is otherwise intact and not subject to movement. On larger lakes the navigation during winter is often most restricted by the intact ice cover that forms in the harbors and connecting channels and by accumulations of ice resulting from wind movement. Icebreakers are often used to assist other vessels and to maintain a broken channel area.

Occasionally ice covers have been dusted to decrease the albedo and take advantage of the radiation component of the energy budgets. A light snowfall over such a dusted area largely neutralizes the effect of the dust application.

VI. SUMMARY

A general review of freshwater ice growth, motion, and decay was given here for river ice and lake ice. Considerably more attention was devoted to river ice because there seems to be more information available on it. A number of subjects have not been covered, including the development of icings [Carey (1973)] on small streams and shallow rivers, forces due to ice [Korzhavin (1971); Michel (1970); Kerr (1976); Neill (1976)], intake design for ice conditions [Hayes (1974); Logan (1974)], sediment transport during ice conditions and ice crystallography [Michel and Ramseier (1971)]. Much systematic field work must yet be done to improve present analyses and define their range of applicability.

ACKNOWLEDGMENTS

This work was substantially completed while visiting the Hydraulics Laboratory of the Swedish State Power Board in Alvkarleby, Sweden. Use of their facilities and discussions with Dr. Peter Larsen there were most helpful. The support of the ice engineering program of the U.S. Army Cold Regions Research and Engineering Laboratory is also greatly appreciated.

LIST OF SYMBOLS

A_o	Area of open water (m^2)	R	Hydraulic radius (m)
B	Width of river (m)	Re	Reynolds number, $UD\rho/\mu$
C_n	Constant in Eq. (7)	S	Friction slope of a river
C_i	Surface concentration of ice	S_w	Flux of water surface (m^2 sec^{-1})
C	Chezy coefficient, $(8g/f)^{1/2}$ (m$^{1/2}$ sec^{-1})	S_d	Degree-days of freezing (day °C)
D	Depth of flow (m)	T_a	Air temperature (°C)
E_z	Dispersion coefficient (m^2 sec^{-1})	T_m	Ice point of water, 0°C (°C)
Fr	Froude number, $U/(gD)^{1/2}$	T_s	Top surface temperature (°C)
Fr$_d$	Densimetric Froude number, $\mathrm{Fr}/(\Delta\rho/\rho)^{1/2}$	T_w	Water temperature (°C)
		$T_{w,0}$	Water temperature at an initial point
F	Force per unit width (N m^{-1})	U	Mean flow velocity (m sec^{-1})
H	Total head (m)	U_c	Critical velocity for submergence (m sec^{-1})
K	Form coefficient	U_u	Velocity beneath ice cover (m sec^{-1})
Nu	Nusselt number, $\phi_{wi}D/(2k_w(T_w - T_m))$	U_*	Shear velocity, $U(f/8)^{1/2}$ (m sec^{-1})
Pr	Prandtl number, $\mu C_p/k_w$	V_a	Wind velocity (m sec^{-1})
Q	Volume flux of water (m^3 sec^{-1})	V	Volume (m^3)

$X_{0.2}$ Mixing distance (m)
  e Porosity
  e_s Vapor pressure at surface (N m^{-2})
  e_a Vapor pressure in air (N m^{-2})
  f Darcy–Weisbach friction coefficient
  g Gravitational constant (m sec^{-2})
  h_{ia} Heat transfer coefficient, ice to air (W m^{-2} °C^{-1})
  h_{wa} Heat transfer coefficient, water to air (W m^{-2} °C^{-1})
  h_{wi} Heat transfer coefficient, water to ice (W m^{-2} °C^{-1})
  k_i Thermal conductivity of ice (W m^{-1} °C^{-1})
  k_s Thermal conductivity of snow (W m^{-1} °C^{-1})
  k_w Thermal conductivity of water (W m^{-1} °C^{-1})
  k_z Dispersion coefficient
  l Floe length (m)
  n Manning coefficient (ft$^{1/6}$)
  x Longitudinal coordinate (m)
  t Time (sec)
  α Empirical coefficient in Eq. (14) (m sec$^{-1/2}$ °C$^{-1/2}$)
  η Ice thickness (m)
  η_a Accumulation ice thickness (m)
  η_c Accumulation ice thickness (m)

  η_i Ice thickness (m)
  η_s Snow thickness (m)
  λ Heat of fusion (J kg^{-1})
  μ Viscosity of water (kg m^{-1} sec^{-1})
  μ Coefficient in Eq. (26)
  ρ_i Density of ice (kg m^{-3})
  ρ_w Density of water (kg m^{-3})
  τ_c Cohesive shearing stress (N m^{-2})
  τ_g Force per unit area due to gravity (N m^{-2})
  τ_w Shear stress due to water flow (N m^{-2})
  ϕ Heat flux (W m^{-2})
  ϕ_{ia} Heat flux from ice surface to air (W m^{-2})
  ϕ_{wa} Heat flux from water to air (W m^{-2})
  ϕ_{wi} Heat flux from water to ice (W m^{-2})
  ϕ_i Heat flux through ice (W m^{-2})
  ϕ_s Short-wave radiation flux to water surface (W m^{-2})
  ϕ_e Evaporative heat flux at water surface (W m^{-2})
  ϕ_c Sensible heat flux at water surface (W m^{-2})
  ϕ_{sn} Heat flux associated with snow falling on water (W m^{-2})
  ϕ_l Long-wave radiation flux to water surface (W m^{-2})

REFERENCES

Altberg, W. J. (1936). *Proc. Int. Union Geod. Geophys. Int. Assoc. Sci. Hydrol.*, pp. 373–407.

Ashton, G. D. (1972). *Int. Assoc. Hydraul. Res., Symp. Ice Its Action Hydraul. Struct., 1972* pp. 123–129.

Ashton, G. D. (1974a). *J. Glaciol.* **13,** 307–313.

Ashton, G. D. (1974b). *Int. Assoc. Hydraul. Res., Symp. River Ice, 1972* pp. 83–89.

Ashton, G. D. (1978a). *Annu. Rev. Fluid Mech.* **10,** 369–372.

Ashton, G. D. (1978b). *Can. J. Civ. Eng.* **5,** 231–238.

Ashton, G. D., and Kennedy, J. F. (1970). *Int. Assoc. Hydraul. Res., Symp. Ice Its Action Hydraul. Struct., 1970* Paper 2.4, p. 12.

Ashton, G. D., and Kennedy, J. F. (1972). *J. Hydraul. Div. ASCE* **98,** 1603–1624.

Barnes, H. T. (1928). "Ice Engineering." Renouf, Montreal.

Bates, R. E., and Bilello, M. A. (1966). "Defining the Cold Regions of the Northern Hemisphere," TR. 178. U.S. Army Cold Reg. Res. Eng. Lab., Hanover, New Hampshire.

Bilello, M. A. (197). *Proc. Can. Assoc. Geol., Annu. Meet., 1977* (in press).

Bilello, M. A., and Bates, R. E. (1966). "Ice Thickness Observations, North American Arctic and Subarctic, 1962–63, 1963–64," SR 43, Pt. III. U.S. Army Cold Reg. Res. Eng. Lab., Hanover, New Hampshire.

Billfalk, L., and Desmond, R. M. (1978). *Int. Assoc. Hydraul. Res., Symp. Ice Probl., 1978.*
Bolsenga, S. J. (1968). "River Ice James—A Literature review." U.S. Lake Surv., Detroit, Michigan.
Bronstad, J. (1965). "The Vikings." Penguin Books, Baltimore, Maryland.
Calkins, D. J., and Ashton, G. D. (1975). *Can. J. Civ. Eng.* **2,** 392–399.
Carey, K. L. (1966). *U.S., Geol. Surv., Prof. Pap.* **550-B,** B192–B198.
Carey, K. L. (1967). *U.S., Geol. Surv., Prof. Pap.* **575-C,** C200–C207.
Carey, K. L. (1973). "Icing Developed from Surface water and Groundwater," Monogr. III-D3. U.S. Army Cold Reg. Res. Eng. Lab., Hanover, New Hampshire.
Carstens, T. (1966). *Geophys. Publ.* **26,** 1–18.
Chalmers, B., and Williamson, R. B. (1965). *Science* **148,** 1717–1718.
Chizov, A. N., ed. (1974). "Investigations and Calculations of Ice Jams," TL473. U.S. Army Cold Reg. Res. Eng. Lab., Hanover, New Hampshire.
Cold Regions Research and Engineering Laboratory (1951–1977). "Bibliography on Cold Regions Science and Technology," CRREL Rep. 12. U.S. Army Cold Reg. Res. Eng. Lab., Hanover, New Hampshire.
Dean, A. (1977). "Remote Sensing of Accumulated Frazil and Brash Ice in the St. Lawrence River," CR 77-8. U.S. Army Cold Reg. Res. Engl. Lab., Hanover, New Hampshire.
Deryugin, A. G. (1975). *Sov. Hydrol. Sel. Pap.* **4,** 278–280.
Devik, O. (1942). *Geophys. Publ.* **13,** 1–10.
Dingman, S. L., Weeks, W. F., and Yen, Y. C. (1968). *Water Resour. Res.* **4,** 349–362.
Donchenko, R. V. (1978). "Conditions for Ice Jam Formation in Tailwaters," TL 669. U.S. Army Cold Reg. Res. Eng. Lab., Hanover, New Hampshire.
Dorsey, N. E. (1940). "Properties of Ordinary Water-Substance." Van Nostrand-Reinhold, Princeton, New Jersey.
Engmann, E. O. (1977). *J. Hydraul. Res.* **15,** 327–335.
Foulds, D. M. (1974). *Can. J. Civ. Eng.* **1,** 137–140.
Frankenstein, G. E., and Assur, A. (1972). *Int. Assoc. Hydraul. Res., Symp. Ice Its Action Hydraul. Struct., 1972* pp. 153–157.
Gow, A. J. (1977). "Growth History of Lake Ice in Relation to its Stratigraphic, Crystalline, and Mechanical Structure," CR 77-1. U.S. Army Cold Reg. Res. Eng. Lab., Hanover, New Hampshire.
Hayes, R. B. (1974). "Design and Operation of Shallow River Diversions in Cold Regions," Rep. REC-ERC-74-19. U.S. Bureau of Reclamation, Denver, Colorado.
Hoyt, W. G. (1913). *U.S., Geol. Surv., Water-Supply Pap.* **337,** 1–77.
International Association for Hydrological Research (1977). "Multilingual Ice Terminology." Res. Cent. Water Resour., Budapest, Hungary.
Kennedy, J. F. (1975). *Int. Assoc. Hydraul. Res., Proc. Int. Symp. Ice Probl., 3rd, 1975* pp. 143–164.
Kerr, A. D. (1976). *J. Glaciol.* **17,** 229–268.
Korzhavin, K. N. (1971). "Action of Ice on Engineering Structure," TL 260. U.S. Army Cold Reg. Res. Eng. Lab., Hanover, New Hampshire.
Larsen, P. A. (1969). *J. Boston Soc. Civ. Eng.* **56,** 45–67.
Larsen, P. A. (1973). *J. Hydraul. Div. ASCE* **99,** 111–119.
Larsen, P. A. (1975). *Int. Assoc. Hydraul. Res., Proc. Int. Symp. Ice Probl., 3rd, 1975* pp. 305–314.
Logan, T. H. (1974). "The Prevention of Frazil Ice Clogging of Water Intakes by Application of Heat," Rep. REC-ERC-74-15. U.S. Bureau of Reclamation.
Mellor, M. (1977). *J. Glaciol.* **19,** 15–66.
Michel, B. (1967). *Phys. Snow Ice, Conf., Proc., 1966* Vol. 1, pp. 129–137.

Michel, B. (1970). "Ice Pressure on Engineering Structures," Monogr. III B1b. U.S. Army Cold Reg. Res. Eng. Lab., Hanover, New Hampshire.

Michel, B. (1971). "Winter Regime of Rivers and Lakes," Monogr. III B1a. U.S. Army Cold Reg. Res. Eng. Lab., Hanover, New Hampshire.

Michel, B., and Ramseier, R. O. (1971). *Can. Geotech. J.* **8**, 36–45.

Michel, B., and Triquet, C. (1967). "Bibliography of River and Lake Ice Mechanics," Rep. S-10. Université Laval, Quebec City, Canada.

Neill, C. R. (1976). *Can. J. Civ. Eng.* **3**, 305–341.

Nezhikovskiy, R. A. (1964). *Sov. Hydrol. Sel. Pap.* 127–150.

Osterkamp, T. E. (1975). "Frazil Ice Nucleation Mechanisms," Rep. No. UAGR-230. University of Alaska, College.

Osterkamp, T. E. (1978). *J. Hydraul. Div. ASCE* **104**, 1239–1256.

Paily, P. P., and Macagno, E. O. (1976). *J. Hydrual. Div. ASCE* **102**, 255–274.

Paily, P. P., Macagno, E. O., and Kennedy, J. F. (1974). *J. Hydraul. Div. ASCE* **100**, 531–551.

Pariset, E., and Hausser, R. (1961). *Trans. Eng. Inst. Can.* **5**, 41–49.

Pariset, E., Hausser, R., and Gagnon, A. (1966). *J. Hydraul. Div. ASCE* **92**, 1–24.

Pivovarov, A. A. (1973). "Thermal Conditions in Freezing Lakes and Rivers." Wiley, New York.

Rimsha, V. A., and Donchenko, R. V. (1957). *Tr. Leningr. GOS. Gidrol. Inst.* **65**, 58–83 (in Russian).

Ryan, P. J., Harleman, D. R. F., and Stolzenbach, K. D. (1974). *Water Resour. Res.* **10**, 930–938.

Schiller, E. J., and Sayre, W. W. (1975). *J. Hydraul. Div. ASCE* **101**, 749–761.

Scholklitsch, A. (1937). "Hydraulic Structures," Vol. I. Am. Soc. Mech. Eng., New York.

Starosolsky, O. (1969). "Ice in Hydraulic Engineering," Rep. No. 70-1. Norw. Inst. Technol., Trondheim.

Stewart, K. M. (1973). *Proc. Conf. Great Lakes Res.* **16**, 845–857.

Stigebrandt, A. (1978). *Nord. Hydrol.* **9**, 219–244.

Tatinclaux, J. E. (1977). *J. Hydraul. Div. ASCE* **102**, 959–974.

Tatinclaux, J. C., and Lee, C. L. (1978). *Can. J. Civ. Eng.* **5**, 202–212.

Tsang, G. (1975). *Int. Assoc. Hydraul. Res., Proc. Int. Symp. Ice Probl., 3rd, 1975* pp. 93–110.

Uzuner, M. S. (1975). *J. Hydraul. Res.* **13**, 79–102.

Uzuner, M. S., and Kennedy, J. F. (1972). *J. Hydraul. Div. ASCE* **98**, 2117–2133.

Uzuner, M. S., and Kennedy, J. F. (1976). *J. Hydraul. Div. ASCE* **102**, 1365–1383.

Weeks, W. F., and Dingman, S. L. (1972). *IHD/UNESCO/WMO Symp. Role Snow Ice Hydrol., 1972* Vol. 2, pp. 1427–1435.

Williams, G. P. (1965). *Can. Geotech. J.* **11**, 313–326.

Williams, G. P. (1973). "Seminar on Ice Jams in Canada," Tech.Memo. No. 107. Natl. Res. Counc., Ottawa, Ontario, Canada.

6 THE SEASONAL SNOWCOVER

D. H. Male
Division of Hydrology
University of Saskatchewan
Saskatoon, Saskatchewan, Canada

I. INTRODUCTION

The seasonal snowcover is a major environmental factor over much of the northern hemisphere and in the Antarctic. It is highly variable in time and space, and to gain an appreciation of its significance on a global scale it is necessary to consider its duration and maximum depth on the ground. Such factors as the rate of accumulation and variation of depth during the winter season are also important. Studies of the geographical distribution of snow by Rikhter (1960) and Bates and Bilello (1966) for the northern

DYNAMICS OF SNOW AND ICE MASSES

hemisphere show a general increase in both snowfall and the duration of snowcover with altitude and latitude. Some indication of the extent of the snowcover can be seen by examining Fig. 1, which is based on the work of Rikhter.

The various forms of snow crystals that constitute newly deposited snow result from variations in the temperature and humidity of the atmosphere at the time of their formation and the action of the wind during their descent to the ground. Once on the ground, snowflakes undergo a rapid metamorphism that reduces their surface area and brings them to a more stable thermodynamic state. This process is accompanied by an increase in the strength of bonds between grains. Blizzards or drifting snow conditions can also alter the shape of individual snow grains, breaking the crystals and abrading them into roughly equidimensional particles. Blowing snow is the major mechanism for the redistribution of masses of deposited snow.

Seasonal snowpacks normally develop from a series of winter storms and are modified by the action of freezing rain, the formation of wind crusts, and diurnal melting and refreezing at the surface. As a result, both the deep mountainous snowpacks and the shallow cover found in areas of low relief develop a characteristic layered structure in which "ice" layers or relatively impermeable fine-textured layers of high density alternate with layers of a coarser texture, lower density, and a higher permeability.

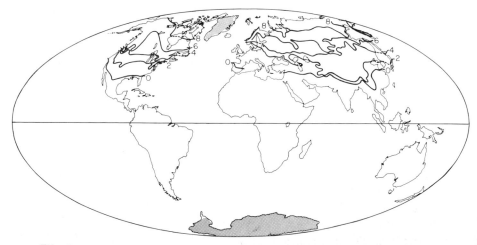

Fig. 1. Seasonal snowcover. The numbered lines represent the duration of the snowcover in months. The major ice sheets are shaded. [Adapted from "Snow and Ice on the Earth's Surface" (1964). CRREL Rep. II-Cl. U.S. Army Cold Reg. Res. Eng. Lab., Hanover, New Hampshire. CRREL map adapted from Rikhter, G. D. (1960). "Geography of the Snow Cover." Moscow.]

During each snowfall the type of snow deposited and the wind direction may change resulting in a stratification that is highly variable.

Snow is a good insulator, and the movement of heat through the snow-cover will cause a temperature gradient across any layer that has been on the ground for more than a few days. The movement of water vapor associated with such gradients produces new crystals that take many shapes but have a layered structure or a stepped or ribbed surface on some crystal facets. These crystals, known as depth hoar, are weakly bonded together.

The introduction of water into the snowcover, in the form of either rain or surface melt, causes a rapid metamorphism that eliminates the smaller particles and reduces the larger particles to a more or less spheroid shape. Water movement through wet snow is a highly complicated process. Ice layers initially inhibit the downward percolation of water but after a few hours vertical drainage channels develop [Gerdel (1954)], which draw off an increasing proportion of the water.

The processes mentioned above are common to the seasonal snow-cover as a whole. However, they do not necessarily occur in the same sequence in all areas of the globe, and many of them will occur simultaneously within a short distance of each other. In a review of the mechanical behavior of the snowcover Mellor (1975) states ". . . there is no material of broad engineering significance that under normal conditions displays the bewildering complexities found in snow." This statement could well be generalized to include all aspects of the seasonal snow-cover. In this chapter a selection of some of the important dynamic characteristics of the snow are examined from its deposition through to its final melt. The emphasis is on the underlying physical principles governing the snow's behavior.

II. METAMORPHISM OF DRY SNOW

Natural snow crystals nucleate and begin their growth in clouds where the temperature is below 0°C. A bewildering variety of crystal types are possible, depending on the temperature and humidity conditions at the time of their formation. During their fall through the atmosphere the crystals can change form and grow or disintegrate as they encounter changes in temperature and humidity. Furthermore, individual crystals may agglomerate into multicrystal flakes. Magono and Lee (1966) classify more than 75 snow crystal types which form as a result of the variety of meteorological conditions that can be encountered during their descent.

Snow can be deposited under a wide variety of wind conditions and at higher speeds can be badly broken either in the highly turbulent boundary layer near the ground surface or by the process of saltation discussed later. Crystals of blowing snow are about 0.1 mm in diameter while unbroken crystals or crystal aggregates range in size from about 0.1 to 4 mm.

The initial state of the deposited snow depends on the particular combination of wind conditions and crystal type present during deposition. At one extreme, aggregates of dendritic crystals deposited during a calm can form a highly porous layer having a bulk density of about 10 kg/m³. At the other extreme, equidimensional aggregates of frozen water droplets or graupel can form a layer having a bulk density as high as 500 kg/m³, particularly if they have been distributed by surface winds during deposition.

After the snow is deposited it can be considered a matrix of ice particles. Following Sommerfeld and LaChapelle (1970) these particles are called grains. Grains may consist of single crystals or be polycrystalline in nature. If bonds have not formed between grains, then grain boundaries are easily identified. Once bonds form it becomes increasingly difficult to define the grain boundaries, and the snow matrix becomes a complex three-dimensional network of connected particles.

Snow particles on the ground continue to change their shape and size, much as they did while falling through the atmosphere. These changes are largely governed by the continually changing temperature distribution in the snowcover which, in turn, is a product of the energy exchange between the snow and the atmosphere. Thus, the shape and size of snow particles found on the ground at any time depends on the thermal history of snowcover from the time of its deposition. This thermal history can be considered as a combination of two idealized conditions: a uniform temperature throughout the snowcover and a constant temperature gradient. In both cases, the temperature remains below 0°C. (Snowpacks at 0°C undergo rapid changes in grain shape and size due to the presence of liquid water. This process is considered in a later section.). Using the notation first suggested by Sommerfeld and LaChapelle (1970) the processes occurring under conditions of uniform temperature are called "equitemperature metamorphism," while those occurring under a constant temperature gradient are known as "temperature gradient metamorphism." Different physical processes are important under these two idealized conditions.

A. Equitemperature Metamorphism

Freshly deposited snow is made up of many dendrites, needles, and columns. The changes experienced by such crystals in a uniform temperature field have been carefully documented by Bader *et al.* (1939) and Yosida *et al.* (1955). Figure 2 illustrates the initial stages of this process for a

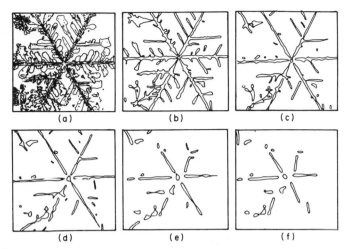

Fig. 2. Equitemperature metamorphism of a dendritic snow crystal at $-7°C$. The crystal is resting on a net woven of single silk fibres. (a) Original crystal; (b) after 2 days; (c) after 10 days; (d) after 15 days; (e) after 30 days; (f) after 36 days. [Drawn from photographs in Yosida *et al.* (1955).]

dendritic crystal and are drawn from photographs in Yosida *et al.* (1955). In general, the lower the temperature the slower the changes.

The first evidence of equitemperature metamorphism is the rounding of sharp corners and the decrease in thickness at the base of individual branches. In badly shattered blown snow this process is difficult to detect. In the next stage the slender necks disappear and the snow crystal breaks up into smaller grains. In naturally deposited snow where grains are close to each other, this breaking-up process is accompanied by the formation of bonds at the points of contact between grains. If this bond growth, known as sintering, is allowed to proceed further, the grains tend to become more equidimensional; the smaller grains disappear and the larger ones grow. Thus the snow at this stage is characterized by an increasing grain size and a decrease in grain number. As the grains become larger and more nearly spherical, there is a continuous decrease in the rate at which changes take place. Sommerfeld and LaChapelle (1970) suggest that the practical limit of grain size that can be reached under uniform subfreezing temperatures is approximately 1 mm.

If the snow is considered as a thermodynamic system, then the metamorphic processes described above can be viewed as the movement of the snow grains towards a state of equilibrium. The thermodynamic criterion for the equilibrium of several snow particles in contact is that the free energy be a minimum. Since the surface energy is important on the scale of individual crystals, the crystals move in a direction which decreases both the surface energy per unit area and the surface area per unit volume. There is considerable experimental evidence to suggest that the mechanism by which these changes are accomplished is the movement of water molecules through the vapor phase [Yosida *et al.* (1955); Hobbs and Mason (1964); Hobbs and Radke (1967)]. Surface diffusion and diffusion through the ice crystal also occur but are of minor importance during the initial stage of metamorphism. The vapor movement is the result of local differences in vapor pressure in the vicinity of the grains, which, in turn, depends on the curvature of the grain surfaces, local surface stress, and crystal structure. Assuming the vapor flux j in the pores separating the snow grains can be described by Fick's law, the relative importance of these factors can be examined. In the general case, Fick's law is

$$j = -D \text{ grad } c, \tag{1}$$

where D (m²/sec) is the diffusion coefficient and c (kg/m³) is the concentration of water vapor. In the simplest case of one-dimensional diffusion between the surface of adjacent grains separated by a space Δx, Eq. (1) becomes

$$j = -D(c_1 - c_2)/\Delta x, \tag{2}$$

where c_1 and c_2 are the vapor concentrations at the surfaces. Water vapor in air near atmospheric pressure obeys the ideal gas law and Eq. (2) can be written as

$$j = -(D/R \, \Delta x)|P_1/T_1 - P_2/T_2| \tag{3}$$

where R, the specific gas constant for water, is 0.462 kJ/kg °K. Hobbs and Mason (1964) express the vapor pressure over a curved surface as

$$P = P_0 \exp(\alpha\delta^3/kTr), \tag{4}$$

where P_0 represents the saturation vapor pressure over a flat ice surface obtained at a given temperature from the Clausius–Clapeyron equation. δ is the mean intermolecular distance, α the surface energy, k the Boltzman constant (1.38×10^{-23} J/°K), and r is measured from a center of curvature within the ice (r is positive if the surface is concave toward the solid). Equation (4) can be approximated as

$$P = P_0(1 + \alpha\delta^3/kTr). \tag{5}$$

Equations (3) and (5) give the following expression for the vapor flux:

$$j \simeq (D\gamma\delta^3 P_0/RkT^2 \, \Delta x)(1/r_1 - 1/r_2), \tag{6}$$

if r_1 and r_2 are both convex surfaces. If one of the surfaces is concave toward the vapor, then its radius will be negative. In order to calculate the range of radii over which curvature is important, it is necessary to determine D and P_0. Following Perla (1978) the diffusion coefficient for water vapor in still air is given by

$$D = D_0(101.3/P_a)(T/273)^n. \tag{7}$$

P_a (kPa) is the atmospheric pressure and D_0 is the diffusion coefficient measured at 273°K and 101.3 kPa (approximately 2.2×10^{-5} m^2/sec). The exponent n has values in the range 1.5–2.0. A convenient approximation for the saturation vapor pressure $p(T)$ over a flat surface in the range of temperature encountered in the snowcover is [de Quervain (1973)]

$$p(T) = 0.611 \exp[0.0857(T - 273)], \tag{8}$$

where $p(T)$ is measured in kilopascals.

Using Eq. (6) Perla (1978) considered the rate at which a small crystal of radius r would shrink in the presence of a flat ice surface. He shows that if the two ice surfaces are separated by a distance of 1 mm, then at a temperature of 273°K and an atmospheric pressure of 70 kPa (elevation approximately 3000 m) a sphere having an initial radius r_0 (m) would disappear in a time t (sec) of the order

$$t \sim (r_0^2/3)10^{16}. \tag{9}$$

A particle of radius 1 μm would disappear in about 1 hr while a particle ten times this radius would survive for about four days and a particle with a radius of 100 μm would survive for over a year. Thus ice surfaces with characteristic radii of the order of 100 μm are essentially flat, and curvature effects can only determine the metamorphism of the smallest crystal features such as the rounding of sharp corners and the disappearance of very thin dendrites (Fig. 2). This conclusion is supported by the work of Yosida et al. (1955), who showed that a rodlike ice needle having a diameter of 0.1 mm would survive for approximately 6 years. Yosida worked with a more complicated two-dimensional model than the one developed by Perla and concluded that the reduction in root radius of most dendrites takes place several orders of magnitude faster than can be accounted for by curvature alone.

By contrast Hobbs and Mason (1964) show that the dominant mechanism for growth of the area of contact, or neck, between two sintering spheres is vapor transport caused by the difference in curvature between the neck and the surface of the ice particles. This sintering process leads to an increase in snow density and strength. Radke and Hobbs (1967) suggest that once the density reaches a value of 400 kg/m^3, volume diffusion through the ice crystal becomes the dominant mechanism for the growth of bonds between ice spheres. Furthermore random packing of the ice grains can increase the snow density to a maximum value of about 600 kg/m^3. In the absence of a significant overburden pressure (surface stress), any further density increase must occur as the result of volume diffusion through the crystal lattice, which allows a decrease in the distance between the centers of adjacent spheres.

The influence of surface stress on the vapor pressure above the ice surface and hence on the flux of vapor from one ice grain to another is difficult to assess. Yosida (1963) outlines a detailed analysis of the pressure difference to be expected above an ice surface as the result of elastic stresses due to surface geometry or the weight of adjoining crystals but suggests that such stresses are not likely to be large enough to have any noticeable effect on the equitemperature metamorphism. Other factors such as crystal imperfections and the quasi-liquid structure of the ice surface near 273°K could influence vapor movement, but Yosida (1963), de Quervain (1973), and Perla (1978) suggest that all such surface energy effects have a much smaller influence on vapor diffusion than the influence of temperature gradients, particularly for grain sizes in the range of 1 mm.

B. Temperature Gradient Metamorphism

Temperature gradients occur in a naturally deposited snowcover as the result of a difference in temperature between the ground surface and the atmosphere near the snow surface. These gradients are related to the energy transfer processes taking place at the upper and lower surface of the pack. Heat transfer to the snowpack is discussed in a later section where it is shown that large variations in energy fluxes are possible over a 24-h period. As a result, the temperature gradient across a given layer of the snowcover is continually changing, although the temperature near the ground is generally higher than the temperature near the upper surface. If the gradient is maintained at a constant value for a sufficiently long period, distinctive crystal shapes will grow in the snowcover depending on the magnitude of the gradient and the grain size and density of the original snow. Based on a series of laboratory and field experiments, Akitaya (1975) classifies these crystals, commonly known as depth hoar, into two categories: "solid-type" depth hoar crystals and "skeleton-type" depth hoar crystals. The solid-type crystals have the general shape of a plate or a column with sharp edges, corners and flat surfaces. A layer of these crystals is relatively hard and appears as a fine-grained compact snow. The skeleton-type depth hoar has larger grains than the solid type and is characterized by a stepped or ribbed surface. Such grains form in a variety of shapes which resemble cups, needles, scrolls, and plates (Fig. 3). A layer of this type of crystals has a very low strength.

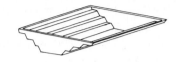

Fig. 3. Depth hoar crystals. [From Müller (1968)].

Akitaya (1975) found that skeleton-type depth hoar predominated when the average temperature gradient was greater than 0.25°C/cm and the solid-type depth hoar when the gradient was less than 0.25°C/cm. The size of the crystals seemed to be controlled by the size of the voids between crystals in the original snow. Akitaya distinguishes between three snow textures according to the size of the void space: type A has large voids and includes new snow, lightly compacted snow, and coarse-grained granular snow of low density; type C has small voids and includes fine-grained compact snow and coarse grained granular snow of medium and high density; and type B has medium-sized air spaces and consists of some fine-grained snows and coarse-grained granular snow of medium density. A qualitative graphical representation of the relationship between grain size, density, size of voids, temperature gradient, and type of depth hoar as suggested by Akitaya is shown in Fig. 4.

The primary mechanism responsible for the formation of depth hoar is vapor transport resulting from the temperature gradient. Equation (8) shows the relation between vapor pressure and temperature: the higher the temperature the higher the vapor pressure. Water vapor moves down

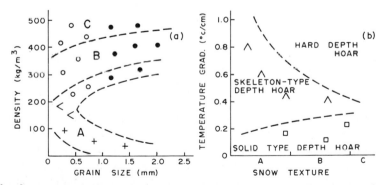

Fig. 4. Relation between the texture of snow and metamorphism under a negative temperature gradient (bottom of sample warmer than top). (a) Types of metamorphosed snow (A, B, and C) and properties of original snow: +, new snow; <, lightly compacted snow; ○, fine-grained compact snow; ●, coarse grained granular snow. (b) Relation between snow texture, temperature gradient, and shape of depth hoar. [From Akitaya (1975)].

the pressure gradient in a "hand-to-hand" process first described by Yosida et al. (1955). If it is assumed the grains in a layer near the ground have a higher temperature, then water will sublime from the top of these grains, move across an air space, and condense on the grains immediately above. This process repeats itself between adjacent layers of grains as long as the temperature gradient exists. The relevant temperature gradient governing this process is the gradient in the pores. This could be higher than the average gradient across several grains because the thermal conductivity of the ice matrix is about one hundred times that of the air and convection currents in the air spaces are unlikely [Akitaya (1975)].

An analysis of temperature gradient metamorphism has been developed by Yosida et al. (1955), Giddings and LaChapelle (1962), and de Quervain (1973). Assuming a one-dimensional temperature gradient in the vertical z direction within the snow the rate of diffusion of water vapor j (kg/m^2 sec) is

$$j = (-fD/RT_m)(\partial p/\partial z), \tag{10}$$

where T_m is the mean absolute temperature and f is a structural parameter that accounts for the fact that only a fraction of the unit area perpendicular to the temperature or pressure gradient is in a pore space where diffusion is possible. Now $\partial p/\partial z = (\partial p/\partial T) \cdot (\partial T/\partial z)$ and using Eq. (8) for p the mass flux becomes

$$j = -(fD/RT_m)0.611 \cdot 0.0857 \exp[0.0857(T - 273](\partial T/\partial z). \tag{11}$$

For this one-dimensional situation the continuity equation for a horizontal layer of snow with thickness dz is

$$\partial\rho/\partial t = -\partial j/\partial z$$

and, if T_m is assumed constant,

$$\partial\rho/\partial t = 0.0524(fD/RT_m)[\partial^2 T/\partial x^2$$
$$+ 0.0857(\partial T/\partial z)^2] \exp[0.0857(T - T_0)]. \tag{12}$$

Equation (12) was first derived by de Quervain (1973), although the basic equations with some variation have been used by Giddings and LaChapelle (1972) and Yosida et al. (1955). Note that for a constant positive temperature gradient $\partial\rho/\partial t$ is positive; the density increases throughout the snow layer. However, for a given temperature gradient, j increases rapidly with mean temperature because of the exponential function. Giddings and LaChapelle (1962) suggest the vapor flux is reduced by a factor of 2 for every 8-degree drop in temperature. Thus the rate of formation of depth hoar is much larger in the lower parts of the snowcover where the temperature is generally higher.

Qualitative evaluations of density changes based on Eq. (12) require some consideration of the factor f. Giddings and LaChapelle argue that for low-density snow f is approximately unity. They recognize that the ice matrix will obstruct the diffusion process but suggest that this effect is canceled by the increased temperature gradient found in the pore space and the fact that vapor can condense on the ice particles rather than go around them. Their calculations of the formation rate of depth hoar are in fair agreement with field measurements considering the range of initial snow types and climatic conditions encountered during their tests which were conducted over a period of from 12 to 55 days.

de Quervain (1973) examines two basic structural models which may be considered limiting cases for vapor diffusion. These models are illustrated in Fig. 5. The first model consists of a block of ice perforated by a network of cylindrical pores with axis aligned parallel to the temperature gradient. For this model the factor f will be equal to the porosity ϕ. The second model consists of alternating layers of air and ice arranged perpendicular to the temperature gradient. In this case f is equal to $1/\phi$ and vapor movement occurs exclusively by the "hand-to-hand" process mentioned earlier. However, several aspects of temperature gradient metamorphism cannot be accounted for by simple models. In particular the first model yields values for changes in density which are considerably lower than observed changes [de Quervain (1973)]. This has led to speculation that the diffusion coefficient in snow is higher than values predicted by Eq. (7). The major shortcoming of the second model is that it neglects the heat sources and sinks created by the condensation and evaporation processes occurring on alternate ice surfaces. As a result of these phase changes the local temperature gradients both in the air gaps and in the ice will be altered. de Quervain suggests that the conditions encountered in natural snow lie somewhere between the extremes of Fig. 5 but points out that a combined model would still have shortcomings. In particular there

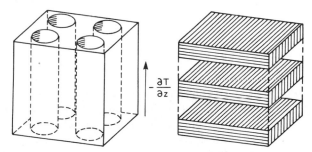

Fig. 5.   Basic models considered by de Quervain (1973) for the calculation of vapor movement through dry snow.

is a selective growth of some crystals and an increasing intrinsic permeability of the snow as metamorphism advances. He suggests that selective growth may be a function of the pore geometry. It is possible crystals with relatively higher and lower temperatures may be in the immediate neighborhood of each other, the local temperature being determined by the location of their point of attachment to the ice matrix. Cold surfaces will encourage growth by condensation while the warm surfaces will act as vapor sources by evaporation and tend to shrink into the matrix.

III. PROPERTIES OF DRY SNOW

In this section the important bulk properties of snow that influence the accumulation and melt of the snowcover are described. In particular those thermal properties that affect the movement of heat through snow and hence the rate of melt are discussed along with the mechanical properties that govern the behavior of snow under a uniaxial stress.

Metamorphism controls the properties of snow. Changes in density, crystal size and shape, and the growth and decay of bonds between grains influence all snow properties to a greater or lesser extent. Much of the published data is from measurements on homogeneous snow samples on a scale of several centimeters which contain many hundred grains. The relation of these average or bulk properties to the crystalline nature of the snow grains and intergranular bonds is imperfectly understood. The thermal properties of specific heat capacity and latent heat of fusion, which depend only on density, are well defined. Properties such as thermal conductivity and intrinsic permeability are influenced by particle size and shape and the growth of bonds between grains. Elastic properties such as Young's modulus and the shear modulus are strongly dependent on the sintering process while optical properties such as the reflectivity or albedo depend on porosity, grain size, and shape. The accuracy with which the various properties can be determined is directly related to their dependence on the metamorphic process and our understanding of these processes.

A. Thermal Properties

The latent heat of snow per unit mass is equivalent to that of ice. The latent heat of fusion for ice at 0°C and atmospheric pressure is 333.66 kJ/kg [Dorsey (1940)]. This value decreases by approximately 0.6% per degree drop in temperature, but this is of no practical importance in the calculation of water yield from the melting snowcover. The latent heat of sublimation is also temperature dependent as shown in Table 1. The

TABLE 1

Latent Heat (Enthalpy) of Sublimation for Ice[a]

Temperature (°C)	Latent heat (kJ/kg)
0	2834.8
−10	2837.0
−20	2838.4
−30	2839.0
−40	2838.9

[a] From Keenan *et al.* (1969).

values listed are for pure ice in equilibrium with its vapor at the given temperature. In the presence of air these values will change slightly.

The specific heat capacity of dry snow is also considered to be that of an equivalent mass of ice since the contribution of the air in the pores is negligible. For pure ice in the range −40 to −5°C, Dickinson and Osborne (1915) suggest the following expression:

$$C_p = 2.117 + 0.0078T \quad \text{(kJ/kg °C)}, \tag{13}$$

where T is in degrees centigrade. Impurities can influence C_p, particularly near the melting temperature, but for hydrological applications an approximate value of 2.1 kJ/kg °C is adequate.

The thermal conductivity of snow is not known to the same accuracy as other thermal properties. It is normally measured as the proportionality constant in one-dimensional heat flow situations where the Fourier heat conduction equation has the form

$$q = -k \, dT/dz, \tag{14}$$

where q is the heat transfer per unit area (in kilowatts per squared meter), dT/dz the temperature gradient, and k is thermal conductivity (in kilowatts per meter-degree). In snow, conduction will take place in the matrix of ice grains and across the air spaces if a temperature gradient is present. However, as mentioned in the previous section, the temperature gradient will also cause the diffusion of vapor through the air spaces. This movement represents a transfer of energy not considered in Eq. (14). In the measurement of k it is virtually impossible to separate the effects of the conduction and diffusion process, and it is customary to express the measured values as an effective conductivity k_e, which accounts for both energy transfer mechanisms. Because of the nature of the diffusion process, k_e depends on the magnitude of the temperature gradient and whether the gradient is constant or changing with time. Yosida *et al.* (1955) suggest that the movement of water vapor contributes 37% to the apparent thermal conductivity of snow having a density of 100 kg/m³. At a density of 500 kg/m³ this value is reduced to 8%. Measured values are normally plotted as a function of density and comparisons of these values can be found in Mellor (1977). Anderson (1976) suggests an expression of the form

$$k_e = 2.09 \times 10^{-5} + C\rho_s^2, \tag{15}$$

where ρ_s is the snow density (in kilograms per cubic meter) and C is about 2.5×10^{-9} for a seasonal snowcover.

B. Mechanical Properties

Snow responds to stress in a variety of ways depending on the magnitude and rate of application of the applied load. Snow deforms elastically when subjected to small loads for short periods of time but also deforms continuously and permanently if a load is applied over longer periods. Over the last 40 years or so attempts have been made to apply the laws of continuum mechanics to snow, but the formulation of relationships between stress, strain, and time are still in the initial stages of development. A comprehensive review of the basic mechanics of dry snow may be found in Mellor (1975). What follows is a discussion of some of the simpler aspects of snow mechanics as they relate to its elastic behavior and its densification or creep resulting from an overburden pressure.

Snow that has been deposited for a sufficient time to develop intergranular bonds acquires a mechanical strength and offers resistance to deformation under an applied load. Under moderate loadings for short periods of time (a few seconds) snow responds elastically with strains that are proportional to stress and that can be recovered once the load is removed. Under sustained loads the initial elastic deformation is followed by a continuous strain or creep. Mellor (1966a) states that the rheological behavior of snow is similar to that of polycrystalline ice with the added complication of a highly irreversible compressibility. Snow that has undergone sintering can therefore be classified as a compressible viscoelastic material in which the viscosity is nonlinear with respect to the stress. In particular, snow exhibits the phenomenon known as tertiary creep. That is, if the stress is above a limiting level and the load is maintained for a sufficiently long period, the rate of straining will increase. This condition normally marks the onset of failure, which is important in the release of avalanches (see Chapter 7).

The elastic behavior of snow is confined to loads of a short duration in which the strains are small enough to leave the bonds between grains undisturbed. Mellor (1975) suggests that the most consistent measure-

ments of the elastic modulus or Young's modulus for snow are derived
from snow samples vibrating at acoustic or ultrasonic frequencies. Traditional uniaxial load tests also give good results if tests are run at high
strain rates. Fig. 6 is a summary of measurements of Young's modulus E
made by various investigators and compiled by Mellor (1975). The tests
were conducted on dry well-bonded snow samples. The influence of temperature on E is shown in the inset. It is noteworthy that within the range
of densities normally encountered in the field (200–500 kg/m³), E varies
by approximately three orders of magnitude.

The sintering process has an important influence on Young's modulus.
Nakaya (1961) has examined this factor by determining E as a function of

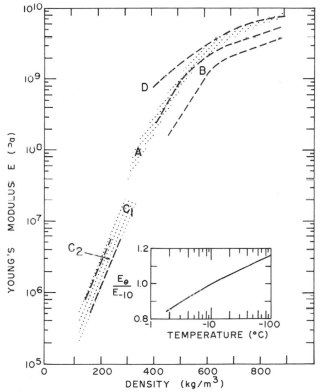

Fig. 6. Young's modulus for dry well-bonded snow: (A) Pulse propagation on flexural
vibration at high frequencies, −10 to −25°C [Smith (1965); Nakaya (1959a,b); Bentley *et al.*
(1957); Crary *et al.* (1962); Lee (1961); Ramseier (1966)]. (B) Uniaxial compression strain
rate approximately 3×10^{-3} to 2×10^{-2} sec⁻¹, temperature −25°C [Kovacs *et al.* (1969)].
(C₁) Uniaxial compression and tension, strain rate approximately 8×10^{-6} to 4×10^{-4} sec⁻¹,
temperature −12 to −25°C. (C₂) Static creep test, −6.5 to −19°C [Kojima (1954)]. (D) Complex modulus, 10^3 Hz, −14°C [Smith (1969)]. [From Mellor (1975).]

time for snow that has been passed through a snowblower. For snow densities in the range 400–600 kg/m³, E increases exponentially with time during sintering. Typically, values for E double in about three weeks and reach a value approximately three times the initial value after one year.

A comprehensive theoretical treatment of the viscoelastic behavior of snow has yet to be developed, although numerous measurements reported in the literature are related to standard rheological models such as the Maxwell, Kelvin–Voigt, or Burgers model [Mellor (1975)]. One of the most commonly observed characteristics of snow and the principal mechanical process occurring in snow deposited on relatively flat terrain is the settlement or densification of a given layer under the action of the overburden pressure. Bader (1953) was the first to suggest that this process could be characterized by the "compactive viscosity" η_c, which is defined as the ratio of the overburden pressure σ_v to the vertical strain rate ε_v at a given depth z. That is,

$$\eta_c = \sigma_v / \varepsilon_v.$$

Following Bader (1960, 1962) the overburden pressure is given as

$$\sigma_v = \int_0^z \rho g \, dz = \dot{A} t_s,$$

where ρ is the snow density, t_s the time since the layer was deposited, and $\dot{A}$ the rate of accumulation of snow. The velocity V at which the layer sinks below the surface is given by

$$V = dz/dt = (dz/d\sigma_v)(d\sigma_v/dt) = \dot{A}/(\rho g).$$

Therefore

$$\varepsilon_v = dV/dz = [-A/(\rho^2 g)] \, d\rho/dz$$

or

$$\varepsilon_v = -1/\rho(d\rho/dt). \tag{16}$$

The rate $\dot{A}$ can be measured directly or from an analysis of the snowpack stratigraphy and therefore depth–density profiles of snow deposits along with measurements of density enable η_c to be determined from field observations. Values of compactive viscosity as a function of density are plotted in Fig. 7. There is considerable difference between the value measured by various investigators, although the concept has been used successfully in the analysis of the settlement of foundations and the closure of tunnels in Antarctica and Greenland and has been applied to the settlement of the seasonal snowcover.

Mellor (1975) emphasizes the nonlinear relationship between viscosity,

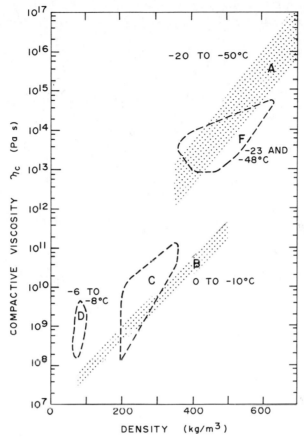

Fig. 7. Compactive viscosity plotted against density. (A) Greenland and Antarctica, −20°C to −50°C [Bader (1962)]. (B) Seasonal snow, Japan, 0 to −10°C [Kojima (1967)]. (C) Alps and Rocky Mountains [Keeler (1969)]. (D) Uniaxial-strain creep tests, −6 to −8°C [Keeler (1969)]. (F) Uniaxial-strain creep tests, −23 to −48°C [Mellor and Hendrickson (1965)]. [From Mellor (1975).]

stress, and density and suggests that the concept of compactive viscosity is difficult to interpret, particularly for seasonal snowpacks. When new snow is deposited the underlying layers experience rapid densification for a few days, which eventually decreases to a relatively slow rate. The process is one of quasi-plastic collapse rather than a steady creep. In this respect the difference between measurements made on seasonal snowpacks where snowfalls are frequent and polar snowfields is significant.

There have been a few investigations of the physical processes occurring within and between individual grains during the deformation of snow, although a comprehensive theory of the process has yet to be achieved. Wakahama (1967a,b) gives a good description of what happens when snow is deformed at a relatively slow constant rate of from 1 to 10% per hour. Initially there is an elastic yield with little relative motion between grains. Eventually slip planes form within individual ice grains and fracture occurs along grain boundaries. Once grains become detached intergranular slip is the dominant mechanism for continued movement. A one-dimensional theory of the deformation process using Wakahama's concepts has been developed by St. Lawrence and Bradley (1975). The analysis shows very good agreement with measurements of changes in stress of snow samples deformed at a constant rate for both tension and compression. The authors are careful to point out that the theory at present is not general and further development is necessary. Nevertheless, it is clear that it is possible to describe the deformation of snow by physical considerations occurring on the scale of individual grains.

The behavior of snow under multidirectional or multiaxial stress fields is dominated by the effect of compression on the strain rate and the strain components are separated into pure shear or uniaxial strains at constant volume and volumetric strains [Westergaard (1964)]. Unlike many materials, compressive volumetric strains for snow are significant and largely irreversible, and the uniaxial components are strong functions of compressibility. The volumetric stress–strain relationship is complicated and a suitable theoretical framework within which measurements may be interpreted is lacking. Mellor (1975) discusses the major difficulties in this area.

IV. BLOWING SNOW

The major process for the redistribution of deposited snow is transport by the wind. On the open plains of the northern hemisphere blowing snow occurs as much as 30% of the time during the winter months. In mountainous regions drifting snow is a major water source for glaciers and an important factor in the formation of avalanches. On the polar ice sheets it could form an important part of the mass balance.

Particles of blowing snow may be supplied from precipitation or they may be lifted from the snow surface by the wind. During precipitation, snow particles may have a significant horizontal velocity even at low wind speeds, but during periods when no snow is falling blowing snow will occur only when the wind velocities are strong enough to produce surface shear and lift forces capable of dislodging particles from the surface. Obviously, the velocity needed depends on the condition of the snow surface. Mellor (1965) suggests that wind speeds (at a height of 10 m) in the range 3–8 m/sec are sufficient in loose unbonded snow while snow grains bonded by sintering or the freeze–thaw process may require wind speeds greater than 30 m/sec to initiate snow transport. This section describes the blowing process in the absence of precipitation.

Two modes of transport have been distinguished in blowing or drifting snow. In a layer approximately 10 cm high, just above the surface, particles bound along traveling in a curved trajectory under the influence of gravity and the drag force resulting from the relative velocity between the wind and the particle. This process is known as saltation by analogy with the motion occurring in low-level sandstorms or near the surface of river beds. When the wind velocity reaches a certain level, particles are held in suspension in the air stream by turbulent diffusion.

These two modes of snow transport have led to the development of two theories of the drifting snow process. In one theory, developed principally by Dyunin (1959, 1967), saltation is viewed as the dominant drift process, and the momentum equation for a two-phase (snow–air) mixture is applied to a layer near the snow surface. The second theory considers diffusion as the dominant mechanism and concentrates on processes that occur in the air layer up to tens of meters above the snow surface. This theory results from studies of blowing snow in Antarctica [Budd (1966a), Budd et al. (1966)].

A. Turbulent Diffusion

Historically, the theory of turbulent snow drift developed from observations of the variation of wind velocity with height above the surface during blowing snow conditions. Liljequist (1957), Dingle and Radok (1961), and Budd et al. (1966) all show that during snowdrift the mean wind profile is proportional to the logarithm of height above the surface, or

$$u = (u_*/k) \ln(z/z_0), \tag{17}$$

where u is the mean wind speed at a height z above the surface, z_0 a roughness height characteristic of the snow surface at which the wind velocity approaches zero, u_* a reference velocity known as the friction velocity, and k von Kármán's constant (approximately 0.4).

The form of Eq. (17) was first explained by Prandtl (1925), who proposed that the shear stress τ near the surface is constant with height and can be expressed as

$$\tau = \rho K \, \partial u / \partial z, \tag{18}$$

where ρ is the air density and K the eddy diffusivity or eddy viscosity by analogy to the viscosity of a Newtonian fluid in laminar flow. For a neutrally stable atmosphere (decrease in air temperature with height of approximately 1°C/100 m) Prandtl proposed a mixing length l such that

$$\tau/\rho = l^2(\partial u/\partial z)^2, \tag{19}$$

where l is assumed proportional to height z so that

$$l = kz. \tag{20}$$

The proportionality constant k is von Karman's constant. If Eq. (20) is substituted into Eq. (19) then

$$\partial u/\partial z = u_*/kz, \tag{21}$$

where $u_*^2 = \tau/\rho$ is the friction velocity. Equation (17) is the integrated form of Eq. (21). From Eqs. (18) and (21) it is seen that the eddy diffusivity K is

$$K = ku_*z. \tag{22}$$

Thus the logarithmic wind profile can be explained in terms of an eddy diffusivity K that is proportional to height above the surface. This result is applicable to any flat surface, and the fact that a logarithmic profile is commonly present during blowing snow conditions has led to the basic assumption that the turbulent mechanisms that produce the momentum transport and shear stress in the surface layer are also responsible for the steady upward movement of snow particles.

The diffusion theory considers the mass flux of snow particles in a given direction to consist of a mean component and a fluctuating component. For example, in the horizontal direction the mass flux of snow s_x at any height is given by

$$s_x = un - (un)', \tag{23}$$

where n is the mass of snow particles per unit volume (in kilograms per cubic meter), u the mean particle velocity, un the mean flux averaged over a period of time, and $(un)'$ the deviation from the mean due to the turbulent eddy motion. The fluctuating component is described by a flux-gradient relation of the form

$$(un)' = -K_{sx}\, \partial n/\partial x. \tag{24}$$

Thus, at any point (x, z) in a two-dimensional stream of blowing snow the mass flux components are

$$s_x = un - K_{sx}\, \partial n/\partial x, \qquad s_z = wn - K_{sz}\, \partial n/\partial z. \tag{25}$$

On the basis of Eqs. (25) the mass fluxes entering and leaving a differential volume dx by dz by unit depth centered at the point (x, z) in the air stream is shown in Fig. 8.

If the net mass flux is equated to the accumulation in the element $(\partial n/\partial t)\, dx\, dz$ then a continuity equation of the following form results:

$$-\partial(un - K_{sx}\, \partial n/\partial x)/\partial x - \partial(wn - K_{sz}\, \partial n/\partial z)/\partial z = \partial n/\partial t. \qquad (26)$$

For flow across a flat plane it can be assumed $\partial n/\partial x$ is much smaller than $\partial n/\partial z$ and diffusion in the x direction is neglected. Furthermore, if the concentration n at any level is constant with time (steady state) then $\partial n/\partial t = 0$. Under these conditions Eq. (26) reduces to the one-dimensional form

$$wn \quad K_{sz}\, \partial n/\partial z = 0. \qquad (27)$$

As mentioned earlier, the fact that the wind profile in blowing snow is logarithmic leads to the assumption that $K_{sz} = K$, where K is given by Eq. (22). It is further assumed that there is no relative horizontal velocity between the snow particles and the air stream and that the vertical particle velocity w is equal to the terminal fall velocity in still air, $-\omega$. Equa-

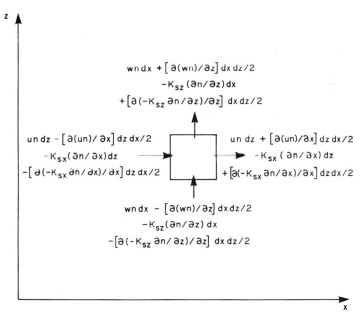

Fig. 8. Mass flux of snow particles entering and leaving a differential element of the air stream having a unit depth and area dx by dz.

tion (27) now becomes

$$\omega n + k u_* z \, \partial n / \partial z = 0. \tag{28}$$

This expression was developed independently by Shiotani and Arai (1953) and Loewe (1956). If the fall velocity is constant with height, which implies the average particle size is constant with height, Eq. (28) can be integrated to give

$$n/n_1 = (z/z_1)^{-\omega/k u_*}. \tag{29}$$

The subscript 1 denotes any reference level.

An expression for the drift density n at a given height as a function of wind velocity u at the same height can be obtained by eliminating u_* between Eqs. (29) and (17) [Dingle and Radok (1961)]. This expression has the form

$$n = n_1 e^{a/u}, \tag{30}$$

where $a = -(\omega/k^2) \ln(z/z_0) \ln(z/z_1)$. Tests of Eqs. (29) and (30) require very careful field measurements under difficult conditions. To date, the technique used most commonly involves the mechanical metering of the mass flow of snow particles by placing an orifice perpendicular to the flow for a timed interval. Snow particles that pass through the orifice are trapped and weighed. The characteristics of these gages are discussed by Mellor (1965) and their performance has been evaluated by Budd et al. (1966). Gages of this type measure the mass flow rate un and the concentration n at a given level is obtained by dividing the flow rate by the wind speed u at the same level.

Both Eqs. (29) and (30) have been tested against a large volume of data from Byrd Station, Antarctica, by Budd et al. (1966). Fig. 9, reproduced from this study, shows the overall mean deviation from the linear log density profile expected from Eq. (29). Also shown is the average deviation from the logarithmic wind profile plotted on a comparable scale. There is a systematic and continuous change in the deviation with height. Equation (29) is based on the assumption of steady-state conditions and a constant particle fall velocity ω, both of which must be questioned in the light of Fig. 9. However, based on a consideration of the difference in density profiles during periods of net accumulation and net erosion, Budd et al. conclude that the steady-state assumption does not explain the deviations of Fig. 9 but that there is a systematic change in particle fall velocity with height.

Equation (30) has also been used to test the simple diffusion theory. It suggests that $\ln(n)$ should vary linearly with u^{-1}, assuming the roughness height z_0 and the reference drift density n are constant. Budd et al. suggest that near the 1-cm level a layer exists in which the drift density is largely

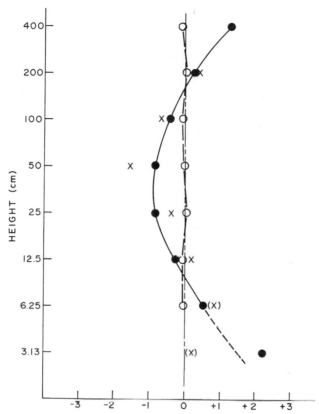

Fig. 9. Test of the linearity of ln n versus ln z [Eq. 29)]. ●, Byrd; ×, Wilkes (Dingle and Radok, 1961); ○, average deviation from log wind profile; solid line, log density ratio (dB); dashed line velocity deviation (m sec^{-1}). [From Budd *et al.* (1966), copyrighted by American Geophysical Union.]

independent of wind speed and on this basis have produced the comparison shown in Fig. 10. The solid lines represent a least-squares regression and, although the scatter is considerable, the figure confirms the general validity of Eq. (30). However, Budd *et al.* also examine the mean drift densities as a function of wind velocity and show deviations from the theoretical straight-line relationship at heights from 3 to 12.5 cm and above 100 cm.

The steady-state diffusion theory has been extended to the case of a variable fall velocity by considering the distribution of particle sizes in the blowing snow [Budd (1966b)]. This analysis is based on partical size distributions at various heights reported by Budd *et al.* (1966). Measurements of particle size were made indirectly on Formvar replicas produced on glass slides enclosed by specially designed collectors. The results of

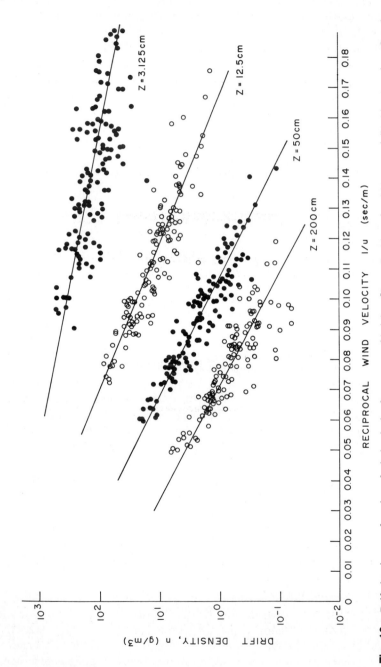

Fig. 10. Drift density as a function of wind velocity for a range of heights. [From Budd et al. (1966), copyrighted by American Geophysical Union.]

these measurements are summarized in Fig. 11 [Budd (1966b)]. The gamma distribution function used by Budd to represent this data is also shown on the figure.

Budd's analysis proceeds from the assumption that the particle fall velocity is proportional to the effective diameter. He further assumes that particles of a given size are distributed with height according to Eq. (29) and that particles of different fall velocities do not influence each other. On this basis and by applying the gamma distribution function, Budd shows that the drift density as a function of height is given by the following series solution:

$$\ln(\Lambda/\Lambda_1) = (\omega/ku_*) \ln(z/z_1) + [\alpha\beta^2/2(ku_*)^2] \ln^2(z/z_1)$$
$$-[\alpha\beta^3/3(ku_*)^3] \ln^3(z/z_1) + \cdot \cdot \cdot, \tag{31}$$

where Λ is the drift density at a given height and Λ_1 the corresponding value at a reference level. The terms α and β are parameters in the gamma distribution. The product $\alpha\beta$ represents the mean fall velocity of which $\alpha\beta^2$ is the variance. The first term on the right-hand side is the logarithm of the drift density ratio as predicted by Eq. (29) on the basis of a constant fall velocity. The second term on the right-hand side may be considered a correction factor that depends on the variance $\alpha\beta^2$ of the particle distribution. Budd shows that the third term is an order of magnitude less than the squared term. A plot of Eq. (30) in the form of a deviation from the linear log-density–log-height relation predicted by Eq. (29) is shown in Fig. 12. Three wind speeds are considered in the calculations. Note from Eq. (17) that u_* is proportional to the wind speed at a given height and, since u_*^2 appears in the denominator of the first correction term, the deviation from the linear profile is reduced considerably at higher wind speeds. Budd also analyzes the mean particle size with height and the mean fall velocity, and shows that both quantities decrease with height. Fig. 12 is a clear indication of the important influence the particle size distribution plays in the mechanics of blowing snow. Radok (1977) mentions that the direct measurement of particle sizes along with their velocity and quantity is important to the future development of research in blowing snow. A photo electric gauge [Schmidt and Sommerfeld (1967)] that measures light attenuation as snow particles pass between a source and a detector shows promise of providing this information.

The total amount of snow transported through a vertical surface of unit width perpendicular to the wind direction can be obtained by taking the product of the drift density and velocity as a function of height and integrating from the snow surface. Using the notation of Eq. (29) for a con-

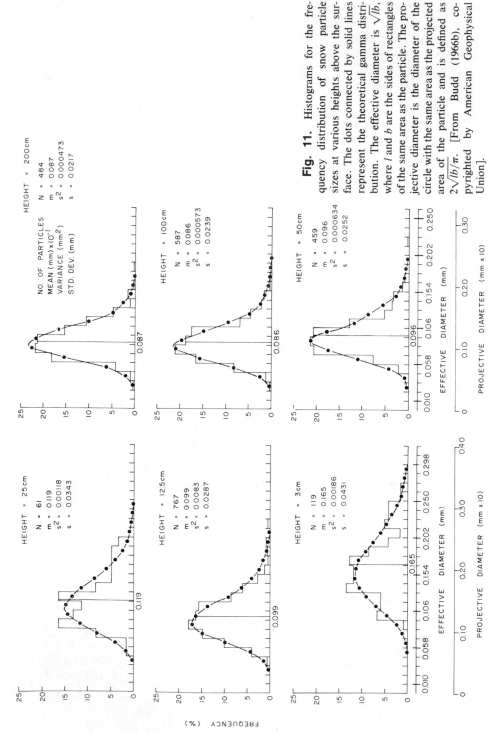

Fig. 11. Histograms for the frequency distribution of snow particle sizes at various heights above the surface. The dots connected by solid lines represent the theoretical gamma distribution. The effective diameter is $\sqrt{lb}$, where l and b are the sides of rectangles of the same area as the particle. The projective diameter is the diameter of the circle with the same area as the projected area of the particle and is defined as $2\sqrt{lb/\pi}$. [From Budd (1966b), copyrighted by American Geophysical Union].

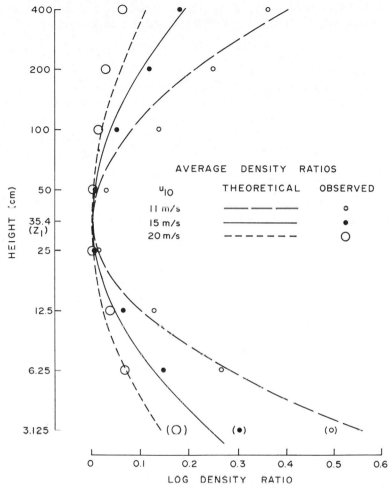

Fig. 12. Average deviation in measured drift density values from the linear log-density–log-height relation for each level and for three wind speeds. The curves are calculated from Eq. (31) with the reference level at 50 cm. [From Budd (1966b), copyrighted by the American Geophysical Union].

stant fall velocity this expression is

$$Q = \int_0^\infty nu \; dz.$$

Even for the simple case of a constant fall velocity the integration results in a complex expression. Simple expressions have been developed using combinations of calculated and measured fluxes at various levels. For example, Budd *et al.* (1966) suggest that in the layer from 1 mm to

300 m the total drift transport Q (g/sec) through a vertical surface 1 m wide perpendicular to the wind direction can be expressed as

$$\log Q = 1.18 + 0.089 u_{10}, \qquad (32)$$

where u_{10} is the wind velocity at the 10-m level.

Typical profiles of the mass flux are shown in Fig. 13. In general, the mass flux decreases slightly with height above the 1-m level while below 1 m there is a very strong increase as the surface is approached. Figure 13 suggests that under blizzard conditions the majority of the mass flux occurs in a shallow layer near the surface. However, measurements of drift concentrations at high levels are not available, and the relative contributions of the saltation and diffusion layers remains in doubt. Nevertheless, the figure indicates that the saltation process is important, and the mechanics of this process are considered in the next section.

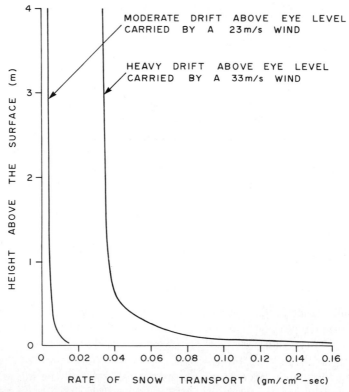

Fig. 13.   Typical profiles of the horizontal mass flux [from Mellor (1965)].

B. Saltation

The nature of saltation was first examined in detail by Bagnold (1941) in a study of blowing sand in the desert. A more recent publication by Bagnold (1973) and a study by Owen (1964) make it possible to clarify the major dynamic processes occurring in this layer.

In the saltation zone particles of snow move under the influence of gravity and an aerodynamic drag force resulting from the relative velocity between the air stream and the particle. Saltation has been observed in laminar flow [Francis (1973)] and hence can be assumed to occur independently of any fluid turbulence or vertical component of velocity although the laminar flow regime is not encountered in blowing snow. Furthermore, Owen (1969) shows that for two-phase flows such as snow particles in air, if the the ratio of solid to fluid density is very high (approximately 720 for ice grains in air), the aerodynamic lift acting on the particles is negligible compared to the drag force.

Saltation begins when the shear stress exerted by the wind on the snow surface, τ_0, exceeds a certain threshold value. Bagnold (1941) describes this threshold in terms of the dimensionless ratio of the hydrodynamic force on a particle to the particle weight

$$\beta = \tau_0/\rho_s g d, \tag{33}$$

where ρ_s is the density of the solid particle of diameter d. For sand grains in air β is of the order 10^{-2}. Bagnold further recognized that if particles were falling on the surface the impact would lower the shear stress required to sustain saltation. He termed this the impact threshold and found that for this condition $\beta = 0.0064$. In snow the metamorphic state of the snow surface and the strength of the bonds between surface gains will influence the threshold conditions. Various workers have published values of the friction velocity u_* at which drifting is initiated ($u_* = \sqrt{\tau_0/\rho}$). Values of u_* range from 0.07 m/sec for very light dry snow [Rikhter (1945)] to 0.4 m/sec for snow which has a wind-hardened surface and a density of approximately 350 kg/m³ [Kotlyakov (1961)]. Budd et al. (1966) suggest that the 10-m wind velocity (u_{10}) can be related to u_* by

$$u_{10} = 26.5u_* \tag{34}$$

for conditions at Byrd Station, Antarctica. If this expression is accepted as representative of snow fields in general, then the range of u_* quoted above corresponds to a range of u_{10} from 1.9 to 10.5 m/sec. Kind (1976) suggests that a representative friction velocity for dry uncompacted snow is 0.15 m/sec, which corresponds to a u_{10} of 4 m/sec.

Once saltation is initiated the moving particles are borne along the surface with trajectories that reach a maximum height of approximately 3 cm. Individual particles rise from the surface in a more or less vertical direction and receive a horizontal acceleration from the aerodynamic drag force. The gravitational force eventually overcomes the initial vertical momentum of the particle and it descends towards the surface. When the particle strikes the surface one or two grains may be splashed upward to continue the saltation process. The saltation process in snow has been photographed by Kobayashi (1973) and a typical particle trajectory is shown in Fig. 14. The increased kinetic energy acquired by the particle as a result of the aerodynamic drag force is dissipated when the particle strikes the bed. In effect the saltation process extracts horizontal momentum from the flow and creates a shear stress between the fluid and the saltation surface.

Owen (1964) proposes two working hypotheses to explain the mechanism of saltation. The first of these is that the saltating layer influences the air flow above it in a manner similar to a surface having a roughness proportional to the height of the saltating layer. For an initial vertical velocity v_1, the maximum height in the trajectory of a saltating particle is $v_1^2/2g$. The vertical velocities in a flow with a friction velocity u_* are of the order of u_* and therefore the saltation height h is proportional to $u_*^2/2g$. The logarithmic wind profile above the saltation surface [Eq. (17)] can then be expressed as

$$u/u_* = k^{-1} \ln(2gz/u_*^2) + D. \tag{35}$$

Owen (1964) has plotted wind profile measurements above saltating layers of sand and soil to test the validity of Eq. (35), and Kind (1976) has extended the comparison to field data for blowing snow taken from Oura *et al.* (1967). The result, shown in Fig. 15, confirms Owen's first hypothesis. Accepting that the height of the saltation layer is of the order $u_*^2/2g$ and using the measurements of Oura *et al.* (1967), it can be suggested that the trajectories of most saltating particles reach a maximum height in the range 3 mm–3 cm.

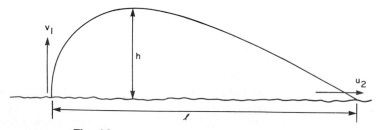

Fig. 14.   Trajectory of a particle during saltation.

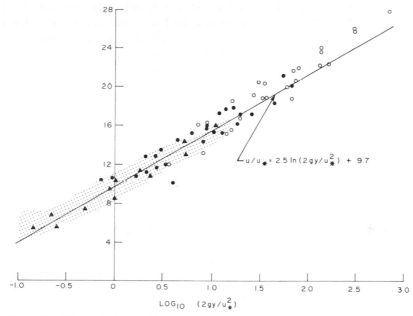

Fig. 15. Wind velocity profile above the saltation layer. Snow: ○, Oura *et al.* (1967); Uniform sand. shaded, Zingg (1953); ▲, Bagnold (1941); nonuniform soil: ●, Chepil (1945) [Reprinted with permission from *Atmos. Environ.* **10** (Kind, 1976). Copyright 1976, Pergamon Press, Ltd.].

The second hypothesis proposed by Owen is that the concentration of particles within the saltation layer is governed by the shearing stress at the surface and that this value is maintained close to the threshold value which ensures that the surface grains are mobile. In other words, the saltating layer is self-governing. An increase in surface stress will produce a momentary increase in the number of saltating particles. As a result there is a momentary increase in the transfer of momentum to the surface layer and the surface stress is restored to its initial value. Using this hypothesis as a starting point, Owen develops an expression for the mass flux in the saltating layer. The analysis is based on a detailed consideration of the motion of the particles and the velocity inside the saltating layer. Kind (1976) develops the same expression by means of a simple argument that clearly shows the implication of Owens second hypothesis. Kind considers a two-dimensional situation only and furthermore assumes the flow is steady; that is, the flow has passed over a sufficiently long upstream fetch so that equilibrium saltation conditions are established. If collisions between saltating particles are neglected then a typical trajectory is shown in Fig. 14. Such a particle will gain an amount of horizontal momentum equal to that lost by the fluid. If it is assumed the initial horizontal velocity of the particle, u_1, is zero and the horizontal velocity just before impact is u_2, then the momentum lost by the fluid per unit length in the downstream direction will be u_2/l. If the total mass flux of particles per unit width perpendicular to the flow is G (g/msec) then the corresponding expression is Gu_2/l. Therefore the shear stress per unit width carried by the saltating particles is

$$\tau_i = Gu_2/l. \tag{36}$$

The total shear stress τ exerted by the two-phase flow on the snow surface consists of a component τ_b exerted directly by the air and τ'. Thus

$$\tau_b = \tau - \tau_i. \tag{37}$$

According to Owen's hypothesis τ_b must remain at the impact threshold value τ_i ($\beta = 0.0064$) so that the snow surface grains are in a mobile state. If τ_b is less than τ_i, no particle can be picked up by the wind to maintain the saltation process. If τ_b is momentarily greater than τ_i, more particles will be picked up by the wind, which will increase τ_i and tend to decrease τ_b. Thus the saltation process is seen to be governed by a self-maintaining mechanism and the ability of a given flow to transport particles by saltation depends on the amount by which the shear stress outside the saltation layer exceeds the threshold value.

If Eqs. (36) and (37) are combined and τ_b is set equal to τ_i then

$$Gu_2/l = \tau - \tau_i$$

or, since $\tau = \rho u_*^2$,

$$Gu_2/l = \rho(u_*^2 - u_{i*}^2) \tag{38}$$

Kind then assumes the saltating particles experience a constant horizontal acceleration and that lift forces and drag in the vertical direction can be ignored. An energy balance on a saltating particle shows that

$$u_2/l = g/v_1, \tag{39}$$

where v_1 is the vertical velocity at the beginning of the trajectory. Kind further assumes that near the surface v_1 is of the same order as u_*, or

$$v_1 = \alpha u_*. \tag{40}$$

If Eqs. (39) and (40) are used in Eq. (38), the following expression for the mass flux G is obtained:

$$Gg/\rho u_*^3 = \alpha(1 - u_{i*}^2/u_*^2). \tag{41}$$

Owen (1964) deduced an expression for the parameter α by using measurements of the mass flux obtained for blowing sand. He found

$$\alpha = (0.25 + \omega/3u_*).$$

Equation (41) has been compared with the measurements of mass flux obtained in blowing snow by Kobayashi (1973) for various combinations of u_{i*} and ω. Figure 16, reproduced from Kind (1976), illustrates the comparison. A fall velocity of 0.75 m/sec corresponds to an average particle size of approximately 0.025 cm. Given the large variation in fall velocity ω and the threshold shear velocity u_{i*} possible in snow, Fig. 16 is a rea-

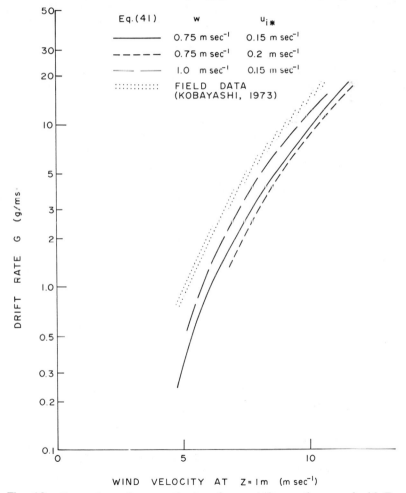

Fig. 16. Comparison of measured rates of snow drift near the ground with Eq. (41) [Reprinted with permission from *Atmos. Environ.* **10** (Kind, 1976). Copyright 1976, Pergamon Press Ltd.]

sonable verification of Owen's second hypothesis for snow. Radok (1968) points out that a further confirmation of Owen's idea that the saltation layer is self-maintaining can be found in Fig. 10. Extrapolation of the curves for various levels to very large wind speeds indicates a trend towards a uniform drift concentration of a magnitude 10^3 g/m³ (similar in magnitude to the air density). Figure 10 also shows that the rate at which drift concentration increases with velocity decreases closer to the surface. Therefore it can be implied that the surface drift concentration is of the same order of magnitude regardless of the wind speed. This is borne out by Eq. (41) and Fig. 16. On the basis of this evidence Radok suggests that the process of net accumulation or ablation of drift snow takes place by a net vertical transport through the self-governing saltation layer which maintains itself on the top of the rising or descending snow surface. Therefore, the factors controlling deposition or erosion must result in changes in the mass flux above the saltation layer which is largely governed by turbulent suspension. Owen's model of saltation implies that the transition from saltation to turbulent suspension will occur when β ($= \tau_0/\rho_s g d$) is approximately equal to unity. In other words, the hydrodynamic force on a particle at the surface is of the same order of magnitude as the weight of the particle. Radok has examined the transition between saltation and turbulent suspension and postulates that the suspension mechanism becomes important once the vertical component of the turbulent air velocity exceeds the particle fall velocities with sufficient frequency to move a substantial number of particles upward against gravity at any time. He suggests that the onset of suspension should be placed in the range of horizontal wind velocity where the rate of increase of upward eddy currents greater than the fall velocity reaches a maximum. Radok expresses this criterion in mathematical form by applying the expression for the vertical velocity w developed by Panofsky and McCormick (1960). These authors show that w can be represented by a normal frequency distribution having a variance w^2 proportional to u_*^2. The criterion is applied to snow, sand, and silt. For snow, Radok uses a fall velocity of 0.3 m/sec and a value of 1.5 m/sec for sand. For silt having a mean diameter of 0.028 mm, a fall velocity of 8 cm/sec is assumed. Figure 17 shows the results of Radok's calculation and gives a clear indication of the difference in behavior of the three materials. In the case of silt when the 10-m velocity reaches 4 m/sec the proportion of eddy currents exceeding the particle fall velocity is 70%. For sand, the increase is very gradual and does not reach the 50% level until u_{10} is near 37 m/sec. Thus, for the most common range of wind velocities, sand is predominantly in saltation. Snow, on the other hand, undergoes a transition from saltation to suspension in a range

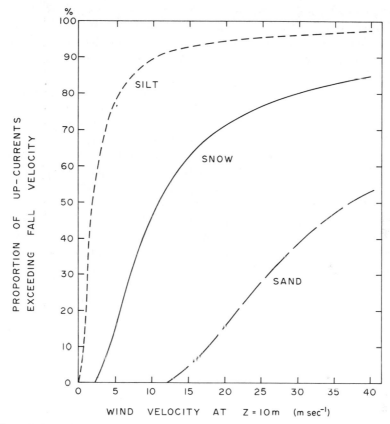

Fig. 17. Cumulative frequency of upward eddy velocities capable of supporting suspended particles of silt, snow and sand. [From Radok (1968)].

of u_{10} from 4 to 12 m/sec, velocities that are frequently encountered in nature.

Figure 17 further emphasizes the need for direct measurements of the particle size distribution near the snow surface. Figure 17 is based on a fall velocity of 0.3 m/sec, which is representative of the measurements in the Antarctic reported by Budd *et al.* (1966). On the other hand, the calculations of Fig. 16 [Kind (1976)] assume fall velocities from 0.75 to 1 m/sec. Clearly, both figures are sensitive to this parameter. However, regardless of the numerical value chosen for fall velocity, Fig. 17 clearly supports Radok's contention that the transition from saltation to suspension is gradual as the wind velocity increases.

C. Dyunin's Dynamical Theory

Beginning in the early 1950s a theory of blowing snow has developed independently in Siberia. The principal work was carried out by Dyunin, who has named his analysis "the dynamical theory" because the starting point for the work is the set of continuity and momentum equations that can be written for a two-phase flow. Much of this work has not been translated from the Russian, but, from what is available [Dyunin (1945, 1959, 1967)], it is evident that a great deal of experimental data has been accumulated on drifting snow and that the dynamical theory has been used as a framework within which successful design criteria have been developed for the control of snow drifting through various types of snow fences and shelter belts.

The dynamical theory considers the processes that occur in the layer up to 40–50 cm above the snow surface and is therefore mainly concerned with the saltation process. Radok (1977) has presented a review of the basic arguments behind this analysis based on a partial translation of Dyunin (1974). Radok suggests that the momentum equations used as a starting point for Dyunin's analysis are of secondary importance and that, for consideration of the density profiles in blowing snow or the total snow transport, the important consideration is the manner in which the concentration of blowing snow at a level above the surface is related to the mean flow parameters. Dyunin puts forward the hypothesis that the entrainment of snow particles takes place as a result of a local reduction in pressure resulting from the formation of small, intense vortices directly above the snow surface. The vortices are unstable, and continuously form, increase in size, and disintegrate in the surface layer. Radok (1977) states this pressure deficit has the form $\rho u_1^2/2$, where u_1 is the horizontal wind velocity and the vertical gradient is $\rho u_1 \, \partial u_1/\partial z$. One of the implications of this approach is that the concentration of drift particles should reach a maximum near the surface at a height less than the roughness length z_0. The detailed measurements of Kobayashi (1973) do not show such a maximum down to the 1-mm level.

Perhaps the most striking feature of Dyunin's analysis is the attention paid to the sublimation of individual snow particles in the air stream. He suggests [Dyunin (1959)] that the rate of sublimation is small when compared to the mass flux of the blowing snow and therefore need not be considered in an analysis of the deposition or erosion process near obstructions to the flow. But total losses that occur during a snowstorm are significant and can be of the order of several millimeters in a 24-hr period depending on the air temperature and humidity. This aspect of blowing snow has been examined in detail in the Russian literature but is, as yet, not translated.

D. Sublimation from Blowing Snow

One major investigation of sublimation from blowing snow is that of Schmidt (1972), which examines all the factors contributing to the sublimation of a single blowing particle. Schmidt begins his analysis from the classical equations describing heat and mass transfer from a sphere which have the form

$$(dm/dt) = 2\pi Dr(\rho - \rho_r)Sh,$$
$$L_s(dm/dt) = 2\pi kr(T_r - T)Nu. \tag{42}$$

In these equations (dm/dt) is the rate at which a spherical particle of radius r loses mass, ρ_r the density of water vapor at the surface of the sphere, and T_r the surface temperature. ρ and T are the corresponding values of these properties in the ambient air stream, D is the diffusion coefficient for water vapor in still air, and k is the thermal conductivity of the air. L_s is the latent heat of sublimation, and Sh and Nu are dimensionless groups, the Sherwood number and Nusselt number, respectively. When these numbers are unity Eqs. (42) reduce to the forms that apply to sublimation in still air. Experiments by Thorpe and Mason (1966) show that, for ice spheres in the range of diameters from 0.6 to 3.6 mm and at a relative velocity V between the air and the particle (ventilation velocity) from 0.25 to 1.0 m/sec,

$$Nu \simeq Sh = 1.88 + 0.58\sqrt{Re}. \tag{43}$$

Re is the Reynolds number (rV/ν), where ν is the kinematic viscosity for air. Equation (43) was developed for a range $10 < Re < 200$. By combining Eq. (42) with the Clausius–Clapyron equation to describe differences in the water vapor density Thorpe and Mason show that

$$dm/dt = 2\pi r\sigma/[f(Nu) - f(Sh)], \tag{44}$$

where

$$f(Nu) = (L_s/kT\ Nu)(L_sM/RT - 1)$$

and

$$f(Sh) = 1/D\rho_{sT}\ Sh$$

In these expressions R is the universal gas constant, M the molecular weight of water, and ρ_{sT} the saturation density of water vapor at the temperature T. σ is the quantity $\rho/\rho_{sT} - 1$ and represents the undersaturation or vapor deficit of the ambient air with respect to water vapor. Schmidt (1972) applies Eqs. (44) and (43) to a single ice particle of varying diameter and his results are shown in Fig. 18. The sublimation rate doubles for a 10°C rise in temperature for the temperature range examined and increases by more than a factor of two when the particle size is doubled. Equation (44) shows that the sublimation rate is directly proportional to σ, the vapor pressure deficit, and therefore a decrease in relative humidity from 90 to 80% will also double the rate of sublimation.

Figure 18 shows a range of particle diameters larger than is normally

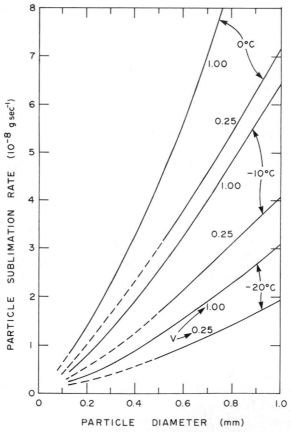

Fig. 18. Effect of ambient temperature on the sublimation rate of a single spherical ice particle. Calculations are made at an atmospheric pressure of 100 kPa and 90% relative humidity. The dashed lines represent extrapolations below a Reynolds number of 10. V is the ventilation velocity in meters per second. [From Schmidt (1972)].

encountered in blowing snow. Budd (1966b) reports mean particle sizes in the range 0.165–0.087 mm, depending on the height above the snow surface. However, the application of Eq. (40) is limited by the range of Reynolds numbers examined by Thorpe and Mason in developing Eq. (43).

Schmidt has extended his analysis to consider solar radiation as an additional energy source for evaporation. In this case the heat transfer Eq. (42) is altered to the form

$$L_s \, dm/dt = 4\pi k r (T_r - T) + Q, \qquad (45)$$

where Q is the rate at which radiation is transferred to the ice particle. Using this equation Schmidt derives an expression for the particle sublimation rate analogous to Eq. (44). He assumes that the total radiation flux intercepted by a particle consists of a direct component from above and a reflected component from the underlying snow surface. Using a snow surface albedo of 0.8 and a particle albedo of 0.5 Schmidt shows that the rate of sublimation increases by 18% for a sphere having a diameter of 0.1 mm under conditions similar to those of Fig. 18. The incident radiation was assumed to be 700 W/m^2, a value near the maximum possible for a clear day during the summer months at an elevation of 500 m.

At higher elevations and lower atmospheric pressure, sublimation rates increase for a given particle size. Two factors lead to this increase. The diffusivity of water vapor increases with a decrease in pressure and Schmidt includes this factor in the model by assuming a variation of the form [List (1966)]

$$D/D_0 = (T/T_0)^n (p_0/p),$$

where D_0 the diffusivity at $p_0 = 100$ kPa was taken as 0.18 cm^2/sec.

The second factor is the increase in solar radiation at higher elevations due to the decrease in the thickness of the atmosphere. Kuz'min (1961) mentions various field studies that show the direct solar radiation on a surface perpendicular to the sun's rays increases between 35 and 140 W/m^2 for every kilometer of rise in elevation. A third factor that tends to counteract the other two is a drop in the Nusselt and Sherwood members with elevation for the same free stream velocity. This is caused by a decrease in the Reynolds number Re brought about by a drop in air density with elevation. However, the net effect is one of increasing sublimation rates with elevation as shown in Fig. 19.

In order to apply the expressions derived for a single ice particle to sublimation in blowing snow it is necessary to calculate the total sublimation rate for a volume of air containing many particles. Schmidt shows that for the particle concentrations or drift densities occurring in blizzards any interaction between particles will be insignificant except in the first

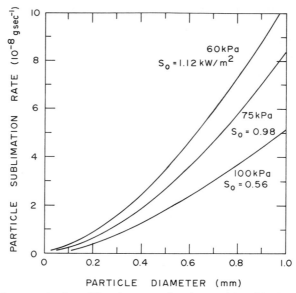

Fig. 19. Increase in the sublimation rate of an ice particle with elevation. For each curve an air temperature of $-20°C$, 90% relative humidity, and a ventilation velocity of 1 m/sec are assumed. The 75-kPa curve corresponds to an elevation of 2.5 km, the 60-kPa curve to an elevation of 4 km. [From Schmidt (1972)].

few millimeters above the snow surface, and therefore the sublimation rates derived from Eq. (44) or the analogous expression that includes the effects of solar radiation can be applied directly to each particle in the volume. Hence, the sublimation rate for all particles having nominal diameters between x and $x + dx$ in a unit volume of air can be expressed by an equation of the form

$$(dM/dt)_x = (dm/dt)_x \, Nf(x) \, dx, \tag{46}$$

where $(dM/dt)_x$ is the total sublimation rate for all particles having a diameter x while $(dm/dt)_x$ is the corresponding rate of sublimation for a single particle; N is the total number of particles in the volume and $f(x)$ represents a frequency distribution function. Schmidt uses the gamma function first used by Budd (1966b) for $f(x)$. The total sublimation rate for the volume can then be found by integrating Eq. (46) over the range of particle diameters,

$$dM/dt = N \int_0^\infty (dm/dt)_x \, f(x) \, dx. \tag{47}$$

In applying this equation to a column of snow Schmidt uses Eq. (29) to express the variation in drift density with height for particles of a given size.

If the above ideas are applied to a vertical column of blowing snow 10 m high and having a 1-m^2 cross-sectional area, the result is a sublimation rate of 0.28% per second for an air temperature of $-20°C$ and a constant relative humidity of 90% throughout the column. The wind velocity at a height of 10 m was assumed to be 15 m/sec.

Calculations of this type are perhaps premature given the many assumptions that still require verification. Nevertheless, they show that sublimation from blowing snow is a subject worthy of future investigation. Recently an extension of Schmidt's work has been completed by Lee (1975). He shows that the assumption of a relative particle velocity equal to the free fall velocity is questionable and, if the effect of atmospheric turbulence is considered, the overall sublimation rate is increased by 15–40% over the range of wind velocities encountered in blowing snow. Furthermore, Lee concludes that the mechanics of the sublimation process is dominated by turbulent processes in that the relative velocity governing sublimation is induced primarily by turbulence rather than the fall velocity.

E. Accumulation and Erosion

Snow deposition or erosion in the vicinity of an obstacle is usually explained in terms of eddies created by the obstacle or variations in the shear stress, although Radok (1968) argues that these processes should be explainable in terms of minor variations in the horizontal drift flux past the obstacle. Theoretical work on boundary layer flow around obstacles is not sufficiently advanced to provide much guidance for the control of drift accumulation in the vicinity of buildings or in other practical situations. The central problem is the distribution of drifting snow with height. If the bulk of the snow particles are near the ground then buildings raised above the surface should allow the snow to pass without major drift formation. However, Radok (1968) shows that the drift that formed upwind of an elevated building in Greenland was influenced by the wind to the top of the building, a height of 30 m above the surface.

The development of techniques for controlling snow erosion and accumulation has proceeded independently of more fundamental studies on the mechanics of blowing snow. Mellor (1965) provides a good summary of this work. The recent work of Tabler (1975) has led to a successful model for calculating annual amounts of drifting snow and is of particular interest in that estimates of evaporative losses are considered. Tabler proceeds from the work of Schmidt (1972) and Lee (1975) and shows that their expressions for the instantaneous rate of sublimation can be approximated by

$$dm/dt = kk_1^{-0.47}m^{0.47}, \qquad (48)$$

where m is the mass of the particle, k is a negative constant, and $k_1 = \rho_s \pi/6$, with ρ_s the density of the particle. This equation can be integrated to give the mass of a particle at any time t and, by introducing a reference time t_{s1}, which is the time required for a reference particle (initial mass m_{01} and diameter x_{01}) to completely sublimate, Tabler shows that the integration leads to an equation of the form (for $x_{0i} \geq x_{01}$)

$$m_i = k_1\{x_{0i}^{1.6} - x_{01}^{1.6}\}^{1.89}, \qquad (49)$$

where m_i is the residual mass at a time t_{s1} for any particle with an initial diameter x_{0i} larger than x_{01}. If the gamma function introduced by Budd (1966b) is used to describe the particle size distribution in the blowing snow then Eq. (49) can be integrated to give the total residual mass of snow particles in the blowing snow at a time t_{s1}. This leads to an expression for the ratio of residual mass M to initial mass M_0 in a volume of air. Tabler shows that this expression can be approximated by

$$M/M_0 = e^{-2(R/R_u)}, \qquad (50)$$

where R is the distance over which the blowing snow has been transported and R_u the distance required for a particle of average size to sublimate. The equation suggests the residual mass is about 39% for $R/R_u = 0.5$, and 11% for $R/R_u = 1.0$.

Equation (50) is used as a weighting factor to determine the relative importance of an increment of distance dr with respect to its contribution to the total amount of blowing snow passing some point downwind. dQ is defined as the water equivalent of blowing snow per unit width perpendicular to the wind contributed by a surface of width dr located at a distance R upwind of the point of interest, and dQ_0 is the initial volume of snow blown off dr and is expressed as

$$dQ_0 = P_r \, dr, \qquad (51)$$

where P_r is the water equivalent precipitation blown from dr. For example, in the calculation of annual transport, P_r is the total winter precipitation less the melt and the amount held in place by vegetation and terrain irregularities. Using the above ideas, Eq. (50) becomes

$$dQ = P_r \, e^{-2(R/R_u)} \, dr$$

or

$$dQ = P_r(0.14)^{(R/R_u)} \, dr. \qquad (52)$$

The total snow transport contributed by some finite distance ΔR can then be found by integration of Eq. (52). For relatively simple situations where P_r and environmental conditions are uniform over the entire contributing distance R_c downwind of a barrier across which there is a negligible snow transport (for example, the margin of an extensive forest), Tabler shows that the integrated form of Eq. (52) is

$$Q = (P_r R_u/2)[1 - (0.14)^{R_c/R_u}].\qquad(53)$$

The total evaporation loss Q_L would be

$$Q_L = P_r P_c - Q.\qquad(54)$$

Based on extensive field tests in Wyoming, Tabler concludes that the best estimate of R_u is 3050 m. For situations in which the upwind condi-

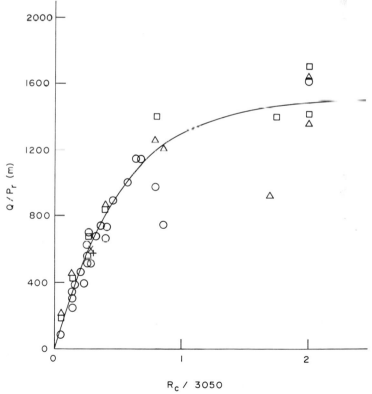

Fig. 20.   Measured annual snow transport water equivalent Q normalized with respect to the relocated precipitation P_r, for R_u = 3050 m. The curve is a plot of Eq. (55): +, 1969–1970; ×, 1970–1971; □, 1971–1972; △, 1972–1973; ○, 1973–1974. (From Tabler, 1975).

tions are not uniform the distance R_c is divided into increments $\Delta R = R_i - R_{i-1}$ that are relatively uniform, and the expression corresponding to Eq. (53) is

$$Q = \sum_{i=1}^{n} \int_{R_{i-1}}^{R_i} (P_r)_i (0.14)^{\Delta R/R_u} \, dr. \tag{55}$$

Figure 20 summarizes the field data collected by Tabler and compares the measurements with this expression. Tabler attributes the scatter with increasing distance to cumulative errors in determining the terrain variables that govern the increments ΔR and errors in determining the wind direction that can lead to significant changes in P_r. Given the extreme difficulty of acquiring accurate field data, Fig. 20 represents a striking confirmation of Tabler's model. Of equal interest is the loss of snow due to sublimation predicted by this analysis. Figure 21 is a plot of Eq. (54) and suggests that such losses are significant. For the Wyoming sites from which the data of Fig. 20 were collected, Tabler suggests the evaporation loss represents approximately 33% of the total winter precipitation.

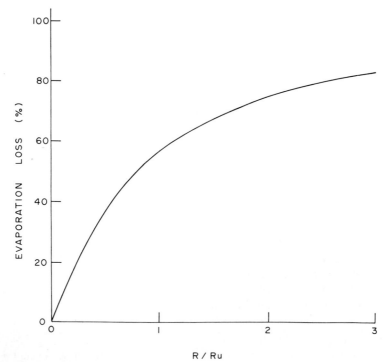

Fig. 21. Percentage of relocated precipitation P_r lost to evaporation. [From Tabler (1975)].

V. SNOWMELT

The rate at which snow melts is governed by the energy balance at the upper surface of the snowpack. Two methods are used in the description of this balance. One approach considers the vertical components of the energy and mass fluxes at the snow–air interface and has the following general form [Kraus (1973)]:

$$\Sigma \, Q_i + \Sigma(mh)_i = 0, \tag{56}$$

where Q_i [kW/m²] represents an energy flux due to radiation, sensible and latent heat transfer (convection), and heat transfer from the lower layers of the snowpack, which is normally assumed to be governed by the effective conductivity k_e. The product $(mh)_i$ represents an energy transfer due to precipitation m (kg/m² sec) at the upper surface, which has associated with it a specific enthalpy h (kJ/kg) determined by the temperature of the precipitation.

In the second approach the entire snowpack is considered as a control volume to which energy can be transferred by radiation, convection, and conduction and across whose boundaries (snow–air and snow–ground interface) mass fluxes in the solid, liquid, or vapor phases are possible. In this case the energy balance takes the form

$$\frac{dU}{dt} = \Sigma \, Q_i + \Sigma(mh)_i. \tag{57}$$

The symbols have the same meaning as in Eq. (56), although Q_i now includes heat transfer between the ground and the snow and $(mh)_i$ includes melt water draining from the bottom of the pack. dU/dt is the change in internal energy or stored energy in the pack and here is written on the basis of a unit area (kJ/sec m²). Kraus (1973) develops the relationship between Eqs. (56) and (57) and shows that internal energy changes result from changes in snowpack temperature or, in the case of an isothermal pack at 0°C, from changes in the relative proportions of ice, liquid water, and, to a minor extent, water vapor.

Equation (56) has been applied to the deeper snowpacks found in mountainous areas and on glaciers since it only requires measurements at or near the upper snow surface. Equation (57) is required to describe completely the thermal regime of a snowpack and is of practical use for the shallow snowcover (less than 40 cm in depth) found in most temperate and arctic regions. For shallow snow, internal energy can be described in terms of average temperature and density in the pack. Equations (56) and (57) can be applied only where horizontal advection is negligible. The snowcover must be continuous and away from open water or significant areas of bare ground. A discontinuous snowcover is considered later.

The remainder of this section contains a description of the major factors that influence each of the energy fluxes important to metamorphism and melt.

A. Radiation

The radiation balance at the earth's surface has been discussed by many authors [e.g., Sutton (1953); Geiger (1966); Kondratyev (1969)]. Traditionally, the net all-wave radiation balance Q_n is considered to consist of a short-wave and a long-wave component and may be written

$$Q_n = S_n + L_n = S_i - S_0 + L\downarrow - L\uparrow, \qquad (58)$$

where S_n is the net shortwave flux and is the difference between the incident or global radiation S_i and the portion of S_i reflected at the snow surface S_0. The global radiation in turn consists of direct solar radiation I reaching the earth's surface and a diffuse component D consisting of solar radiation that has been scattered by air, dust, and other particles in the atmosphere. D also consists of solar radiation reflected from the snow surface and backscattered by the atmosphere. Global radiation is present only during the daytime and practically all of its energy is in the range of wavelengths from 0.3 to 3.0 μm.

The downward atmospheric radiation $L\downarrow$, or counterradiation, originates from ozone, water vapor, carbon dioxide, and nitrogen compounds in the atmosphere that have absorbed solar radiation. $L\uparrow$ is the terrestrial radiation emitted at the earth's surface according to the Stephan–Boltzman law. The net long-wave radiation L_n is the difference between $L\downarrow$ and $L\uparrow$ and falls in the wavelengths from 3.0 to 80 μm.

1. Short-Wave Radiation

The daily amount of direct-beam radiation reaching the snow surface is a function of latitude, time of year, slope, orientation, and cloud cover. One of the simplest methods of analyzing the interactions between these factors has been developed by Garnier and Ohmura (1970). They express the direct clear-day radiation I (kW/m²) falling on a slope as

$$I = \int (I_0/r^2)p^m \cos(X\Lambda S) \, dH, \qquad (59)$$

where I_0 is the solar constant, r is the radius of the earth's orbit, p is the mean or overall transmissivity of the atmosphere along the zenith path and is a measure of the fraction of the solar radiation that reaches the earth's surface without being absorbed, m is the optical air mass which is the ratio of the distance the sun's rays travel through the atmosphere to the depth of the atmosphere along the zenith path, $\cos(X\Lambda S)$ is the cosine of the angle of incidence of the sun's rays on the slope, and H is the hour angle measured from solar noon, the integral being taken over the duration of sunlight on the slope.

Integration of this expression to obtain daily totals of direct-beam radiation over a 12-month period yields the well-known distribution shown in Fig. 22. Note the rapid increase in I during March and April, the melt period over much of the northern hemisphere. For example, on 1 March at a latitude of 50° N, I is 10.2 MJ/m². The corresponding value for 1 April is 17.8 MJ/m². Since solar radiation is a major component of the total energy input to snow, time of year has an important influence on the melt rate. This fact is well known to hydrologists involved in flood forecasting. As a rule, the longer the spring melt is delayed, the greater the danger of flooding.

The transmissivity p is highest in winter and lowest in summer because the atmosphere contains more water vapor and dust during the summer months. It also varies somewhat with latitude, increasing towards the

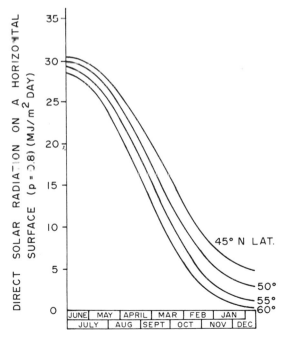

Fig. 22. Daily totals of direct-beam solar radiation on a horizontal surface ($p = 0.8$). [From Male and Granger (1979)].

north. Kondratyev (1969) reports a mean annual value of p at Pavlosk near Leningrad (59°41' N) of 0.745 with a deviation of ± 0.05 for the years 1906–1936. Williams *et al.* (1972) suggest values ranging from 0.85 to 0.75 for February and March at Benson, U.K. (51°30' N), and Paris (48°49' N). Kuz'min (1961) recommends a value of 0.8 ± 0.05 for the snowmelt period within the European portion of the USSR.

Slope and orientation also influence the direct-beam flux reaching the snow surface, the intensity being greatest on south-facing slopes in the northern hemisphere and north-facing slopes in the southern hemisphere. In general, the influence of these factors is a maximum at the winter solstice (in the northern hemisphere) and a minimum at the summer solstice. For example, on 1 March at 50° N lat, a 10° south-facing slope receives approximately 2.5 times as much direct-beam radiation as a 10° north-facing slope. The corresponding ratio on 1 April is 1.5.

In mountainous terrain the computation of direct solar radiation is complicated by the shadows that are present at the point being considered. Methods for considering this effect have been developed by Williams *et al.* (1972) and Dozier (1979). The region being considered is divided into a uniform grid of known elevation at all grid points. By considering the elevation at each grid point and searching along the line of the sun's azimuth for points high enough to block the sun, the grid points in shadow at any hour angle may be found.

The presence of a forest canopy further complicates the computation of solar radiation. The forest canopy is a heterogeneous, anisotropic medium that absorbs, scatters, and reflects the direct-beam radiation and emits long-wave radiation. The amount of solar radiation reaching the snow surface is reduced by the shading effect caused by the trunks and the canopy. This effect is normally considered by multiplying I by the factor of $1 - V$, where V is a beam shading function that changes according to the solar angle. Dozier (1979) describes a complete solar radiation model that includes a shading function.

The diffuse component of the short-wave radiation reaching the snow surface is less amenable to mathematical computation. On clear days it may amount to only 10% of the total short-wave flux, but on cloudy or overcast days the percentage is much higher. The amount of energy received on a horizontal surface because of the scattering of incoming solar radiation depends mainly on the position of the sun, the concentration of aerosol particles, and changes in air density. The simplest expression for calculating this flux is given by List (1966, p. 420) and attributed to Fritz, and estimates the diffuse radiation D_0 on a horizontal surface from the expression

$$D_0 = 0.5(1 - a_w - a_0)I_t - I \quad \text{kW/m}^2, \tag{60}$$

where a_w is the radiation absorbed by water vapor (assumed to be 7%) and a_0 is the proportion absorbed by ozone (assumed to be 2%). The term I_t is the extraterrestrial radiation where

$$I_t = \frac{I_0}{r^2} \int \cos z_s \, dH$$

and z_s is the zenith angle of the sun. The factor of 0.5 in Eq. (60) expresses the assumption that half the radiation is scattered towards the surface and half away from the surface. Empirical expressions for calculating D_0 are also presented by Stanhill (1966) and Liu and Jordan (1960). In both cases it is necessary to have measurements or estimates of the global radiation on a horizontal surface. D_0 can also be calculated from a consideration of the physics of the scattering process. Dozier (1979) presents a computing scheme in which the amount initially scattered out of the beam is determined by aerosol and Rayleigh scattering coefficients, ozone and water vapor absorption coefficients, all of which are functions of wavelength. Models of this type are complicated and require among other things an estimate of the ozone and water vapor distribution with altitude and variations of air pressure with altitude.

A second source of diffuse radiation to the snow surface is the backscattered radiation or the portion of the global radiation reflected at the snow surface which is redirected downward by subsequent scattering and reflection in the atmosphere. Kuz'min (1961) suggests that the increase in scattered radiation over snow as opposed to bare ground ranges from 65% when the sun is near the horizon to 12% at solar elevations in the range 35–55°. Models that account for backscattering have been developed by Hay (1976) and Dozier (1979). The critical parameter in these types of calculations is the reflectance or albedo of the snow surface. Albedo is considered later in this section.

For a sloping surface without surrounding obstructions the diffuse component can be calculated from the expression [Kondratyev (1969)]

$$D = D_0 \cos^2(\theta/2),$$

where θ is the acute angle of the slope in question. This expression assumes the diffuse radiation is isotropic. In mountainous regions a given slope is likely to have a decreased sky dome because of the surrounding topography. The factor by which the downward-scattered radiation is reduced can be calculated by assuming the radiation intensity is constant from all parts of the sky dome and integrating over the unobscured portion. For a given slope the reduction factor is constant and need be calculated only once. Details of this calculation are described by Obled and Harder (1979) and Dozier (1979).

An additional source of diffuse radiation which must be considered in a mountainous environment is the radiation reflected from the obscured portion of the sky dome: radiation that has been reflected to a slope from adjacent terrain of different slope and orientation. This flux can arise from the reflection of direct-beam radiation or from the reflection of the incoming diffuse radiation. For snow the reflection from the direct beam is assumed to be isotropic as specular reflectance is rare. This source of radiation is computed in the model developed by Dozier (1979) but can be considered a special case of the reflection of short-wave radiation from snow. This flux is described in the next section.

Although most calculations of diffuse radiation are made on the assumption that this flux is isotropic, Kondratyev (1969) shows that clear-sky diffuse radiation is significantly anisotropic. The intensity is greatest in the vicinity of the sun and near the horizon because of the greater optical thickness of the atmosphere at large zenith angles. The angular distribution of diffuse radiation has been examined by Steven (1977) and Temps and Coulson (1977). The correction factor developed by Temps and Coulson are used in the computing scheme of Dozier (1979). When one considers that in mountainous regions anisotropic diffuse radiation can be reflected from surrounding surfaces the picture becomes very complicated indeed. A discussion of this phenomenon is given by Obled and Harder (1979).

2. Albedo

A large portion of the short-wave radiation incident to the snow surface is reflected. The ratio of the reflected radiation to the incident flux is referred to as the albedo A. Thus

$$A = S_0/S_i = S_0/(I + D).$$

The short-wave radiation absorbed by a snow surface S_n is often determined as

$$S_n = (1 - A)S_i = (1 - A)(I + D). \tag{61}$$

The reflectance r of snow varies with wavelength and albedo may be considered the integrated reflectance over the short-wavelength spectrum

$$A = \int_{\lambda_1}^{\lambda_2} rI(\lambda) \, d\lambda \Big/ \int_{\lambda_1}^{\lambda_2} I(\lambda) \, d\lambda. \tag{62}$$

In Eq. (62), $I(\lambda)$ and $rI(\lambda)$ are the incident and reflected monochromatic intensities at the wavelength λ. The limits of integration are in the range 0.3–3 μm and are determined by the characteristics of the measuring instrument.

When short-wave radiation strikes a snow surface a fraction is reflected and a fraction passes into the snow. The intensity of the transmitted radiation decreases with distance below the snow surface and is often assumed to attenuate exponentially. The attenuation process is due to absorption and scattering with scattering being the dominant process. Bohren and Barkstrom (1974) have investigated this process in detail. Scattering occurs primarily from the grain boundaries of the snow matrix, although subsurface backscattering makes some contribution to the albedo. Giddings and LaChapelle (1961) show that the reflectance at $0.6 \mu m$ becomes independent of thickness for depths greater than 10–20 mm. Mellor (1966b) suggests that from 20 to 60% of the reflected intensity for snow in the density range 370–620 kg/m³ could be attributed to scattering from below the surface.

Spectral reflectance r has been measured by various investigators [Mellor (1977)] but there are considerable differences among the sets of data. Albedo is known to be a function of grain size and density, whether the surface is illuminated by diffuse radiation or direct sunlight, and the roughness of the surface. Bohren and Barkstrom (1974) show that albedo will be inversely proportional to the square root of the grain size and that there should be a slight decrease of r with increasing wavelength. In general, the spectral reflectance of fresh low-density snow is high and, once the snow has thawed, reflectance decreases throughout the short-wavelength range. Refreezing causes virtually no change in the values. In addition, small amounts of impurities are known to have an influence on the albedo.

In the field, the albedo is a major determinant of the amount of radiant energy absorbed by the snow and hence is an important factor in the determination of the rate at which the pack melts. In general, albedo decreases with time from the last snowfall, although the rate of decay differs for deep and shallow snowpacks. In Fig. 23 the curve for the mountain snowpack is taken from the U.S. Army Corps of Engineers (1956) and the curve for shallow snow from measurements made at an elevation of approximately 1000 m within the wavelength range 0.2–1.2 μm [O'Neill and Gray (1973)]. Clearly, as the snowcover becomes shallow the underlying surface influences the albedo.

3. Long-Wave Radiation

The net longwave radiation L_n at the snow surface is the difference between the downward atmospheric radiation $L\downarrow$ and the upward flux from the snow surface $L\uparrow$. Over a snow surface $L\uparrow$ is normally greater than $L\downarrow$ and the net long-wave radiation represents a loss of energy from the snow.

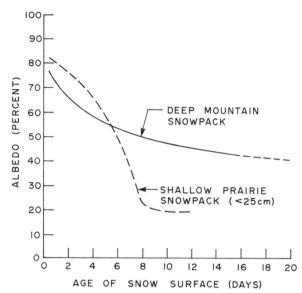

Fig. 23. Decay of the spatial albedo with the age of snow for prairie and mountain snowpacks. [From O'Neill and Gray (1973)].

Downward atmospheric radiation is emitted from all levels but Geiger (1966) shows that under clear-sky conditions the largest portion reaching the earth's surface originates in the lowest 100 m. Radiation is emitted by ozone, carbon dioxide, and water vapor, ozone contributing approximately 2% of the total and carbon dioxide approximately 17%. The remainder comes from water vapor and variations in $L\downarrow$ are largely due to the variations in the amount and temperature of the water vapor. Many investigations have shown there is a correlation between the downward-directed long-wave radiation and the air temperature and vapor pressure at a height of 1.5–2 m above the surface. The most widely quoted expression is that of Brunt (1952), which has the form

$$L\downarrow = \sigma T_a^4[a + b\sqrt{e}], \tag{63}$$

where σ is the Stephan–Boltzmann constant (5.67×10^{-11} kW/m² K⁴), T_a the absolute air temperature at 1.5 or 2 m (°K), e the vapor pressure of the air at 1.5 or 2 m (Pa), and a and b are empirical parameters. On the basis of long-term measurements over snow in the Russian plains and steppe regions, Kuz'min (1961) suggests Eq. (63) should take the form

$$L\downarrow = \sigma T_a^4[0.62 + 0.005\sqrt{e}] \quad \text{kW/m²}. \tag{64}$$

Kuz'min emphasizes the empirical nature of Eq. (64) and points out the parameters a and b will change according to the time of year and location on the earth's surface. Kondratyev (1969) provides convincing evidence that this flux is not a unique function of air temperature and vapor pressure near the surface. Recently, Male and Granger (1979) have shown that considerable scatter is to be expected when Brunt's equation is compared with daily values measured over a continuous snow surface.

Brutsaert (1975) has suggested a means of estimating incoming long-wave radiation which takes into consideration atmospheric temperature and humidity profiles above the surface. Assuming a constant linear temperature decrease with height, he shows that $L\downarrow$ can be estimated by

$$L\downarrow = 0.642(e_0/T_0)^{1/7}[\sigma T_0{}^4], \tag{65}$$

when T_0 is the air temperature near the ground (°K), and e_0 is the corresponding vapor pressure (Pa). In the derivation of Eq. (65) a lapse rate of $-0.006°C/m$ and a standard atmosphere at 100 kPa is used. However, Brutsaert shows that the coefficient 0.642 is relatively insensitive to these assumption. Marks (1979) has modified Eq. (65) for use in alpine areas on the assumption that the relative humidity is constant with height and temperature variations with height follow the standard lapse rate. He reports good agreement between measured values and this expression during the snow season.

The long-wave radiation emitted from the snow surface can be calculated from the Stephan–Boltzman law:

$$L\uparrow = \varepsilon_s\sigma T_s{}^4, \tag{66}$$

where T_s is the absolute temperature of the snow surface (°K) and ε_s is the emissivity. Snow is a near perfect black body in the long-wave portion of the spectrum and values for ε are very close to unity [Kondratyev (1969); Matveev (1965); Anderson (1976)].

In mountainous areas topographical variations can cause significant changes in the net long-wave radiation. In a valley, radiation from the atmosphere is reduced because a portion of the sky is obscured by the valley walls. However, the valley floor will receive radiation from these surfaces in an amount governed by a form of Eq. (66), since reflected long-wave radiation from snow and most other natural surfaces is negligible. Marks (1979) considered this effect by using a thermal view factor V_f to estimate the flux from the unobscured portion of the sky. According to Lee (1962) V_f has the form

$$V_f = \cos^2(90 - H) \tag{67}$$

with H the average horizon angle measured from the zenith. Thus, in areas of high relief the long-wave radiation incident at a point is given by

$$L\downarrow = (\varepsilon_a\sigma T_a{}^4)V_f + (\varepsilon_s\sigma T_s{}^4)(1 - V_f), \tag{68}$$

where ε_a is the effective emissivity of the atmosphere and is calculated using Brutsaert's approach [Eq. (65)]. Marks shows that under clear-sky conditions Eq. (68) gives values that are within 10–15% of the measured mean during the snow season.

4. Radiation Balance over Snow

Measurements of the radiation balance over snow show a wide variation in the individual fluxes because of such factors as cloud cover, forest canopy, and topography. For open areas of low relief the results shown in Fig. 24 are typical. The first two days of this period were clear followed by a more or less continuous cloud cover on the last day. The first day incorporates a period of fog. Because of the high albedo of snow the incoming global radiation is of the same magnitude as the reflected flux. Similarly the incoming long-wave radiation $L\downarrow$ and the radiation emitted by the snow surface $L\uparrow$ are of the same magnitude. The nearly constant surface emission is a reflection of the limited range of surface temperatures possible during the melt period. During periods of no melt, the temperature of the snow surface can drop rapidly at night because the thermal conductivity of snow is low. This prevents heat transfer from below sufficient to compensate for $L\uparrow$.

The presence of a forest canopy will obviously alter the relative magnitudes shown in Fig. 24. Reifsnyder and Lull (1965) show that in a stand of

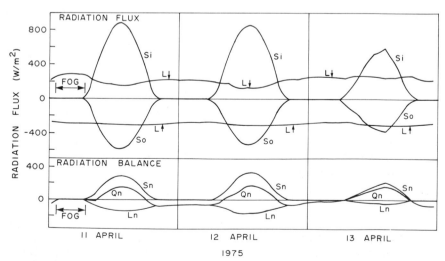

Fig. 24. The radiation balance over melting snow. [From Male and Granger (1979)].

Douglas fir the net short-wave flux to the snow decreases rapidly with an increase in canopy closure while the net long-wave flux from the forest to the snow increases. The net all-wave flux decreases with increasing crown closure until the closure is about 60%. At higher closures there is an increase in Q_n. In general Q_n is smaller in the forest than in the open. However, for a leafless deciduous forest Hendrie and Price (1979) suggest the net all-wave radiation can be significantly greater than that received in an open area.

B.   Convective and Latent Energy Exchange

The turbulent fluxes of sensible (Q_h) and latent heat (Q_e) are of secondary importance in most snowmelt situations when compared to the radiation flux. Nevertheless, on many occasions turbulent exchange plays an important role in determining the rate of melt of the snowcover. Both Q_h and Q_e are governed by processes occurring in the two or three meters of the atmosphere immediately above the snow surface. These fluxes can be defined by the expressions

$$Q_h = -C_p\rho\overline{w'T'}, \qquad Q_e = -L_v\rho\overline{w'q'}, \tag{69}$$

where C_p is the specific heat of the air at constant pressure, L_v the latent heat of vaporization for water, and ρ is the air density. w', T', and q' represent the departure of the quantities w, T, and q from their respective mean values. w is the vertical component of the wind velocity, T the air temperature, and q the specific humidity. In the development of Eqs. (69) it is assumed that the product ρw is zero [Swinbank (1951)]. This is true for a horizontal uniform surface if the period of observation is sufficiently long. Eddy correlation instruments based on Eqs. (69) have been developed to measure Q_h and Q_e [Dyer (1961)]. Hicks and Martin (1972) and more recently McKay and Thurtel (1978) have used this approach to measure the turbulent exchange processes over snow.

More commonly these fluxes are computed from the measurements of wind velocity, temperature, and humidity profiles obtained in the first few meters above the snow surface. Assuming that the sensible heat flux Q_h and the latent flux Q_e can be characterized by the gradients of the mean potential temperature θ and the mean specific humidity q, then

$$Q_h = -C_p\rho K_h \, \partial\theta/\partial z \qquad \text{and} \qquad Q_e = -L_v\rho K_e \, \partial q/\partial z, \tag{70}$$

where K_h and K_e are transfer coefficients or eddy diffusivities for sensible heat and water vapor, respectively. Equations (70) are written on the assumption that the vertical fluxes are constant with height. The region of the atmosphere to which this conditions applies is referred to as the constant flux layer. Panofsky (1974) extends this layer to approximately 30 m above the surface, but points out that this height is highly variable and is affected by the strength of the wind, the temperature gradients, and the upwind surface conditions.

The eddy diffusivities K_h and K_e are computed from the wind profiles using a procedure outlined by many investigators [Kraus (1973); Anderson (1976); Priestly (1959)]. The analysis is based on the assumption that the turbulent shear stress can be expressed as

$$\tau = \rho K_m \, \partial u / \partial z \quad \text{or} \quad K_m = u_*^2/(\partial u/\partial z), \tag{71}$$

where the ratio of shear stress τ to air density ρ is denoted u_*^2, the square of the friction velocity. In the constant flux layer u_* is independent of height and is a scaling parameter. K_m is the eddy diffusivity for momentum flux and $\partial u/\partial z$ is the gradient of the average horizontal wind velocity component. Assuming the wind profiles can be described by a logarithmic law, Q_h and Q_e can be calculated from measured wind speed, air temperature, and humidity ratios taken at corresponding heights using the equations

$$\begin{aligned}
Q_h &= -\rho C_p k^2 (K_h/K_m)(u_b - u_a)(\theta_b - \theta_a)/\ln^2(b/a), \\
Q_e &= \rho L_v k^2 (K_e/K_m)(u_b - u_a)(q_b - q_a)/\ln^2(b/a).
\end{aligned} \tag{72}$$

In these expressions a and b denote the measurement heights and k is the von Kármán constant (approximately 0.4). The potential temperature θ is the temperature that a volume of air assumes when brought adiabatically from its existing pressure to a standard pressure (generally that at the surface). θ at a given height is related to the temperature T by the expression

$$\theta_z = T_z + \Gamma z,$$

where Γ is the adiabatic lapse rate (approximately 1°C/100 m) and z is the height above ground level. The atmosphere is said to be in a neutral condition at this temperature gradient.

The ratios K_h/K_m and K_e/K_m are usually assumed to be unity for a neutral atmosphere. Deviations from neutrality are considered through application of the Monin–Obukhov similarity theory [Monin and Obukhov (1954)]. In this analysis a dimensionless velocity (ϕ_m) and temperature (ϕ_h) gradient are introduced

$$\phi_m = (kz/u_*) \, \partial u/\partial z, \qquad \phi_h = (kz/T_*) \, \partial\theta/\partial z, \tag{73}$$

where T_* is defined under neutral conditions by the expression

$$Q_h = -\rho C_p u_* T_*.$$

Under neutral conditions ϕ_m and ϕ_h are unity. Monin and Obukhov (1954) show that the dimensionless gradients ϕ_m and ϕ_h are universal functions of the dimensionless stability parameter z/L with L a scaling length first introduced by Obukhov (1946) and defined as

$$L = -(u_*^3 C_p \rho T)/(kg Q_h), \tag{74}$$

where T is the mean absolute temperature of the boundary layer and g is the gravitational acceleration. By combining Eqs. (71) and (73) it can be shown that

$$K_h/K_m = \phi_m/\phi_h. \tag{75}$$

The problem of determining the ratio K_h/K_m is equivalent to that of finding the ratio ϕ_m/ϕ_h.

A dimensionless humidity gradient ϕ_e can also be defined [Dyer (1968)], where

$$\phi_e = (kz/q_*)(\partial q/\partial z) \tag{76}$$

with q_* a characteristic humidity ratio analogous to u_*. From Eq. (76) it follows that

$$K_e/K_m = \phi_m/\phi_e. \tag{77}$$

The functional dependence of the ratios ϕ_m/ϕ_h and ϕ_m/ϕ_e on the stability parameter z/L has been the subject of many investigations and there are significant differences among the various sets of data. In the stable range (z/L positive) which is common over a melting snow surface there is general agreement that K_h/K_m is unity [Anderson (1976)]. However, for z/L greater than 0.1 (very light winds), Granger and Male (1978) report that measurements over snow suggest the ratio K_h/K_m decreases with increasing stability and is better described by the relation (Fig. 25)

$$K_h/K_m = [1 + 7(z/L)]^{-0.1}. \tag{78}$$

In arriving at this expression a log–linear profile [Webb (1970)] was fitted to the measured values of wind and temperature.

The greatest variation of results occurs in the unstable range. Figure 25 gives a comparison of various investigations. The curve based on the work of Dyer and Hicks (1970) is recommended by Anderson (1976) for use over snow. The measurements of Granger (1977) are the only data collected over a snow surface and correspond very closely with the equation proposed by Swinbank (1968). In fact Granger's work provides a reasonable extrapolation to the neutral region for Swinbank's equation. Businger et al. (1971) found a much higher value for K_h/K_m under neutral conditions but the difference is largely due to their choice of 0.35 as a

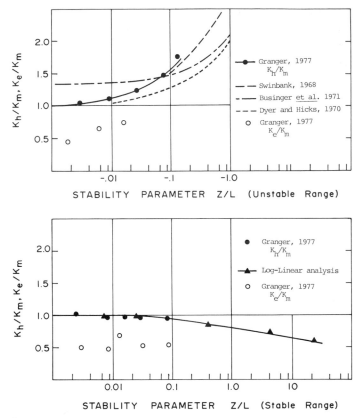

Fig. 25. Comparison of results of various investigations for K_h/K_m and K_e/K_m. [From Granger and Male (1978)].

value for the von Kármán constant k, as opposed to the usual value of 0.4. It is interesting to note that both Dyer and Hicks (1970) and Businger *et al.* (1971) used the eddy correlation technique to measure the shear stress and turbulent heat flux directly, whereas the analysis of Granger (1977) and Swinbank (1968) is based on low-level wind velocity and temperature measurements. It is possible that systematic errors are associated with one or both procedures.

Measurements of the ratio K_e/K_m over snow are relatively scarce, reflecting the greater degree of difficulty associated with the measurement of humidity profiles or fluctuations in the moisture content of the air. Mean values of K_e/K_m calculated from the measurements of Granger (1977) and grouped according to ranges of z/L are plotted in Fig. 25. Comparison of these data with measurements made over short grass and cereal crops [Högström (1974); Blad and Rosenberg (1974)] show large differences. McBean and Miyake (1972) suggest one possible reason for this. In their work fluctuations in velocity, temperature, and humidity were measured directly and analyzed in the frequency domain to show the effects of different scales of motion on the turbulent transfer mechanisms near the surface. Their measurements indicate that moisture is a passive scalar; that is, the moisture content of the air does not affect buoyancy (in contrast to the active scalar temperature which is directly related to the convective motion). They suggest that universal relationships pertaining to the transfer of passive scalars are unlikely and that the transfer mechanisms will depend on both the surface boundary conditions and the large-scale circulations. If this is the case more measurements are needed to establish the trend in K_e/K_m over snow.

In many situations where an estimate of water production from snowmelt is required, sufficient data to use Eqs. (72) is not available. In such cases both the sensible and latent heat transfer are described by expressions of the form

$$Q_e = D_e u_a (e_a - e_0) \quad \text{and} \quad Q_h = D_c u_a (T_a - T_0), \tag{79}$$

where D_e and D_c are known as bulk transfer coefficients and e is vapor pressure. The subscript "a" denotes values at some level above the snow surface, while the subscript zero denotes surface conditions. Both Kuz'min (1961) and Anderson (1976) review the use of equations of this type while Male and Granger (1979) give some indication of the loss of accuracy to be expected from such methods. Their data suggests that there is a correlation between daily totals of sensible heat as estimated by Eqs. (79) and calculated from Eq. (72), although significant differences can exist on any given day. The bulk transfer formula is a poor estimator of the latent heat flux Q_e. It would seem that vapor pressure at one level above the snow surface is insufficient to describe the vapor flux.

In operational snowmelt forecasts the use of simplified equations similar to Eqs. (79) are possible because in many field situations the sensible and latent heat fluxes are of secondary importance compared to the radiation flux. An indication of the relative importance of these fluxes in an open, relatively flat area over which the snowcover is continuous is given in Fig. 26. Each term in the energy balance [Eq. (56)] has been measured independently. The difference between the change in internal energy dU/dt and the summation of the heat flux values ΣQ_i is an indication of the accuracy of the data.

During the daytime (Q_n positive) both Fig. 26a and b show that the sensible heat flux is small compared to the radiation flux. Nevertheless Q_h can have an important influence on the melt rate. Note that in Fig. 26a Q_h

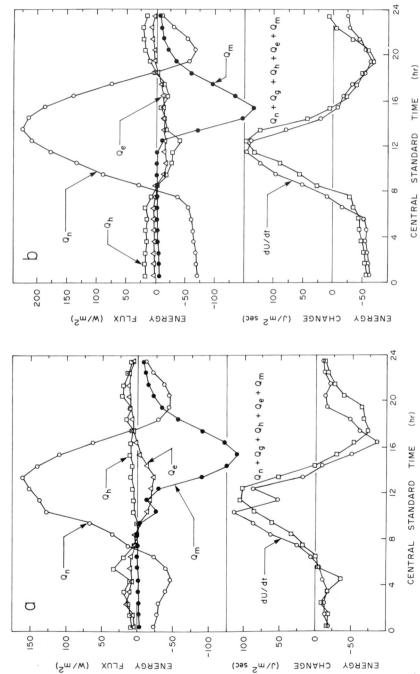

Fig. 26. Hourly averages for the energy budget over snow at Bad Lake, Saskatchewan: (a) 10 April 1974, (b) 14 April 1975. Q_n, net radiation; Q_g, ground heat flux; Q_h, sensible heat flux; Q_e, latent heat flux; Q_m, melt. [From Granger and Male (1978)].

is positive except for 1 hr while in Fig. 26b Q_h is positive at night and negative during the daytime. This suggests that the sensible heat flux has the ability to either assist or counteract the radiation flux. Granger and Male (1978) show that this ability is a major factor in determining whether runoff will occur during the early melt period and is governed by the energy content of the air mass above the snow. For a continuous snowcover of great areal extent, in the absence of appreciable bare soil, forest cover, or nearby body of water over which the air can be warmed, any significant contribution of energy by convective heat transfer must be the result of advection on an air mass scale. A positive sensible flux over the 24-hr period (Fig. 26a) suggests the presence of a relatively warm air mass aloft, while the diurnal cycle in Q_h (Fig. 26b) suggests the presence of a cooler air mass. Details of this argument are presented by Granger and Male (1978) who show a correlation between the daily totals of Q_h and the mean daily upper air temperature at the 85-kPa level.

Generally, the latent heat flux Q_e responds to the radiation flux following a cycle of evaporation during the day and condensation at night. Male and Granger (1979) report daily net evaporation rates ranging from 0.30 to 0.02 mm of water equivalent per day with a mean value of 0.10 mm/day. Net daily condensation is possible if nighttime radiation losses are excessive. For the semidesert region of West Kazakhstan, Rylov (1969) reports net daily evaporation rates ranging from 0.08 to 0.27 mm of water equivalent per day with an average of 0.11 mm/day during the period of continuous snowcover. These calculations are based on 12 years of measurements with snow evaporation pans. Despite these small mass fluxes, Q_e can be a significant portion of the daily energy budget. Male and Granger (1979) suggest that evaporation accounts for from 14 to 22% of the incoming energy to the snow surface in an open area.

Under a forest cover the relative magnitudes of the energy fluxes at the snow surface will be significantly different from measurements made in the open. However, data comparable to Fig. 26 are not available for this situation. In general the sensible heat transfer should be related to the air temperature gradient above the snow surface and the latent heat transfer should be determined by the corresponding vapor pressure gradient. The functional relationship between these gradients and their respective fluxes is not known. Equations (72) can only be used in relatively flat open areas where the assumption of one-dimensional heat flow implicit in these expressions is valid. Furthermore, the open areas must be sufficiently large to allow the constant flux boundary layer adjacent to the snow surface to develop to a thickness that will allow accurate measurement of the wind temperature and humidity profiles. In the case of the eddy correlation measurements the boundary layer must have a sufficient height to avoid the smaller scales of turbulence near the surface which are beyond the response of present-day sensors. An upstream fetch of several hundred meters is usually required for this purpose.

In a forest environment wind velocities are lower than in the open and this is often taken as an indication that the turbulent exchange rates are small. In coniferous forests Miller (1955, 1959) shows that the radiative flux is the dominant term in the energy balance. Hendrie and Price (1979) argue that the same is true of a deciduous cover. They show a strong correlation between the measured net radiation above the snow and daily melt volumes.

Turbulent exchange processes in a mountainous environment are also difficult to determine and, as with a forest environment, the functional relationship between the wind velocity temperature and moisture content has yet to be developed except for isolated terrain features. Unlike the forest environment, turbulent exchange processes can have a significant influence on the melt rate. The most spectacular evidence of this is the chinook wind, which results from the airmass losing moisture while rising on the windward side of a mountain range and then being heated by compression on the descent down the lee side. The result is a strong, warm, dry wind that can produce rapid melt and high rates of evaporation on the lee side of a ridge. A detailed review of the major large-scale and local wind patterns in a mountainous terrain is given by Obled and Harder (1979). They emphasize the difficulties that must be overcome in modeling such patterns and hence in arriving at estimates of the sensible and latent heat flux in these regions.

A third implication, which must be considered when examining the turbulent exchange processes over snow, is the discontinuous or patchy snowcover that is characteristic of the relatively flat prairie and tundra regions of the northern hemisphere over much of the melt period. Bare ground within the snowcover significantly alters the energy balance at the surface. Local advection from the bare patches to the snow must be considered and becomes increasingly important as the snowcover dwindles. Gray and O'Neill (1974) estimate that during a 6-day interval 44% of the energy supplied to an isolated melting snowpatch was by sensible heat transfer. During the period of continuous snowcover the corresponding figure for the same location was 7%.

When the terrain is a patchwork of bare ground and snowcovered areas, Eqs. (72) must be used with caution since they assume a vertical, one-dimensional flux, which is not valid near the edge of the patches. The correct theoretical approach to this problem is to calculate the sensible and latent heat transfer on the basis of a two- or three-dimensional model that considers the boundary layer development from the upwind edge of the snowpack. An analysis of this type has been developed by Weisman (1977) and gives some insight into the variation in fluxes from the leading edge of the pack. In the analysis it is assumed the radiation flux is constant over the snow and variations in the melt rates produced by latent and sensible heat transfer are considered. The specific problem analyzed is one of a warm moist air mass moving over the leading edge of a snowpack. The results suggest that the melt rate is at a maximum near the leading edge and decreases by one-third in approximately 15–25 m. Beyond this point melt decreases according to a power law relationship. Weisman also shows that temperature changes relatively gradually several meters from the leading edge and suggests that one-dimensional expressions such as Eqs. (72) or (79) should give reasonable results if temperatures and vapor pressures are measured near the center of the patch. This conclusion is supported by Cox and Zuzel (1976), who show that the sensible heat to a snowpatch 250 m long by 100 m wide can be estimated using a one-dimensional formula. Kraus (1973) in his review of the energy exchange at an air–ice interface suggests that one-dimensional expressions or expressions that assume horizontal homogeneity can be used to study variations in space "if the effects of horizontal inhomogeneity do not dominate over the processes regarded in the (one-dimensional) model." On the basis of Weisman's analysis it would seem this restriction holds as long as the snowpatch has a length of approximately 1000 m in the prevailing wind direction.

C. Soil Heat Flux

The ground heat flux Q_g is a negligible component in the daily totals of the energy balance for a snowpack when compared to the radiation, sensible and latent heat flux. In Fig. 26, Q_g is too small to register in the energy ordinate although its value has been included in the summation of fluxes. However, Q_g does not normally change direction over a 24-hr period and its cumulative effect can be significant over a season. This flux influences the temperature regime of the snow near the ground surface and is an important factor in the formation of depth hoar.

The vertical flux of heat at the soil surface is described by the one-dimensional Fourier heat conduction equation

$$Q_g = -k \; \partial T/\partial z|_s, \tag{80}$$

where k is the thermal conductivity (in watts per meter-degree centigrade) and $\partial T/\partial z|_s$ is the temperature gradient in the ground at the surface (in degrees centigrade per meter). Thermal conductivity values for silt and clay soils range from 0.4 to 2.1 W/m°C and for sand from 0.25 to 3

W/m°C, depending on moisture content and density. The moisture content, in particular, has a strong influence on conductivity values. Since the thermal conductivity of ice is approximately 4 times that of water, frozen soils exhibit somewhat higher k values. A comprehensive review of the factors influencing the thermal conductivity of a soil is given in De Vries (1963). Penner (1970) shows the variation of k that may be expected as the soil freezes.

In snowmelt calculations, average values of Q_g over a 24-hr period are used. For shallow snowpacks Kuz'min (1961) suggests maximum values of ± 400 kJ/m² per day, the negative sign expressing a loss of heat from the melting snow when the underlying surface is frozen and the positive sign representing a heat gain by the snow when the soil is not frozen. Granger (1977) has measured values in the range 0–260 kJ/m² per day. Gold (1957) measured an average flux of 860 kJ/m² per day from the soil to the pack at Ottawa, Canada, while Yoshida (1962) observed melt rates equivalent to 260–360 kJ/m² per day at the snow ground interface of a deep pack in Japan.

VI. WET SNOW

A. Metamorphism

The presence of liquid water in a snowpack causes very rapid metamorphic changes. According to the observations of Wakahama (1968a), small ice grains are eliminated and the larger grains grow rapidly until they reach diameters in a range from 1 to 2 mm. This process is accompanied by a loss of strength between snow particles and an increase in density. Colbeck (1978) suggests that the disappearance of the smaller ice grains partly explains the frequently noted observation that snow can store liquid water and then release it suddenly, particularly during the early stages of melt. Water flowing into a snowcover for the first time is held in the snow matrix by capillary attraction to which the smaller grains make an important contribution. Once the smaller grains disappear this water is released.

The rapid changes that take place in wet snow can be understood by considering the temperature differences between individual snow grains. Such differences are caused by local differences in the mean stress and in the radius of curvature between the solid, liquid, and vapor phases of the water which, in turn, determine the local equilibrium temperature. A thermodynamic analysis of the grain growth phenomenon, developed by Colbeck (1975a), shows the radius of curvature of the phase boundary is the most important parameter for natural snowpacks.

Colbeck considers two modes of saturation. At saturations greater than about 14% of the pore volume, air occurs as individual bubbles trapped in the pores and the water phase consists of continuous paths completely surrounding the snow grains. This saturation condition, known as the funicular regime, occurs in a natural snowpack immediately above relatively impermeable boundaries such as ice layers or frozen ground. Wet snow normally exists at lower values of saturation known as the pendular regime. In this condition the gaseous air–vapor phase is present in more or less continuous paths throughout the snow matrix. The water volume is greater than can be held by capillarity and the water pressure is much less than the atmospheric pressure. Water saturation during gravity drainage of homogeneous snow generally lies within the pendular regime.

In the funicular regime only two phase boundaries exist: the solid–liquid and the liquid–vapor boundary. The solid–liquid boundary can be described by the radius of the ice particle r_p and the liquid–vapor boundary by the radius of the air bubble r_a. Following Colbeck (1975a) these parameters can be related to the phase equilibrium temperature T (°K) by applying the Gibbs–Duhem equation to each phase i:

$$-s_i\, dT + v_i\, dp_i = du_i. \tag{81}$$

Here s_i is the specific entropy of the ith phase (kJ/kg °K), u_i the chemical potential (kJ/kg), and v_i is the specific volume. The pressures p between the two adjacent phases are related to the mean radius of curvature r_{ij} and the interfacial tension σ_{ij} (N/m) by the expression

$$p_i - p_j = 2\sigma_{ij}/r_{ij}. \tag{82}$$

The mean radius of curvature is related to the two principle radii (R_1 and R_2) at each point by the expression

$$2/r_{ij} = 1/R_1 + 1/R_2. \tag{83}$$

Equations (81) and (82) can be combined [Colbeck (1973a)] to give

$$\ln\!\left(\frac{T}{T_0}\right) = \frac{2}{L}\left(\frac{1}{\rho_1} - \frac{1}{\rho_s}\right)\frac{\sigma_{lg}}{r_a} - \frac{2}{L\rho_s}\frac{\sigma_{sl}}{r_p}, \tag{84}$$

where T is the melting temperature (°K), T_0 the corresponding melt temperature over a flat surface, and ρ is the density. L is the latent heat of fusion (kJ/kg) and the subscripts s, l, and g refer to the solid, liquid, and gaseous phases, respectively. Using the approximation

$$\ln(T/T_0) \simeq T_m/T_0,$$

where T_m is the melt temperature in degrees centigrade, Eq. (84) can be written

$$T_{\mathrm{m}} = \frac{2T_0}{L} \left(\frac{1}{\rho_1} - \frac{1}{\rho_s} \right) \frac{\sigma_{\mathrm{lg}}}{r_{\mathrm{a}}} - \frac{2T_0}{\rho_s} \frac{\sigma_{\mathrm{sl}}}{r_{\mathrm{p}}}. \tag{85}$$

The expression suggests that in the funicular regime there are two opposing effects related to the geometry of the snow matrix which govern the local melt temperature. The melt temperature increases as the size of the air bubble (r_{a}) becomes smaller and decreases with a decrease in the size of the ice grain. If it is assumed that smaller air bubbles escape from the snow matrix due to buoyancy effects and only larger bubbles remain then the temperature distribution among grains is largely determined by the grain size. The larger grains exist at a higher temperature than the smaller grains. Figure 27 shows equilibrium temperatures for a range of snow grain size for various radii of air bubbles based on Eq. (85). The temperature difference between the larger and smaller grains is the driving potential for heat flow by conduction to the smaller grains causing them to melt. Note that as the smaller grains decrease in size the controlling temperature gradient increases and the melt process is accelerated.

Colbeck (1973a) calculates the life expectancy of a smaller ice grain in the presence of a neighboring larger particle by considering the rate of heat transfer by conduction through the liquid phase between the grains. When this expression is combined with the restriction that the total volume of the two particles remain constant and the heat transferred from the larger particle must equal the energy required for the phase change at the surface of the smaller particle, it is possible to calculate the life expectancy of the smaller grains. It is shown that this life expectancy increases with the size of the neighboring grain. For example a particle with a radius of 0.1 mm has a life expectancy of approximately 0.6 days in the presence of a particle of radius 0.5 mm and approximately 1.1 days if the neighboring particle has a radius of 1 mm.

Wakahama (1968a) observed the metamorphism of snow immersed in water. His results show that grains smaller than 0.2 mm radius disappeared within three days and those smaller than 0.28 mm disappeared within 6 days. These results are in fair agreement with the predictions of Colbeck when it is considered that two grains in relative isolation are a rarity in natural snow where multigrain contacts and complicated heat flow patterns exist.

In the pendular regime the snow is unsaturated and a third interface can exist, the vapor–solid interface. By applying Eq. (82) to each of the three phases it can be shown that the requirement for mechanical equilibrium is

$$\sigma_{\mathrm{sg}}/r_{\mathrm{sg}} + \sigma_{\mathrm{gl}}/r_{\mathrm{gl}} = \sigma_{\mathrm{sl}}/r_{\mathrm{sl}}. \tag{86}$$

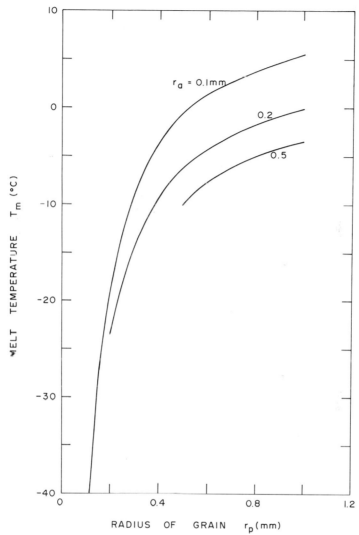

Fig. 27. Equilibrium grain temperature as a function of the radius of the grains and of the air bubbles. [From Colbeck (1973a)].

In the pendular regime r_{sg} is equal to the particle radius r_p and r_{gl} is related to the capillary pressure p_c by the expression

$$p_c = 2\sigma_{gl}/r_{gl}, \tag{87}$$

where $p_c = p_g - p_l$.

By combining Eqs. (86) and (87) with the Gibbs–Duhem equation [Eq. (81)] Colbeck (1973a) shows that the phase equilibrium temperature is given by the expression

$$T_m = -(T_0/\rho_1 L)p_c - (2T_0/\rho_s L)\sigma_{sg}/r_p. \tag{88}$$

The equation indicates that in the pendular regime the melting temperature is influenced by both the radius of the ice grains and the capillary pressure. To determine the relative influence of these two factors Colbeck (1975a) measured the capillary pressure in wet snow samples as a function of liquid water saturation S_w (S_w represents the fraction of the pore volume occupied by water). The results are reproduced in Fig. 28. In the pendular regime capillary pressure increases rapidly with decreasing saturation. During free drainage of homogeneous snow water saturation is less than 20% and capillary pressures are greater than 700 Pa. Thus, the effect of capillary pressure on the equilibrium temperature is dominant and the size of the ice grain has a second-order influence.

The observations of metamorphism in the pendular regime made by Wakahama (1968a) support the above analysis of the relative importance of capillary pressure (water saturation) and grain size on the rate of grain growth. He studied two situations. In the first case no water was added to the snow and the capillary pressure was high. The observed rate of grain growth was very slow. In the second case a small amount of water was added to the snow sample, which caused a reduction in capillary pressure. An increase in the rate of grain growth was observed. This increase can be attributed to the increased influence of the smaller ice grains with a lowering of capillary pressure as predicted by Eq. (88). An additional factor influencing grain growth in the pendular regime is the reduced area available for heat flow between grains due to the reduced volume of water (air is a good insulator compared to water). Colbeck (1975a) suggests that this effect is secondary compared to the influence of capillary pressure.

The grain-to-grain bonding strength in the pendular regime is high compared to that of the funicular regime. Colbeck (1975a) attributes this difference to the influence of water tension at low saturation drawing particles together. By considering the mechanical equilibrium between two neighboring particles and the temperature difference between the particle surface and the bond, he is able to show that if the cross-sectional area through which water tension acts A_w is much larger than the cross-sectional area of the bond A_b, then bond growth occurs by melting at the contact area. On the other hand, if A_b is much greater than A_w, freezing occurs at the contact area preventing further bond growth. A third influence is that of large overburden pressures which can also cause the contact area to grow by melting. At thermal equilibrium in the pendular

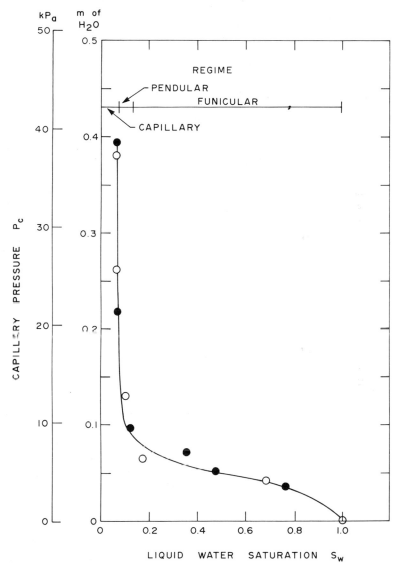

Fig. 28. Capillary pressure in wet snow as a function of water saturation. The two saturation regimes are shown. Density: ●, 550 kg/m³; ○, 590 kg/m³. [From Colbeck (1973a)].

regime, A_w is an order of magnitude larger than A_b and, for the geometrical configurations he analyzed, the effect of liquid tension on bond stength is high compared to the effect of ice-to-ice adhesion.

B. Water Movement through Snow

On the scale of the grains water flow through a snowpack which has experienced the grain growth and densification associated with melt metamorphism occurs in a highly irregular manner. At the melting temperature a thin film of water can cover the individual ice grains, and de Quervain (1973) shows that, for snow having a grain size of 2 mm, much of the water flow could occur through this film. A second possible means of water movement is that of vertical flow through isolated saturated pores. Wakahama (1968b) shows that once the pores are filled with water a Hagen–Poiseuille-type flow may occur, and this is an extremely efficient mechanism for draining the snowpack. For fine-grained snow Colbeck (1974b) suggests that a combination of film flow and droplet movement is possible. He further suggests that pressure increases in the pore space are unlikely during drainage since air is free to move through the spaces between grains in response to the movement of liquid water.

In naturally deposited snow, numerous dye studies have shown that percolation occurs in a highly irregular manner. Water from film or droplet flow tends to collect at individual capillary pores. On a larger scale Gerdel (1954) describes isolated vertical flow channels or "glands" of coarse-grained snow that act as drains or preferred paths for the melt water. The coarse grains are the result of the higher water saturation in these channels which induces a rapid grain growth. This, in turn, allows more water to drain through the channels.

A second large-scale feature of a naturally deposited snow is the presence of relatively impermeable high-density layers often referred to as "ice" layers. Water can accumulate above such layers producing a saturated condition. However, dye studies suggest these layers are not impermeable but rather are characterized by a variable permeability which forces the meltwater to take numerous sideways steps on its route to ground level [Langham (1974a); Gerdel (1949)]. This pattern is further complicated by the fact that ice layers are seldom formed in the horizontal plane but commonly have a variable slope.

1. Water Flow through Homogeneous Snow

For homogeneous snow (that is in the absence of vertical drainage channels and ice layers) the vertical flow of water on a scale approximately equal to that of the snow depth can be described by an application of Darcian flow theory [Colbeck (1972); Colbeck and Davidson (1973)]. Assuming an unsaturated snowpack, the air and water fluxes are balanced. Darcy's law, which relates the capillary pressure gradient $\partial p_c/\partial z$ and the acceleration of gravity g to the volume of water flowing per unit area per unit time u, is

$$u = (k_l/\mu_l)[\partial p_c/\partial z + \rho_l g], \qquad (89)$$

where k_l is the permeability of liquid water, μ_l water viscosity, and z is measured in the vertical direction. The pressure gradient $\partial p_c/\partial z$ is generally of secondary importance compared to the gravity term. Colbeck (1974b) shows that at low water fluxes, which correspond to low values of water saturation and large water tension, the pressure gradient is significant. It can also be important at the leading edge of meltwater moving downward through the pack. However, in this case the pressure gradient acts as a diffusion term and does not influence the rate of frontal movement to any degree [Colbeck (1974b)]. In the subsequent development, which follows Colbeck and Davidson (1973), this term is neglected.

In the application of Eq. (89) to snow it is necessary to have a relationship between the permeability available to the liquid phase k_l and the more easily measured intrinsic permeability of the snow matrix k. Colbeck and Davidson (1973) relate these two quantities to the effective liquid water saturation S^* by means of the expression

$$k_l = kS^{*3}, \qquad (90)$$

where S^* is the fraction of the pore volume containing moving water. Equation (90) implies the permeability k_l increases very rapidly with this fraction.

The combination of Eqs. (89) and (90) yields the following relationship between the vertical liquid flux u through the unsaturated snow and the water saturation S^*:

$$u = \alpha kS^{*3}, \qquad (91)$$

where α is a constant and equals $\rho_l g \mu_l^{-1}$.

The continuity equation for the liquid water phase applied to a differential volume of the snowpack having unit area and thickness dz is

$$\partial u/\partial z + \phi_e(\partial S^*/\partial t) = 0, \qquad (92)$$

where ϕ_e represents the effective porosity, defined as

$$\phi_e \equiv \phi(1 - S_{wi}). \qquad (93)$$

ϕ is the porosity or ratio of pore volume to total volume; S_{wi} is termed the "irreducible water saturation" by Colbeck and is the water retained by capillary forces. Accurate measurements of S_{wi} are difficult to obtain and will vary with the metamorphic state of the snow. Using dielectric probes Lemellä (1973) measured values of 2–3% per unit volume or approximately 5–7% of the pore volume for a seasonal snowpack that had experienced several hours of active melt. Equations (91) and (92) can be combined to give the following differential equation which describes the flux of water:

$$3\alpha^{1/3}k^{1/3}u^{2/3}(\partial u/\partial z) + \phi_e(\partial u/\partial t) = 0. \tag{94}$$

The simplest solution of this equation is obtained by the method of characteristics on the assumption that the snow density is constant with depth. This solution is of the form

$$dz/dt|_u = 3\alpha^{1/3}k^{1/3}\phi_e^{-1}u^{2/3} \tag{95}$$

and relates the flux of water u to the rate of downward movement of that flux, symbolized as $dz/dt|_u$. The important feature of this solution is that $dz/dt|_u$ increases for a given value of u to the two-thirds power. Thus, larger values of flux move more quickly than smaller values and at any given depth the passing wave of meltwater is characterized by a rapid rise to a peak value followed by a slower recession.

Experimental verification of Eq. (95) is reported by Colbeck and Davidson (1973) for isolated columns of repacked, homogeneous snow which were sealed and allowed to drain for several days after a period of intense surface melting. The results of these tests are reproduced in Fig. 29; dz/dt was determined as the ratio of the column length to the travel time. Note that plots of this type can be used to determine the product $k^{1/3}\phi_e^{-1}$. Furthermore, if measured surface fluxes and fluxes at the bottom of the pack are used, then the value of $k^{1/3}\phi_e^{-1}$ represent an integrated average of any inhomogeneities in the snowcover.

Equation (95) can be applied to hydrological forecast schemes, although for such purposes information in the form of Fig. 29 is not likely to be available and it would be necessary to rely on measured values of the intrinsic snow permeability k. This parameter is a function of snow density, grain size, and shape. Colbeck (1978) suggests that the most useful measurements of k have been made by Shimizu (1970), who provides the following relationship:

$$k = 0.077d^2 \exp[-7 \cdot 8(\rho_s/\rho_l)], \tag{96}$$

where d is the grain diameter and ρ_s the snow density.

Measured changes in the shape of a diurnal wave of meltwater with depth are shown in Fig. 30 reproduced from Colbeck and Davidson

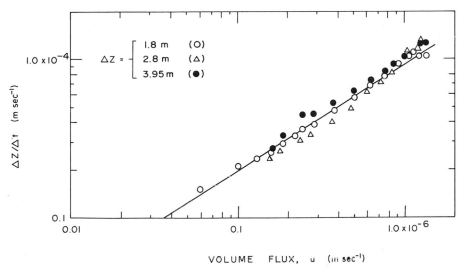

Fig. 29. The measured relationship between the speed of descent (dz/dt) of values of u for 3 repacked columns with the lengths dz indicated. The solid line through the points is a plot of Eq. (95). [From Colbeck and Davidson (1973)].

(1973). If it is assumed that the melt wave originates at the surface and the values of this flux are known as a function of time, then a step-by-step application of Eq. (95) from the surface to any desired depth will show many of the features of Fig. 30. In particular the maximum flow rate decreases with depth, the minimum flow rate increases, and the trailing edge lengthens. Of particular interest is curve D and E which show an almost vertical leading edge. This is evidence of what Colbeck, by analogy to the behavior of a pressure wave in a compressible gas flow, calls a shock wave. As mentioned earlier the rate of propagation of values of volume flux $dz/dt|_u$ increase as the volume flux u increases. Slower-moving smaller values of the flux produced by melt early in the morning are overtaken by the faster-moving larger values of u produced later in the day. Once two values of flux intersect the shock front is formed. In the absence of capillary effects the water flux is discontinuous across the shock front and Eq. (95) is no longer sufficient to describe the flow. Colbeck (1975b) combines Eq. (95) with the continuity requirement that the volume of liquid water must be conserved as the front moves downward to determine the following expression for the rate of propagation of the front $d\xi/dt$:

$$d\xi/dt = \alpha^{1/3}k^{1/3}\phi_e^{-1}(u_+^{2/3} + u_+^{1/3}u_-^{1/3} + u_-^{2/3}),\qquad(97)$$

where u_+ and u_- are values of the faster and slower mass fluxes, respec-

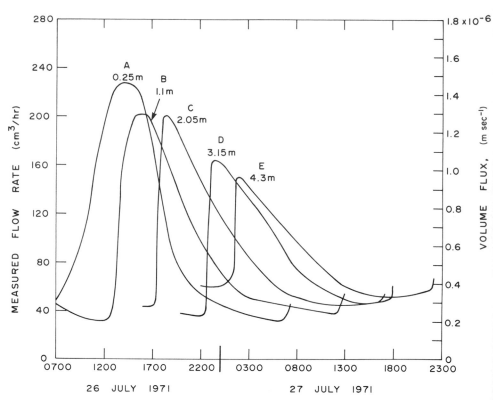

Fig. 30. Characteristics of a diurnal wave of meltwater at various depths below the snow surface. [From Colbeck and Davidson (1973)].

tively, which have intersected to form the shock front. These values change continuously as larger values of flux overtake the front.

When the diurnal melt rate reaches the soil the vertical motion is altered and several situations are possible. If the soil is unsaturated some of the water will infiltrate. The remainder will move over the land surface through a relatively shallow saturated region. A highly idealized schematic representation of this situation is shown in Fig. 31. It must be emphasized that several reasonably common factors can alter this very simple situation. In particular vegetation protruding into the bottom layers of the snowcover and local, small-scale terrain features will considerably complicate the geometry of the saturated layer. In some areas temperature gradient metamorphism can cause significant increases in grain size (depth hoar) at the base of the pack prior to melt. Nevertheless the situation illustrated in Fig. 31 is a useful model for considering the features of the interaction between the unsaturated and saturated zones.

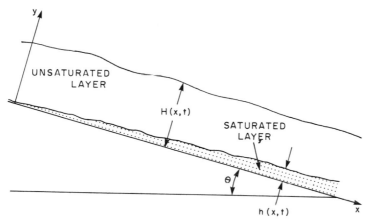

Fig. 31. An idealized cross-sectional view of the snowcover showing the unsaturated vertical flow regime and the saturated flow over the base.[From Colbeck (1974a)].

Flow in the saturated layer has been described by Colbeck (1974a) beginning with a straightforward application of Darcy's law for small values of slope θ in the form

$$u = -(k/\mu_1)(\partial p_1/\partial x - \rho_1 g \theta) \tag{98}$$

In this case the presence of air bubbles trapped in the snow is ignored and the snow permeability to the water phase is assumed to be the intrinsic permeability determined from Eq. (96). Because of rapid metamorphism in the saturated layer the average grain size is larger than in the unsaturated layer and the intrinsic permeability will be higher. In addition, according to Eq. (90) the permeability to liquid water in the unsaturated zone varies as the third power of the effective water saturation S^{*3}. As a result of these two influences permeability in the saturated zone is likely to be several orders of magnitude greater than in the unsaturated zone. Therefore, although Eq. (98) implies a lower driving potential gradient for the basal flow and despite the greater distances water travels in the basal layer, time spent in this mode of flow is often less than the time spent in vertical unsaturated flow.

The continuity equation for the basal layer of height h (Fig. 31) is

$$\partial(uh)/\partial x + \phi\, \partial h/\partial t = I(x, t), \tag{99}$$

where $I(x, t)$ is the net flow to the layer and consists of the inflow from the unsaturated zone less the infiltration to the underlying soil. Infiltration from snow to soil, particularly frozen soil, is an important and largely unexamined aspect of snow hydrology but lies outside the scope of this chapter.

When Eqs. (98) and (99) are combined the expression governing the flow in the saturated layer is obtained [Colbeck (1974a)]:

$$\alpha k_s \theta(\partial h/\partial x) - \alpha k_s \, \partial/\partial x(h \, \partial h/\partial x) + \phi(\partial h/\partial t) = I. \qquad (100)$$

If it is assumed that the increase in the thickness of the saturated layer with downslope distance dh/dx is small compared to the slope θ, then the second term of Eq. (100) can be ignored and a solution to the flow is found using a coordinate system that moves downslope at the velocity c of the water particles [Colbeck (1974a)]. This velocity is

$$c = \alpha k \theta \phi^{-1}. \qquad (101)$$

For a saturated layer of length L the solution to Eq. (100) shows that the total discharge q_L per unit width of the saturated layer can be determined by integrating the input to the basal layer I over a time interval Δt given by

$$\Delta t = L c^{-1}. \qquad (102)$$

It is important to note that this integration does not involve directly the thickness of the saturated layer h because the flux of water through the saturated layer varies linearly with h. Thus the timing of the runoff is independent of the geometrical complexities of the surface over which the water flows.

A test of this theory is reported by Dunne *et al.* (1976) at sites in the Canadian subarctic where the slopes were steep and uniform, the snowcover homogeneous, and infiltration to the soil negligible. Beginning with standard micrometeorological measurements, measurements of snow density and grain size, flow in both the unsaturated and saturated layers was calculated to yield estimates of discharge from the snowpack. In general measured runoff compared reasonably well with observed runoff from the study plots, which were of various sizes, aspect, and slope and had a variety of vegetative cover. One consistent difference between the calculated and measured discharges was the overestimate of the lag time between peak rates of melt and runoff. The authors suggest that the difference may be due to the presence of water channels underneath the snowcover. The rate at which these channels form and grow is a problem that requires detailed examination.

2. Water Flow through Layered Snow

The stratified nature of naturally deposited snow can exert an important influence on the water movement through the pack. "Ice layers" are formed at the upper surface of the pack by freezing rain and are buried by subsequent snowfalls. Relatively impermeable layers of high-density grains can be formed by refreezing of percolating meltwater, especially at buried wind or radiation crusts. Measurements by Gerdel (1954) suggest that once large amounts of meltwater are introduced to the pack, disintegration of such layers can occur in a matter of hours with the formation of vertical drainage channels similar to those present in homogeneous snow that has experienced meltwater runoff.

Langham (1974b) states that ice layers in the snow are polycrystalline and consist of crystals having a diameter of a few millimeters. At the junction of these crystals are veins that at 0°C contain liquid water. Figure 32, taken from Langham (1974b), illustrates the growth of these veins which results in a loss of contact area between crystals, a decrease in strength, and an increased permeability. Assuming the veins are completely filled with water, their phase equilibrium temperature can be calculated from a modified version of Eq. (85) in the form [Langham (1974b)]

$$T_m = -0.137(T_0/\rho_i L)(\sigma_{sl}/r),\qquad(103)$$

where r is the diameter of the largest circle that can fit inside the vein. Obviously, as the temperature T_m increases so does the size of the vein. Sources of energy for this growth are provided by meltwater percolation, heat conduction and the absorption of solar radiation. On the basis of a detailed set of numerical calculations Langham (1975) shows that solar radiation is particularly important and can provide sufficient energy to make ice veins permeable an hour or two after sunrise.

Based on Eq. (103) it is possible to determine that once ice veins grow to a radius in the range 0.01–0.1 mm they have reached thermal equilibrium with the surrounding snow, and are capable of passing meltwater at a rate that prevents the formation of a saturated layer above the ice. Presumably, at some point in their growth the entry of air into the vein is possible and the thermal regime changes to one analogous to that of unsatu-

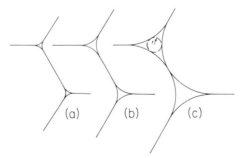

Fig. 32.   Stages in the growth of veins in polycrystalline ice which result in an increased permeability and a decrease in strength. [From Langham (1975)].

rated snow. A detailed study of this phenomenon is necessary to provide a more complete understanding of the changing properties of ice layers in the presence of meltwater.

Our understanding of the influence of ice layers on runoff over large areas is also incomplete. The permeability, thickness, and areal extent of these layers have never been measured in the field. Large variations in these properties are likely both spacially and in time once meltwater is present in the snow. Dye studies by Gerdel (1954) and Langham (1974a) show that even gently sloping layers can divert an appreciable amount of the flow. However, ice layers are generally sufficiently permeable to allow some flow through the pores. The overall flow pattern is further complicated by the vertical channels or drains mentioned earlier. Colbeck (1978) suggests the two-dimensional schematic diagram shown in Fig. 33 to illustrate the flow pattern in the vicinity of a single ice layer. An approximate analysis of this situation is developed by Colbeck (1973b). Flow along the ice surface is saturated and is assumed to obey Darcy's law in a form similar to Eq. (98). Vertical flow through the ice is assumed to obey a form of Darcy's law,

$$w_i = \rho_1 g \mu_i^{-1} k_i (h/h_i + 1), \tag{104}$$

where w_i is the volume flux through the layer and k_i the permeability of the ice layer. Note that according to this equation flow will only occur if ponded water is present above the layer. The ratio of the flow through the pores W to the flow through the drains is [Colbeck (1973b)]

$$W/Q = (k_i/k)(2l/h_i)[h_i/h + 1] \cot \theta, \tag{105}$$

where $2l$ is the length of the ice layer. This equation suggests that the flow

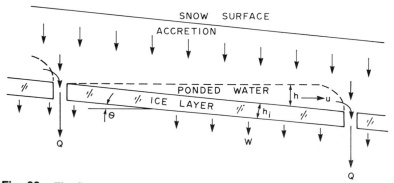

Fig. 33. The flow pattern around and through a sloping ice layer. Flow past the ice layer is by a combination of flow to the drains Q and flow through the pores W. [From Colbeck (1978)].

along the ice layer is strongly dependent on θ at small slopes. The relative flow along the ice layer is about an order of magnitude greater at a slope of $10°$ than it is for a horizontal layer. The permeability of the layers is also an important factor governing the overall flow patterns. Colbeck (1973b) shows that the ponding of water above ice layers will not occur unless k_i is less than 0.5% of that for the surrounding snow. At values of k_i 50 times less than this, vertical flow through a horizontal ice layer will dominate the lateral mode of flow.

The ultimate aim of any analysis of water flow through a structured snowpack is to relate the discharge from the pack to the flow over and through the individual layers. Given a complete understanding of the parameters governing the flow patterns in the vicinity of an individual layer it would be necessary to calculate the interaction between layers. This would require a step-by-step calculation of the flow from one layer to the next, it being recognized that the number, extent, and slope of these layers vary considerably. In general, the computational problems would be formidable.

A second type of analysis, which avoids the necessity of step-by-step calculations, is to represent the layered snowpack as a homogeneous but anisotropic porous medium [Bear (1972)]. This approach must be applied on a scale sufficiently large to average variations in the flow field and in the long term may develop into a useful forecasting tool in snow hydrology. The details of this approach have been developed by Colbeck (1975b), although, at present, the analysis has not been tested in the field.

In general, the permeability of an anisotropic porous medium can be represented by the following symmetrical matrix:

$$k_{ij} = \begin{bmatrix} k_{11} & k_{12} & k_{13} \\ k_{12} & k_{22} & k_{23} \\ k_{13} & k_{23} & k_{33} \end{bmatrix} \tag{106}$$

where the subscripts 1, 2, and 3 represent the three coordinate directions. Colbeck (1975b) applies this matrix to the coordinate axes shown in Fig. 34. He assumes the maximum value of permeability is in the x_1 direction and the minimum permeability is in the x_2 direction. For this system of coordinates the three nondiagonal elements (k_{12}, k_{13}, k_{23}) are zero. It is further assumed that the flow properties in the x_1 and x_3 direction are identical, which reduces the problem to one of two dimensions. Thus, the two values of permeability relevant to the problem are k_{11} parallel to the layers and k_{22} perpendicular to the layers. k_{11} is approximately equal to the permeability of the homogeneous snow between layers; k_{22} is an average value which includes the effect of the alternating layers of ice and homogeneous snow.

Bear (1972) derives a general expression for Darcy's law for an unsaturated anisotropic porous medium which gives the volume flux in any coordinate direction as a function of the components of the tension gradient and gravity. If the angle θ is significant and tension gradients are neglected, Colbeck (1975b) shows that the two components of the flux for a layered snowpack in the coordinate system of Fig. 34 are

$$u_1 = \alpha S^{*3}k_{11} \sin \theta \quad \text{and} \quad u_2 = -\alpha S^{*3}k_{22} \cos \theta, \quad (107)$$

where u_1 is the flow along the slope and u_2 is the flow perpendicular to the slope.

The direction of the flow with respect to a line perpendicular to the slope is given by the angle β, where

$$\tan \beta = (k_{11}/k_{22}) \tan \theta. \quad (108)$$

The travel time through the snowpack is important for predicting runoff and is given by Colbeck (1975b) as

$$T = h(3\alpha\phi_e^{-1}S^{*2})^{-1}k_{11}^{-1} \csc \theta \tan \beta \quad (109)$$

for a given value of S^* or a corresponding value of u; h is the depth of snow perpendicular to the slope.

He further shows that travel times for the anisotropic snowpack and an isotropic pack of comparable depth are related by

$$T = (k_{11}/k_{22})T_z, \quad (110)$$

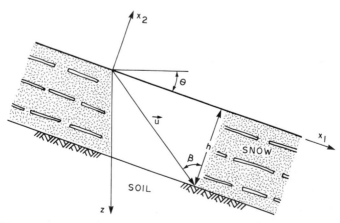

Fig. 34. A sloping snowpack consisting of alternating layers of homogeneous snow and discontinuous ice layers. The average volume flow u is shifted to an angle $\beta-\theta$ from the vertical as a result of ice layers. [From Colbeck (1975b)].

where T_z is the travel time through an isotropic pack having a permeability k_{11}. In a layered snowpack the travel time increases with the ratio of principal values of the permeability matrix k_{11}/k_{22}.

C. Mechanical Properties

The introduction of meltwater into the snowcover can weaken the bonds between grains resulting in a material that has a reduced mechanical strength and is more easily compressed than snow at subfreezing temperatures. In general, the compressibility of wet snow increases with increasing liquid water content [Wakahama (1975)] but decreases with the addition of small amounts of salt [Yarkin et al. (1972)].

Wet snow deforms predominantly by regelation [Colbeck et al. (1978)] but the pressure melting mechanism is a complex one involving the impurity or salt concentration in the snow, the double charged layer at each solid–liquid boundary between individual grains, the small temperature differences between the stressed and stress-free areas of the grain surface, and the rate at which meltwater is discharged from between the grains. Colbeck (1976) shows that the local temperature differences on the grain surfaces are particularly important and that the response of wet snow to an applied load can be dominated by this factor which arises because the equilibrium temperature at a solid–liquid interface decreases with increasing pressure; hence, the contact surfaces between neighboring particles have a temperature that is lower, in general, than the stress-free surface in the pore spaces.

The importance of these small, local temperature differences to the compressibility of wet snow makes this material an extremely difficult one to examine in the field or in the laboratory. Wakahama (1975), in particular, emphasizes the very small amounts of heat that are required to initiate melting at the grain boundaries of ice at 0°C. This energy can be supplied by the small amounts of solar radiation that penetrate below the snow surface. For example, at a depth of 20 cm the daily amount of snow melted by absorbed solar radiation is of the order of 2×10^{-4} g/cm^3. However, as explained by Wakahama, this energy concentrates at the grain boundaries where the temperature is slightly lower and the free energy slightly higher than that of the ice grains. Thus only a small amount of solar radiation that has penetrated below the surface may result in melting of the grain boundaries and a loss in mechanical strength.

The compression process in wet snow has been studied in detail by Colbeck et al. (1978) in a laboratory situation in which care was taken to eliminate all external heat flow to the samples by conduction or radiation. Thus, their measurements isolate the importance of the local flow of heat and the regelation or pressure melting mechanism on the compressibility of wet snow.

Tests were conducted for a range of capillary pressures and various salinities. Capillary pressure was maintained constant for each test and as the compression proceeded the snow sample discharged water to maintain the imposed capillary pressure. The results of these measurements are summarized in Figs. 35 and 36. Figure 36 is a plot of ice density (ice mass per unit volume) as a function of time for various capillary pressures. The curves clearly show the well-known decrease in the rate of densification with decreasing liquid water content (higher capillary pressure). Similarly, Fig. 36 shows the decrease in compressibility with an increase in salinity measured in this case as the product of the gram-moles of NaCl per kilogram of solution and the dissociated ions per molecule).

The investigations of Colbeck *et al.* show that it is possible to describe the compression of high-density wet snow by considering the physical processes of heat and mass transfer occurring on the scale of individual particles. Grain growth plays an important role in the compression process. Refreezing of meltwater on the stress-free surface of the larger particles is a source of heat for both the regelation process and the melting of the smaller particles. Furthermore, Colbeck *et al.* found that over a two-week period seven out of eight of the particles melt away, which suggests that many vacancies must be created temporarily in the ice matrix as particles melt away from their neighbors. Wakahama (1975) observes that once a particle begins to melt it disappears quickly and therefore it can be assumed that only a small fraction of particles is shrinking at any time. However, these particles reduce the average number of contacts per particle and increase the average stress per contact. Some indication of the complexity of the mechanisms involved at the contact between particles can be obtained by considering Fig. 37a, which is an idealized representation of two particles at 0°C being pushed together by a force F. The particles melt at the interparticle stressed surface. The meltwater is removed as a thin film and refreezes on the stress-free surface, thus providing a source of heat for the melting process at the interparticle contact. The thickness of the liquid film between the particles is of the order of 10^{-5} mm and hence the repulsion forces arising from the double charged layers at each solid–liquid boundary are significant and counteract the applied load, especially at low salinities. Furthermore, concentrations of dissolved air and ionic impurities in the meltwater moving from the thin film between the particles play an important role in governing the rate of flow and hence the rate of compression.

A quantitative model of this process has been developed by Colbeck *et al.* (1978), which explains the major characteristics of the measurements

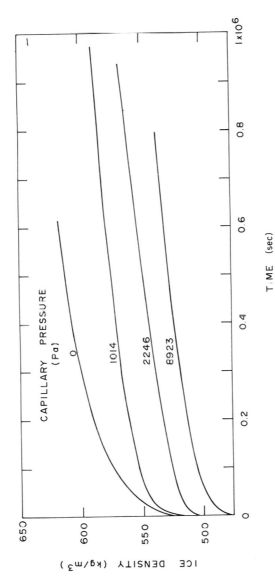

Fig. 35. Measured changes in ice density with time for various capillary pressures. [From Colbeck *et al.* (1978)].

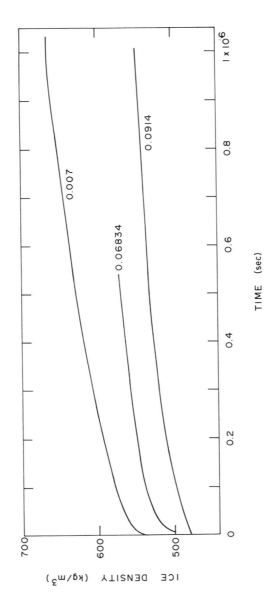

Fig. 36. Measured changes in ice density with time for various sodium chloride contents (gram-moles NaCl per kilogram of solution times dissociated ions per molecule). [From Colbeck *et al.* (1978)].

388

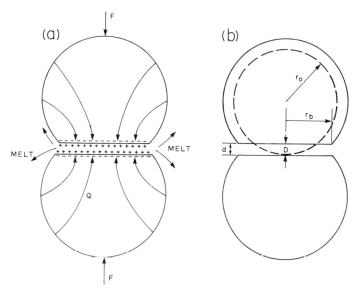

Fig. 37. Schematic representation of the processes occurring when two particles are pushed together and melt at the stressed surfaces. [From Colbeck *et al.* (1978)].

illustrated in Figs. 35 and 36. In their analysis the temperature of the stressed surface at the interparticle contact is given by

$$T_b = \frac{T_0}{L}\left(\frac{1}{\rho_1} - \frac{1}{\rho_s}\right)\sigma_b + 0.0024\left(\frac{C_s - C_b}{C_s}\right) - \frac{RT_0^2}{L}IS_b, \qquad (111)$$

where σ_b is the stress at the bond, C_s and C_b the concentration of dissolved air in the pore water and at the contact between particles, I the number of dissociated ions per molecule, and S_b the concentration of dissolved salt at the bond. The first term of Eq. (111) is the temperature decrease resulting from the stress exerted by the liquid at the bond, the second term is the temperature increase resulting from the relative absence of dissolved air in the melt, and the third term is the decrease in temperature due to the presence of dissolved ionic impurities at the bond. The temperature difference between the stress-free and stressed surfaces at low water saturation (pendular regime) is calculated from Eqs. (88) and (111) where a correction is added to Eq. (88) similar to the third term in Eq. (111) to account for dissolved salt in the pore water. Thus, for a given level of impurities and a given level of stress at the bond the temperature difference between the stress-free and stressed surface will be reduced with an increase in capillary pressure (decrease in liquid water content).

This results in a reduced heat flow and a lower rate of densification as il-lustrated in Fig. 35.

The influence of the double charged layer illustrated in Fig. 37 is most important at low salinities. Colbeck *et al.* (1978) suggest that the stress in the liquid at the bond σ_b can be expressed as

$$\sigma_b = \sigma_{bulk} \frac{6}{n} \frac{4(\bar{r} - D)^2}{\pi r_b^2} - \frac{0.42}{d} \exp(-2.3 \times 10^6 \sqrt{S_b}\, d). \tag{112}$$

In this equation σ_{bulk} is the bulk stress applied to the snow sample (in pascals), n the number of stressed contacts per particle, $\bar{r}$ the average particle radius, D the displacement of the melt surface (see Fig. 37b), r_b the bond radius, and d the thickness of the liquid film. Equation (112) demonstrates that the average stress at an interparticle contact decreases as the size of the contact area increases and as the number of contacts increases. The second term is the repulsive force between the two ice surfaces separated by the liquid film in which the dissolved salt concentration is S_b. Colbeck *et al.* show that this term plays an important role in decreasing the stress σ_b, particularly as the salinity decreases to 10^{-3}. The reduced stress results in a reduced temperature gradient and heat flow to the melting surfaces, which ultimately leads to a decrease in compressibility.

Although the mechanisms involved in the densification of wet snow are complex, Colbeck *et al.* suggest ice density changes may be expressed as a function of time using the expression

$$\rho = Ct^{0.2} + \rho_0, \tag{113}$$

where ρ_0 is the initial ice density at t the lapsed time and C a parameter that depends on the applied bulk stress, capillary pressure, initial particle size, and the salinity of the pore water. As long as the parameter C can be computed properly this expression is a highly accurate substitute for the more complex computations.

REFERENCES

Akitaya, E. (1975). *IAHS-AISH Publ.* **114**, 42–48.

Anderson, E. A. (1976). *NOAA Tech. Rep., NWS* **19**, 150 p.

Bader, H. (1953). "Sorge's Law of Densification of Snow on High Polar Glaciers," Res. Rep. No. 2. U.S. Army Snow, Ice and Permafrost Res. Estab. Hanover, New Hampshire.

Bader, H. (1960). "Theory of Densification of Dry Snow on High Polar Glaciers," Part I, Res. Rep. No. 69. U.S. Army Snow, Ice and Permafrost Res. Estab. Hanover, New Hampshire.

Bader, H. (1962). "Theory of Densification of Dry Snow on High Polar Glaciers," Part II, Res. Rep. No. 108. U.S. Army Snow, Ice and Permafrost Res. Estab. Hanover, New Hampshire.

Bader, H., Haefeli, R., Bucher, E., Neher, J., Eckel, O., Thams, C., and Niggli, P. (1939). *Beitr. Geol. Schweiz, Geotech. Ser., Hydrol.* No. 3 [Engl. transl. by U.S. Army Snow, Ice and Permafrost Res. Estab., Transl. 14(1954), Hanover, New Hampshire].

Bagnold, R. A. (1941). "The Physics of Blown Sands and Desert Dunes." Methuen, London.

Bagnold, R. A. (1973). *Proc. R. Soc. London, Ser. A* **332**, 473–504.

Bates, R. E., and Bilello, M. A. (1966). "Defining the Cold Regions of the Northern Hemisphere," Tech. Rep. No. 178. U.S. Army Cold Reg. Res. Eng. Lab., Hanover, New Hampshire.

Bear, J. (1972). "Dynamics of Fluids in Porous Media." Am. Elsevier, New York.

Bentley, C. R., Pomeroy, P. W., and Dorman, H. J. (1957). *Ann. Geophys.* **13**, 253–285.

Blad, B. L., and Rosenberg, N. J. (1974). *J. Appl. Meteorol.* **13**, 227–236.

Bohren, C. F., and Barkstrom, B. R. (1974). *J. Geophy. Res.* **79**, 4527–35.

Brunt, D. (1952). "Physical and Dynamical Meteorology." Cambridge Univ. Press, London and New York.

Brutsaert, W. (1975). *Water Resour. Res.* **11**, 742–744.

Budd, W. F. (1966a). "Glaciological Studies in the Region of Wilkes, Eastern Antarctic, 1961," ANARE Sci. Rep., Ser. A (IV), Glaciol. Publ. No. 88.

Budd, W. F. (1966b). *Antarct. Res. Ser.* **9**, 59–70.

Budd, W., Dingle, R., and Radok, U. (1966). *Antarct. Res. Ser.* **9**, 71–134.

Businger, J. A., Wyngaard, J. C., Izumi, Y., and Bradley, E. F. (1971). *J. Atmos. Sci.* **28**, 181–189.

Chepil, W. S. (1945). *Soil. Sci.* **60**, 305–320, 397–411, and 475–480.

Colbeck, S. C. (1972). *J. Glaciol.* **11**, 369–385.

Colbeck, S. C. (1973a). "Theory of Metamorphism of Wet Snow," Res. Rep. No. 313. U.S. Army Cold Reg. Res. Eng. Lab., Hanover, New Hampshire.

Colbeck, S. C. (1973b). "Effects of Stratigraphic Layers on Water Flow through Snow," Res. Rep. No. 311. U.S. Army Cold Reg. Res. Eng. Lab., Hanover, New Hampshire.

Colbeck, S. C. (1974a). *Water Resour. Res.* **10**, 119–123.

Colbeck, S. C. (1974b). *J. Glaciol.* **13**, 85–97.

Colbeck, S. C. (1975a). *IAHS-AISH Publ.* **114**, 51–61.

Colbeck, S. C. (1975b). *Water Resour. Res.* **11**, 261–266.

Colbeck, S. C. (1976). "Thermodynamic Deformation of Wet Snow," CRREL Rep. 76-44. U.S. Army Cold Reg. Res. Eng. Lab., Hanover, New Hampshire.

Colbeck, S. C. (1978). *Adv. Hydrosci.* **11**, 165–206.

Colbeck, S. C., and Davidson, G. (1973). *Role Snow Ice Hydrology, Proc. Banff Symp., 1972* Vol. 1, pp. 242–257.

Colbeck, S. C., Shaw, K. A., and Lemieux, G. (1978). "The Compression of Wet Snow," CRREL Rep. 78-10. U.S. Army Cold Reg. Res. Eng. Lab., Hanover, New Hampshire.

Cox, L. M., and Zuzel, J. F. (1976). *Proc. 44th Annu. Meet. West. Snow Conf., 1976* pp. 23–28.

Crary, A. P., Robinson, E. S., Bernett, H. F., and Boyd, W. W. (1962). "Glaciological studies of the Ross Ice Shelf, Antarctica, 1957–1960," IGY Glaciol. Rep. No. 6. Am. Geographical Soc.

de Quervain, M. (1973). *Role Snow Ice Hydrol., Proc. Banff Symp., 1972* Vol. 1, pp. 203–226.

De Vries, D. A. (1963). *In* "Physics of Plant Environment" (W. R. Van Wijk, ed.), pp. 210–235. North-Holland Publ., Amsterdam.

Dickinson, H. C., and Osborne, N. S. (1915). *Bur. Stand. (U.S.), Bull.* **12**, 49–81.

Dingle, W. R. J., and Radok, U. (1961). *Int. Assoc. Sci. Hydrol., Publ.* **55**, 77–87.

Dorsey, N. E. (1940). "Properties of Ordinary Water Substance." Van Nostrand-Reinhold, Princeton, New Jersey (reprinted by Hafner, New York, 1968).

Dozier, J. (1979). *In* "Modeling Snow Cover Runoff" (S. C. Colbeck and M. Ray, eds.), pp. 144–153. U.S. Army Cold Reg. Res. Eng. Lab., Hanover, New Hampshire.

Dunne, T., Price, A. G., and Colbeck, S. C. (1976). *Water Resour. Res.* **12**, 677–685.

Dyer, A. J. (1961). *Q. J. R. Meteorol. Soc.* **87**, 401–412.

Dyer, A. J. (1968). *J. Appl. Meteorol.* **7**, 845–850.

Dyer, A. J., and Hicks, B. B. (1970). *Q. J. R. Meteorol. Soc.* **96**, 715–721.

Dyunin, A. K. (1954). *Tr. Transp.-Energ. Inst., Akad. Nauk SSSR, Sib. Otd.* **4**, 59–69; *NRC Tech. Transl.* **NRC TT-1101** (1963).

Dyunin, A. K. (1959). *Izv. Sib. Otd. Akad. Nauk SSSR, Ser. Tekh. Nauk* **12**, 11–24; *NRC Tech. Transl.* **NRC TT-952** (1961).

Dyunin, A. K. (1967). *Phys. Snow Ice, Conf., Proc., 1966* Vol. 1, Part 2, pp. 1065–1073.

Dyunin, A. K. (1974). *Tr. Novosib. Inst. Inzh. Zheleznodorozhn. Transp.* **159**, 3–110.

Francis, J. R. D. (1973). *Proc. R. Soc. London, Ser. A* **332**, 443–471.

Garnier, B. J., and Ohmura, A. (1970). *Sol. Energy* **13**, 21–34.

Geiger, R. (1966). "The Climate Near the Ground," rev. ed. Harvard Univ. Press, Cambridge, Massachusetts.

Gerdel, R. W. (1949). *Proc. 17th Annu. Meet. West. Snow Conf., 1948* pp. 81–99.

Gerdel, R. W. (1954). *Trans. Am. Geophys. Union.* **35**, 475–485.

Giddings, J. C., and LaChapelle, E. (1961). *J. Geophys. Res.* **66**, 181–189.

Giddings, J. C., and LaChapelle, E. (1962). *J. Geophys. Res.* **67**, 2377–2383.

Gold, L. W. (1957). *Proc. Int. Geod. Geophys. Gen. Assembly Toronto, 1956* Vol. 4, pp. 13–21.

Granger, R. J. (1977). M. Sc. Thesis, Dept. Mech. Eng., University of Saskatchewan (unpublished).

Granger, R. J., and Male, D. H. (1978). *J. Appl. Meteorol.* **17**, 1833–1842.

Gray, D. M., and O'Neill, A. D. J. (1974). *In* "Advanced Concepts and Techniques in the Study of Snow and Ice Resources," (H. S. Santeford and J. L. Smith, eds.) pp. 108–118. Natl. Acad. Sci., Washington, D.C.

Hay, J. E. (1976). *Atmos.* **14**, 278–287.

Hendrie, L. K., and Price, A. G. (1979). *In* "Modeling Snow Cover Runoff" (S. C. Colbeck and M. Ray, eds.), pp. 211–221. U.S. Army Cold Reg. Res. Eng. Lab., Hanover, New Hampshire.

Hicks, B. B., and Martin, H. C. (1972). *Boundary-Layer Meteorol.* **2**, 496–502.

Hobbs, P. V., and Mason, B. J. (1964). *Philos. Mag.* [8] **9**, 181–197.

Hobbs, P. V., and Radke, L. F. (1967). *J. Glaciol.* **6**, 879–891.

Högström, U. (1974). *Q. J. R. Meteorol. Soc.* **100**, 624–639.

Keeler, C. M. (1969). "Some Physical Properties of Alpine Snow," Res. Rep. No. 271. U.S. Army Cold Reg. Res. Eng. Lab., Hanover, New Hampshire.

Keenan, J. H., Keyes, F. G., Hill, P. G., and Moore, J. G. (1969). "Steam Tables. Thermodynamic Properties of Water Including Vapor, Liquid and Solid Phases." Wiley, New York.

Kind, R. J. (1976). *Atmos. Environ.* **10**, 219–227.

Kobayashi, D. (1973). *Low Temp. Sci., Ser. A* **31**, 1–58.

Kojima, K. (1954). *Low Temp. Sci., Ser. A* **12**, 1–13.

Kojima, K. (1967). *Phys. Snow Ice, Conf., Proc., 1966* Vol. 1, Part 2, pp. 929–952.

Kondratyev, L. Ya. (1969). "Radiation in the Atmosphere." Academic Press, New York.

Kotlyakov, V. M. (1961). *Int. Assoc. Sci. Hydrol., Publ.* **55**.

Kovacs, A., Weeks, W. F., and Michetti, F. (1969). "Variation of Some Mechanical Prop-

erties of Polar Snow, Camp Century, Greenland,'' Res. Rep. No. 276. U.S. Army Cold
Reg. Res. Eng. Lab., Hanover, New Hampshire.

Kraus, H. (1973). *Role Snow Ice Hydrol., Proc. Banff Symp., 1972* Vol. 1, pp. 128–164.

Kuz'min, P. P. (1961). ''Protsess Tayaniya Shezhnogo Pokrova (Melting of Snow Cover).''
Gidrometeorol. Izd. Leningrad (Engl. transl. TT71-50095. Isr. Program Sci. Transl.,
Jerusalem, 1972).

Langham, E. J. (1974a). *In* ''Advanced Concepts and Techniques in the Study of Snow and
Ice Resources'' (H. S. Santeford and J. L. Smith, eds.), pp. 67–75. Natl. Acad. Sci.,
Washington, D.C.

Langham, E. J. (1974b). *Can. J. Earth Sci.* **11**, 1280–1287.

Langham, E. J. (1975). *IAHS-AISH Publ.* **114**, 73–81.

Lee, L. W. (1975). Ph.D. Thesis, Dept. Mech. Eng., University Wyoming, Laramie (unpublished).

Lee, R. (1962). Mon. Weather Rev. **90**, 165–166.

Lee, T. M. (1961). ''Note on Young's Modulus and Poisson's Ratio of Naturally Compacted
Snow and Processed Snow,'' Tech. Note. U.S. Army Cold Reg. Res. Engl. Lab., Hanover, New Hampshire (unpublished).

Lemmelä, R. (1973). *Role Snow Ice Hydrol., Proc. Banff Symp., 1972* Vol. 1, pp. 670–679.

Liljequist, G. H. (1957). *In* ''Norwegian-British-Swedish Antarctic Expedition 1949–1952,''
Sci. Results, Vol. II, Part 1C. Nor. Polarinst., Oslo.

List, R. J. (1966). ''Smithsonian Meteorological Tables,'' 6th rev. ed. Smithson. Inst.,
Washington, D.C.

Liu, B. Y., and Jordan, R. C. (1960). *Sol. Energy* **4**, 1–19.

Loewe, F. (1956). ''Etudes de glaciologie en Terre Adelie 1951–1952.'' Expedition Polaires
Française, IX. Paris.

McBean, G. A., and Miyake, M. (1972). *Q. J. R. Meteorol. Soc.* **98**, 383–398.

McKay, D. C., and Thurtell, G. W. (1978). *J. Appl. Meteorol.* **17**, 339–349.

Magono, C., and Lee, C. W. (1966). *J. Fac. Sci., Hokkaido Univ., Ser. 7* **2**, 321–325.

Male, D. H., and Granger, R. J. (1979). *In* ''Modeling Snow Cover Runoff'' (S. C. Colbeck
and M. Ray, eds.), pp. 101–124. U.S. Army Cold Reg. Res. Eng. Lab., Hanover, New
Hampshire.

Marks, D. (1979). *In* ''Modeling Snow Cover Runoff'' (S. C. Colbeck and M. Ray, eds.), pp.
167–178. U.S. Army Cold Reg. Res. Eng. Lab., Hanover, New Hampshire.

Matveev, L. T. (1965). ''Physics of the Atmosphere. Fundamentals of General Meteorology'' Gidrometeorologiche Skoe Izdatel'stvo (Transl. from Russian by Isr. Program
Sci. Transl., Jerusalem, 1967).

Mellor, M. (1965). *In* ''Cold Regions Science and Engineering'' (F. J. Sanger, ed.), Part III,
Sect. A3C. U.S. Army Cold Reg. Res. Eng. Lab., Hanover, New Hampshire.

Mellor, M. (1966a). *Appl. Mech. Rev.* **19**, 379–389.

Mellor, M. (1966b). *Int. Assoc. Sci. Hydrol., Publ.* **69**, 128–140.

Mellor, M. (1975). *IAHS-AISH Publ.* **114**, 251–291.

Mellor, M. (1977). *J. Glaciol.* **19**, 15–66.

Mellor, M., and Hendrickson, G. (1965). ''Confined Creep Tests on Polar Snow,'' Res. Rep.
138. U.S. Army Cold Reg. Res. Eng. Lab., Hanover, New Hampshire.

Miller, D. H. (1955). *Univ. Calif., Berkely, Publ. Geog.* **11**.

Miller, D. H. (1959). *Mitt. Schweiz. Anst. Forstl. Versuchswes.* **35**, 57–79.

Monin, A. S., and Obukhov, A. M. (1954). *Tr. Inst. Geofiz. Akad. Nauk. SSSR* **151**, 163–
187. Am. Meteorol. Soc. *(Engl. Transl.)* **T-R-174** (1959).

Müller, F. (1968). *In* ''Runoff From Snow and Ice,'' Vol. 2, pp. 33–51. Natl. Res. Counc.,
Ottawa, Ontario, Canada.

Nakaya, U. (1959a). "Visco-elastic Properties of Snow and Ice in the Greenland Ice Cap," Res. Rep. No. 46. U.S. Army Snow, Ice and Permafrost Res. Estab., Hanover, New Hampshire.

Nakaya, U. (1959b). "Visco-Elastic Properties of Processed Snow," Res. Rep. No. 58. U.S. Army Snow, Ice and Permafrost Res. Estab., Hanover, New Hampshire.

Nakaya, U. (1961). "Elastic Properties of Processed Snow with Reference to its Internal Structure," Res. Rep. No. 82. U.S. Army Snow, Ice and Permafrost Res. Estab. Hanover, New Hampshire.

Obled, Ch., and Harder, H. (1979). In "Modeling Snow Cover Runoff" (S. C. Colbeck and M. Ray, eds.), pp. 179–204. U.S. Army Cold Reg. Res. Eng. Lab., Hanover, New Hampshire.

Obukhov, A. M. (1946). Tr. Inst. Teor. Geofiz., Akad. Nauk SSSR No. 1; Boundary-Layer Meteorol. **2**, 7–29 (1971).

O'Neill, A. D. J., and Gray, D. M. (1973). Role Snow Ice Hydrol., Proc. Banff Symp., 1972 Vol. 1, pp. 176–186.

Oura, H., Ishida, T., Kobayashi, D., Kobayashi, S., and Yamada, T. (1967). Proc. Int. Conf. Low Temp. Sci., 1966 Vol. 1, Part 2, pp. 1099–1117.

Owen, P. R. (1964). J. Fluid Mech. **20**, 225–242.

Owen, P. R. (1969). J. Fluid Mech. **39**, 407–432.

Panofsky, H. A. (1974). Annu. Rev. Fluid Mech. **6**, 147–177.

Panofsky, H. A., and McCormick, R. A. (1960). Q. J. R. Meteorol. Soc. **86**, 495–503.

Penner, E. (1970). Can. J. Earth Sci. **7**, 982–987.

Perla, R. (1978). "Temperature-gradient and Equi-température Metamorphism of Dry Snow," Paper presented at Deuxième Rencontre Internationale sur la Neige et les Avalanches. Assoc. pour l'Etude de la Neige et des Avalanches, Grenoble, France.

Prandtl, L. (1925). Angew. Math. Mech. **5**, 136–139.

Priestley, C. H. B. (1959). "Turbulent Transfer in the Lower Atmosphere." Univ. of Chicago Press, Chicago, Illinois.

Radke, L. F., and Hobbs, P. V. (1967). J. Glaciol. **6**, 893–896.

Radok, U. (1968). "Deposition and Erosion of Snow by the Wind," Res. Rep. No. 230. U.S. Army Cold Reg. Res. Eng. Lab., Hanover, New Hampshire.

Radok, U. (1977). J. Glaciol. **19**, 123–139.

Ramseier, R. O. (1966). "Some Physical and Mechanical Properties of Polar Snow," Res. Rep. No. 116. U.S. Army Cold Reg. Res. Eng. Lab., Hanover, New Hampshire.

Reifsnyder, W. E., and Lull, H. W. (1965). U.S., Dep. Agric., Tech. Bull. **1344.**

Rikhter, G. D. (1945). "Snezhnyi pokrov, ego gormirovanie i svoistva (Snow Cover, its Formation and Properties)." Izv. Akad. Nauk SSSR, Moscow (Engl. transl. by U.S. Army Snow, Ice and Permafrost Res. Estab., Transl. No. 6 (1954).

Rikhter, G. D. (1960). "Geografiia snezhvogo pokrova (Geography of the Snowcover)." Izv. Akad. Nauk SSSR, Moscow.

Rylov, S. P. (1969). Tr. Kaz. Nauchno-Issled. Gidrometeorol. Inst. **32**, 64–77; Sov. Hydrol. Sel. Pap. (1969) 258–270.

St. Lawrence, W., and Bradley, C. C. (1975). IAHS-AISH Publ. **114**, 155–170.

Schmidt, R. A. (1972). U.S., For. Serv., Res. Pap. RM **90.**

Schmidt, R. A., and Sommerfeld, R. A. (1969). Proc. 37th Annu. Meet. West Snow Conf., 1969 pp. 88–91.

Shimizu, H. (1970). Low Temp. Sci., Ser. A **22**, 1–32.

Shiotani, M., and Arai, H. (1953). Proc. Jpn. Natl. Congr. Appl. Mech., 2nd, 1952 pp. 217–218.

Smith, J. L. (1965). "The Elastic Constants, Strength and Density of Greenland Snow as Determined from Measurement of Sonic Wave Velocity," Tech. Rep. No. 167. U.S. Army Cold Reg. Res. Eng. Lab., Hanover, New Hampshire.

Smith, N. (1969). "Determining the Dynamic Properties of Snow and Ice by Forced Vibration," Tech. Rep. No. 216. U.S. Army Cold Reg. Res. Eng. Lab., Hanover, New Hampshire.

Sommerfeld, R. A., and LaChapelle, E. (1970). *J. Glaciol.* **9,** 3–17.

Stanhill, G. (1966). *Sol. Energy* **19,** 96–101.

Steven, M. D. (1977). *Q. J. R. Meteorol. Soc.* **103,** 457–465.

Sutton, O. G. (1953). "Micrometeorology." McGraw-Hill, New York.

Swinbank, W. C. (1951). *J. Meteorol.* **8,** 135–145.

Swinbank, W. C. (1968). *Q. J. R. Meteorol. Soc.* **94,** 460–467.

Tabler, R. D. (1975). *In* "Snow Management on the Great Plains," Publ. No. 73, pp. 85–104. Res. Comm., Great Plains Agric. Coun. Univ. of Nebraska, Lincoln.

Temps, R. C., and Coulson, K. L. (1977). *Sol. Energy* **19,** 179–184.

Thorpe, A. B., and Mason, B. J. (1966). *Br. J. Appl. Phys.* **17,** 541–548.

U.S. Army Corps of Engineers (1956). "Snow Hydrology." U.S. Dept. of Commerce, Washington, D.C.

Wakahama, G. (1967a). *Phys. Snow Ice, Conf., Proc., 1966* Vol. 1, Part 1, pp. 291–311.

Wakahama, G. (1967b). *Phys. Snow Ice, Conf., Proc., 1966* Vol. 1, Part 2, p. 908.

Wakahama, G. (1968a). *Int. Assoc. Sci. Hydrol., Publ.* **79,** 370–379.

Wakahama, G. (1968b). *Low Temp. Sci., Ser. A* **26,** 77–86.

Wakahama, G. (1975). *IAHS-AISH Publ.* **114,** 66–72.

Webb, E. K. (1970). *Q. J. R. Meteorol. Soc.* **96,** 67–90.

Weisman, R. N. (1977). *Water Resour. Res.* **13,** 337–342.

Westergaard, H. M. (1964). "Theory of Elasticity and Plasticity." Dover, New York.

Williams, L. D., Barry, R. G., and Andrews, J. T. (1972). *J. Appl. Meteorol.* **11,** 526–533.

Yarkın, I. G., Tyutyunov, I. A., and Sadovskıı, A. V. (1972). *Osn., Fundam. Mekh. Gruntov* **4,** 21–22.

Yoshida, S. (1962). *J. Meteorol. Res.* **14,** 17–37.

Yosida, A. (1963). *In* "Ice and Snow" (W. D. Kingery, ed.), pp. 485–527. MIT Press, Cambridge, Massachusetts.

Yosida, A., *et al.* (1955). *Contrib. Inst. Low Temp. Sci., Hokkaido Univ.* **7,** 19–74.

Zingg, A. W. (1953). *Univ. Iowa Stud. Eng., Bull.* No. 34.

7 AVALANCHE RELEASE, MOTION, AND IMPACT

R. I. Perla
Environment Canada
Canmore, Alberta, Canada

I. INTRODUCTION

Avalanches may be a harmless trickle of loose snow descending to a new angle of repose, or a large devastating mass of snow, ice, and earth moving at high speed down a lengthy slope with enough energy to destroy anything in its path. Table 1 provides a qualitative scale of the effects of snow and ice avalanches versus vertical drop, volume of debris, and impact force.

Avalanche frequency decreases rapidly with increasing size. Sluffs and small avalanches are common and are taken for granted as part of the mountain scenery. In North America, approximately 10,000 avalanches are reported annually to the U.S. Forest Service but only a small portion (~1%) of these are responsible for damage and casualties. [See the following references for further information about U.S. avalanches: Gallagher (1967); Williams (1975a,b); Stethem and Schaerer (1979); Perla (1972a); B. R. Armstrong (1976, 1977, 1978)].

DYNAMICS OF SNOW AND ICE MASSES

TABLE 1

Scale of Snow and Ice Avalanches

Size	Potential effects	Order of magnitude estimates		
		Vertical descent	Volume	Impact pressure
Sluffs	Harmless	10 m	$1-10$ m^3	$<10^3$ Pa
Small	Could bury, injure, or kill a human	$10-10^2$	$10-10^2$	10^3
Medium	Could destroy a wood frame house, or auto	10^2	10^3-10^4	10^4
Large	Could destroy a village or forest	10^3	10^5-10^6	10^5
Extreme	Could gouge landscape, world's largest avalanches (Himalayas, Andes)	$10^3-5 \times 10^3$	10^7-10^8 Includes ice, soil, rock, mud	10^5-10^6

Avalanche casualties are more numerous in the densely settled European Alps where disasters were recorded as early as 218 B.C. [Allix (1925); Flaig (1955); Matznetter (1955); Wechsberg (1958); Fraser (1966)]. In Austria alone, 751 people died in the period 1950–1970 [Aulitzky (1974)]. In Switzerland, the 30-year average (1940–1970) is 25 fatalities and 13 injuries annually [Castelberg *et al.* (1972)]. Swiss and international avalanche statistics are published annually in "Schnee und Lawinen in den Schweizeralpen," by the Swiss Institute for Snow and Avalanche Research, Weissfluhjoch, Davos (1947–1973).

Avalanche disasters have also occurred in Japan [Yosida *et al.* (1963); Shimizu (1967)]; in Iran [Roch (1961)]; in Iceland [Jónsson (1957); Björnsson (1977)]; in Norway [Ramsli (1974)]; in Chile [Atwater (1968, 1970); Leon (1978)]; and generally in any country with a combination of mountains and snow. Avalanches have also caused fatalities in such unlikely places as Lewes, Sussex, England [Southern (1976)], and Toronto, Ontario, Canada [Ommanney (1978)].

In terms of human suffering, the most destructive avalanches on record occurred in the Peruvian Andes in 1962, and again in 1970. In 1962, a large mass of ice and snow from Mount Huascarán triggered a massive avalanche of ice and snow that mixed with soil and rock from the lower slopes. The avalanche traveled 16 km into the Santa Valley, destroying 9 small towns, and killing thousands of livestock and 4000 persons [Morales (1966)].

In 1970, an earthquake released from Mount Huascarán an ice–snow avalanche that coupled to a landslide. Total vertical descent was over

4000 m, and estimated volume (10^7-10^8 m³) was approximately 5–10 times the estimated volume of the 1962 catastrophe. This time the avalanche completely buried Yungay, killing over 18,000. Debris flow in the Rio Santa caused additional problems, affecting settled areas over a distance of several hundred kilometers [Plafker and Ericksen (1978)]. These avalanches demonstrate the combined destructive power of earthquake, snow–ice avalanche, and landslide. The snow–ice mass comprised perhaps only 10% of the total avalanche, but was sufficient to liquefy the flow and greatly extend runout distances.

Although avalanche phenomena include descending snow, ice, mud, soil, and rock in various combinations, both on land surfaces and beneath the ocean (submarine flows, turbidity currents), only avalanches consisting entirely of snow or ice will be treated in detail here.

Snow avalanches commonly involve the failure of snow layers with density ρ less than 300 kg/m³, and typically occur in response to a storm period of heavy precipitation. Firn snow ($500 < \rho < 800$ kg/m³) is extremely stable on inclined slopes unless appreciable amounts of rain or meltwater reduce cohesion. As the density of dry firn increases beyond about 800 kg/m³, the firn transforms into glacial ice; that is, pores no longer interconnect but exist as separate air bubbles. Slope instability of glacier ice (ice avalanche) involved mechanisms distinctly different from those observed for snow slopes. The emphasis here is on snow avalanches rather than ice avalanches, since investigations of the latter are comparatively rare, so there is not a well-developed literature to draw upon.

It is customary to consider an avalanche path as consisting of three zones: the starting zone, the track, and the runout-deposition zone. The failure process begins in the starting zone. The dynamic characteristics of the moving avalanche depend then on the relief of the track and the amount of material that is entrained into the avalanche as it gains momentum. First, we shall describe snow slope instability in the starting zone, i.e., the initiation of avalanches; then we shall deal with avalanche dynamics in the track and runout-deposition zone, including impact forces and runout distances.

It is not possible to include all the theoretical and experimental developments related to avalanche technology here. Topics recently reviewed elsewhere include the following: avalanche forecasting [Salway (1976); LaChapelle (1977); Föhn et al. (1977)]; avalanche warning systems [Judson (1977)]; artificial release of avalanches [Gubler (1977); Perla (1978a)]; avalanche mapping and zoning [Martinelli (1974); Burrows and Burrows (1976); Ives and Bovis (1978)]; and avalanche geomorphology [Luckman (1977)]. The technology of avalanche defense structures (including reforestation) is summarized in several monographs: Frutiger and Martinelli (1966), Mellor (1968), and Castelberg *et al.* (1972). The widely used Swiss design criteria for supporting structures are set out in EIDG (1968). General guidelines for avalanche control, safety, and rescue are provided in Mellor (1968), Perla and Martinelli (1976), and the proceedings of symposia held at Davos, Switzerland [Fondation Internationale "Vannie Eigenmann" (1963)]; at Sulden, Italy [Fondation Internationale "Vannie Eigenmann" (1975)]; and at Banff, Alberta [Perla (1978b)]. Theoretical foundations for avalanche rescue are outlined by Good (1972).

II. MECHANISM OF AVALANCHE RELEASE

A. Loose-Snow Avalanches

On the basis of initiation mechanism and starting-zone failure patterns, avalanches are generally classified according to the relative cohesiveness of the snow: loose-snow avalanches and slab avalanches [de Quervain (1966b)].

Loose-snow avalanches start in cohesionless surface layers of either dry, unsintered snow, or wet snow containing liquid water. Initial failure is analogous to the rotational slip of cohesionless sands or soil, but occurs within a small volume (<1 m³) in comparison to initiation volumes in soil slides (e.g., bank sluffs). The initial failure may involve a mass no larger than a "snowball." As soon as it breaks loose, the failed clump moves down the slope, entraining a widening triangular pattern of snow, which when observed from a distance gives the impression that the avalanche originated at the apex of the triangle; loose-snow avalanches are sometimes called point avalanches for this reason.

During cold, dry snow storms there is a continuous surface sluffing of loose snow from slopes steeper than about 45–50°. Typically, these are very small avalanches (sluffs) that simply transfer the snow load to lower angles of repose, stabilize the slope, and prevent a much larger avalanche. However, loose-snow avalanches sometimes entrain large amounts of snow from the track and grow to hazardous size, especially if thaw conditions (solar radiation, rain) cause wet, cohesionless grains to form deep in the pack. Loose-snow avalanches may also trigger more massive, and potentially more dangerous, slab avalanches (discussed later).

Since the loose-snow avalanche involves relatively cohesionless snow, in a first approximation, failure can be assumed to occur for many cases when the angle of repose is exceeded. Table 2 gives values of the angle of repose of various particles and sands [Carrigy (1970)] and values found

TABLE 2

Approximate Angles of Repose of Snow and Other Particles

Material	Angle of repose (deg)
Polished steel balls ($d \approx 3$ mm)	25
Lead shot, graphite lubricated ($d \approx 3$ mm)	23
Ottawa sand, oven dry ($d \approx 0.5-1.0$ mm)	36
Ottawa sand, 0.95% moisture ($d \approx 0.5-1.0$ mm)	43
Dune sand (N.E. Alberta)	35
Sieved depth hoar ($d \approx 2$ mm)	32–37
Pulverized snow ($d \approx 3.5$ mm, $T = -35°C$)	40
Pulverized snow ($d \approx 1.4$ mm, $T = -35°C$)	41
Pulverized snow ($d \approx 0.6$ mm, $T = -35°C$)	42
Pulverized snow ($d \approx 0.6$ mm, $T = -3.5°C$)	55
Natural dendrites ($T = -35°C$)	63
Natural dendrites ($T = -4°C$)	~90
Pulverized snow ($d \approx 0.6$ mm, $T = 0°C$)	90

from tests on mounds of snow consisting of grains of various shapes, sizes, and at various temperatures [Roch (1955); Kuroiwa et al. (1967)]. The experimental snow mounds were built rapidly from disaggregated, sieved snow to minimize sintering. Nonetheless, angles of repose showed marked increases as temperatures approached 0°C, probably due to the increased rate of sintering at warmer temperatures and the resulting increase in cohesion.

The behavior of loose-snow slopes is similar to the trends shown in Table 2. During snowstorms constant surface sluffing occurs on steep slopes ($>45°$) if temperatures are relatively cold ($<-10°C$) and atmospheric humidity is relatively low, conditions that minimize sintering in surface layers. At warmer temperatures ($-5 < T < 0°C$), snow may cling to steeper slopes, increasing the thickness of the layer until stress exceeds strength at depth, or until strain softening reduces strength. Either mechanism may cause a loose-snow avalanche of substantial thickness (~ 0.5 m).

From field observations it is known that the angle of repose drops drastically at 0°C if continued heating of the surface produces liquid water. For example, Ambach and Howorka (1966) noted initiation of avalanche activity at a liquid water content of 7.5% by volume, and noted a major wet avalanche event correlated with an increase of liquid water to 10%. The lowest angles of repose are observed when snow assumes a slushlike consistency at liquid water contents greater than about 25%. Slush avalanches over 100 m wide and over 1 km long occur on gentle slopes (10°) at arctic latitudes and at the base of glaciers where large amounts of water percolate down to impervious surfaces formed during the previous year. The angle of repose of a slush slope may be as low as 2° [Rapp (1963); Nobles (1966)].

Downslope propagation of both dry and wet loose avalanches involves the transition to the kinetic angle of repose, which is less than the static value because of the almost complete loss of cohesion. Roch (1955) mentions that snow grains set in motion from a static angle of repose ($\sim 40°$) come to rest at approximately 17°. This is a rather low angle compared to sands that come to rest on slopes of about 28–34°, depending on texture, moisture, and grain size [Carrigy (1970)].

B. Slab Avalanche Observations

The second category of avalanches involves the release of a cohesive slab over an extended plane of weakness, analogous to the planar failure of rock slopes rather than to the rotational failure of soil slopes. As shear failure spreads along the slip-plane, high tensile stress develops at the upslope boundary and slab failure appears to the observer to begin with the spectacular propagation of tensile fractures across the slope, sometimes continuing as far as 1 km. Areas of shear fracture along the plane of weakness may range from 10^2 to 10^5 m. Slab thickness also varies considerably; thicknesses of more than 1 m are not unusual (see Fig. 1).

Once set in motion, the slab may entrain additional mass; however, it is the large volume of the original slab that generally makes a slab avalanche more dangerous than a loose-snow avalanche. After slab failure, sharply defined fracture surfaces outlining the slab boundaries remain at the starting zone. Based on nomenclature proposed by Varnes (1958) to describe landslide failure patterns, the fracture surfaces of the slab avalanche are designated as shown on Fig. 2. The bed surface and stauchwall experience shear failures while the crown surface experiences a tensile failure. The flank surfaces experience a combination of these two modes of failure.

Several investigators have climbed into starting zones after slab release to study the geometry of fracture surfaces and to measure properties of slab stratigraphy immediately above the crown surface. From published observations and numerous photos [e.g., Perla (1978c); Perla and Martinelli (1976)] the following qualitative picture of slab geometry emerges:

(1) Bed surfaces are flat to a good approximation. The radius of curvature is rarely less than 10^2 m. This indicates reasonable approximations for the respective shear and normal stress along the bed surface prior to

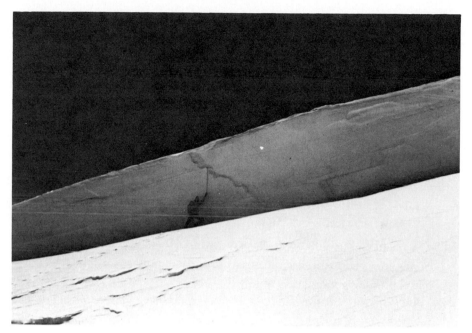

Fig. 1. Example of slab avalanche failure (photo by Knox Williams).

failure are given by

$$B_{xz} \approx \rho g Z \sin \theta \qquad (1)$$

and

$$B_{zz} \approx -\rho g Z \cos \theta, \qquad (2)$$

where θ is the bed surface inclination, Z the slab thickness, and ρ the slab density averaged to depth Z.

(2) The bed surface and the crown surface are approximately perpendicular; the intersecting angle varies along the crown with an overall tendency to slightly exceed 90°. The bed surface of a slab rarely curves up to join the crown surface as in the rotational failure of soil slopes.

(3) Stauchwalls are not always observed, probably because the moving slab obliterates this downslope feature. In a larger number of cases, it is possible that no stauchwall exists, but rather the bed surface tapers gradually to intersect the snow surface.

(4) The crown fracture tends to propagate as a continuous smooth fracture, often as a relatively smooth arc between the flanks, but sometimes will zigzag between natural stress concentrations such as rocks and trees.

(5) The flank fractures often have a peculiar sawtooth pattern that

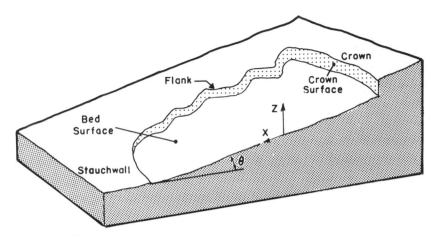

Fig. 2. Slab avalanche nomenclature and coordinate system.

consists of intersecting tension and shear fractures. Roch (1955) calls this pattern *fermetures éclair* ("zipper"). Examples are given by Perla (1978c) and Lang (1977).

(6) The flank-to-flank dimension tends to exceed the crown-to-stauchwall dimension [Brown *et al.* (1972)]. Exceptions are elongated slabs in gullies.

(7) The observed ratio of the flank-to-flank dimension to slab thickness varies between $10-10^3$, and is typically $\sim 10^2$. Plane-strain (or strain-rate) assumptions may be applied along a slab cross section sufficiently centered with respect to the flanks. The slab may also be considered as a shell or plate in a three-dimensional analysis.

(8) An extensive crown fracture ($\sim 10^3$ m) may propagate across the slope, linking several smaller slabs into one avalanche. When this occurs it is likely the bed surface shear failure and crown fracture are closely synchronized [Perla (1975)] because of the observed intersection angle formed by the crown and bed surface.

Table 3 presents statistics of slab measurements compiled by Perla (1977) from various sources. The statistics are based on data taken at the crowns of 205 large slabs. The mean thickness of the sampled slabs was 0.67 m; this probably exceeds the population mean thickness since small slabs were not included in correct proportion.

Figure 3 illustrates the distribution of 194 measurements of bed surface angle (θ). Considering the large number of cases and the small standard deviation (4.79°), high confidence can be placed on the rule of thumb that slab avalanches require a starting zone where a part of the slope has an

TABLE 3

Summary of Slab Measurements[a]

No. of cases	Parameter	Range		Mean	Standard deviation
		Maximum	Minimum		
(193)	Slab thickness Z (m)	4.2	0.08	0.67	0.43
(194)	Slope angle θ (deg)	55	25	38.3	4.79
(121)	Mean density $\bar{\rho}$ (kg/m³)	461	60	206	77.2
(121)	Shear stress at bed surface B_{xz} (Pa)	9050	65	964	1049
(72)	Density at bed surface ρ_B (kg/m³)	400	90	231	75.9
(111)	Temperature at bed surface T_B (°C)	0	−13	−4.58	3.08

[a] From Perla (1977).

inclination of 25° or greater. It is possible that the sample mean (38.3°) is lower than the population mean since small slabs that release from inaccessible steep slopes and cliffs are not included. The bed surface shear stress (B_{xz}) also would be expected to reach a maximum on a 45° slope if slab thickness Z varies as cos θ [Mellor (1968)]. On the other hand, the

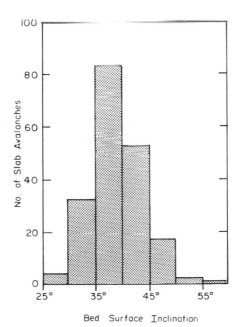

Bed Surface Inclination

Fig. 3. Number of slab avalanches versus bed-surface inclination θ (194 cases).

mean bed surface inclination of about 38° may be significant since failure of certain types of soil slopes are especially frequent at specific angles; for example, Karta till soil fails at 37° [Swanston (1970)].

The mean density $\bar{\rho}$ of 121 slabs was found to be 206 kg/m³, about twice the density of undisturbed newly fallen snow (≈ 100 kg/m³), indicating that slabs have a wind-stiffened or sintered texture.

Bed-surface shear stress B_{xz} computed from Eq. (1) ranges from 10^2 to 10^4Pa, with a high standard deviation around the mean at 10^3Pa. These are extremely low stress levels compared to failure levels of other solid earth materials. Figure 4 shows the scatter of B_{xz} versus density ρ_b measured within a layer at the bed surface level, providing a rough measure of how the shear strength of weak layers varies as a function of density. A large amount of scatter is expected because strength is fundamentally a function of snow texture rather than density [Gubler (1978a)], and also because density samples were 50 mm thick, whereas the failure plane of the bed-surface level may be confined to a discontinuity of crystal dimensions (~ 1 mm). A possible curve, fit to the data of Fig. 4, which explains 65% of the variance, is

$$B_{xz} = 0.01\rho_B{}^2. \tag{3}$$

Equation (3) yields shear strength values considerably less than strengths derived from laboratory tests of small samples, possibly due to scale effects (bed surfaces are many orders of magnitude larger than lab samples) but probably because bed surfaces include the weakest textures possible for a given density.

Figure 5 shows the distribution of 111 measurements of bed-surface temperature. Clearly, the predominant band is from -5 to 0°C; moreover,

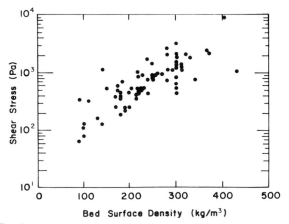

Fig. 4.   Shear stress at bed surface versus bed-surface density.

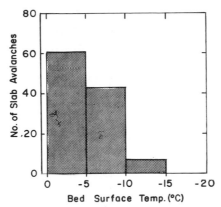

Fig. 5. Number of slab avalanches versus bed-surface temperature.

the population distribution may be further skewed toward 0°C because dry slabs were sampled almost exclusively; only 4 of the 111 measurements were taken after 31 March, and therefore a representative proportion of wet slabs ($T_B = 0$°C) was excluded from the statistics.

Qualitative observations of the snow texture at the crown have also been made. Figure 6 shows one example of crown stratigraphy at Alta, Utah [Perla (1971)]. The relative frequency of a specific slab texture depends on the precipitation, wind, and temperature. Probably the most frequent slab stratigraphy is a layer of newly deposited snow consisting of wind-blown fragments (~ 0.1 mm) that sinter rapidly into a stiff texture

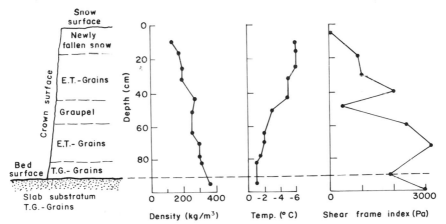

Fig. 6. Example of slab stratigraphy at Alta, Utah [Perla (1971)]. E.T. and T.G. refer respectively to equitemperature and temperature gradient metamorphosed grains. Shear frame index is a relative measure of strength obtained in situ.

[Seligman (1936)]. Due to settling, density may increase somewhat with depth down to the bed surface, which is typically a weak layer, either deposited at colder air temperatures than prevailed during slab deposition or deposited with less wind. The combination of a relatively stiff layer over a weak layer of newly fallen snow also correlates with variations in crystal habit as determined by cloud temperature and supersaturation, or with variations in crystal riming. One example is a heavily rimed slab layer lying on a bed surface of lightly rimed crystals [LaChapelle (1967)]. Occasionally, bed surfaces contain a "ball-bearing" grain known as graupel, a crystal completely encased in rime of a diameter about 1–2 mm [Pahaut (1975)].

In continental ranges such as the Colorado and Canadian Rockies, cloudless dry skies in the winter allow a relatively large portion of terrestrial radiation to escape from the snow surface. The surface temperature drops, and significant temperature gradients (~ 10 deg/m) develop across the snowpack, especially during the fall and early winter when the snowpack is thin. Strong temperature gradients drive recrystallization, as described in Chapter 6, leading to the growth of weak, coarse crystals which may assume very weak, hollow shapes (diameter 1–10 mm). The largest and weakest grains (depth-hoar) are found just above the ground where recrystallization is more rapid due to warmer snow temperatures (ground temperatures are approximately 0°C at temperate latitudes). Coarse-grained layers are often found in the crown substratum, although it is not clear why the bed surface is often located above the stratigraphic plane which demarcates coarse-grained from finer-grained layers. A possible explanation is given by Bradley *et al.* (1977).

Crystallization may also occur at the snow surface due to locally strong temperature gradients. The prime example is surface hoar, an exceptionally weak surface texture composed of feathery crystals [Pahaut (1975)] that are formed by vapor deposition from a moist surface boundary layer of the atmosphere onto the radiation-cooled snow. Buried under a layer of new snow, surface hoar is difficult to detect and thus an important cause of unexpected avalanches [de Quervain (1946)].

Another example is the radiation recrystallized layer [LaChapelle (1970); R. L. Armstrong *et al.* (1974)] formed when outgoing long-wave radiation cools the snow surface at the same time that incoming solar radiation warms the subsurface, causing an intense temperature gradient across a depth of about 10 mm. The subsurface layer melts during the day and refreezes during evening into a thin crust. The net result is a weak assortment of temperature gradient grains (hollow prisms) supported on a thin, friable crust.

The scenario changes in late spring when the most common instability is due to thermal weakening of the bed surface by heat conduction, radiation, and meltwater percolation in various complex combinations. The slab may loosen from a crust, or may slide directly on the ground or grass. In some cases, wet slab fractures may open and propagate very slowly (~ 1 m/day), in contrast to the rapid fracture propagation (~ 100 m/sec) always observed during failure of dry slabs, and very often observed even if the slab is thoroughly wetted.

C. Stress Analysis of the Infinite Slab

The analysis of slab stress t_{ij} properly begins with a study of the simplest case, the inclined slab extending to infinity in the x and y directions, bounded by parallel planes $z = 0$ and $z = Z$. A set of Cartesian coordinates (Fig. 7) is fixed in space at the slab substratum. For this geometry, all gradients in the x and y directions vanish and the components t_{xy} and t_{yx} are zero. Haefeli (1963, 1967) called this a neutral zone problem since it is considered a model of the central region of the slab away from the influences of the slab boundaries.

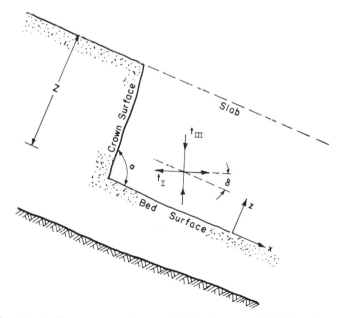

Fig. 7. Principal stress convention used in this chapter. Algebraic maximum principal stress is t_1 and forms an angle δ with the x direction. Angle α formed by crown and bed surface is nearly perpendicular ($\pm 10°$). It is hypothesized that at failure, δ shifts to approximately zero.

Until the time of slab failure, accelerations are negligible so the equilibrium equations for the slowly deforming slab are

$$\partial t_{xz}/\partial z + \rho(z)g \sin \theta = 0,$$
$$\partial t_{zz}/\partial z - \rho(z)g \cos \theta = 0, \tag{4}$$
$$\partial t_{yz}/\partial z = 0,$$

which are easily integrated from the surface ($z = Z$; $t_{xz} = t_{zz} = t_{yz} = 0$) to an arbitrary depth, giving

$$t_{xz}(z) = P(z) \sin \theta,$$
$$t_{zz}(z) = -P(z) \cos \theta, \tag{5}$$
$$t_{yz}(z) = 0,$$

where $P(z)$ is computed from the density profile

$$P(z) = g \int_z^Z \rho(z) \, dz. \tag{6}$$

The remaining task is to find the missing components t_{xx} and t_{yy} from a constitutive assumption that describes the response of snow. It is possible to invest much effort in completing the solution with a complex constitutive law that matches the nonlinear viscoelastic properties of snow [Brown *et al.* (1973); Salm (1975)], but it should be understood that the geometry and boundary conditions of the infinite slab taken together with thermodynamic limitations on material constants impose constraints on the principal stress solutions, irrespective of the precise form of the constitutive law. To illustrate this point, assume first that snow is a linear, isotropic, elastic solid. The stress t_{ij} is therefore related to displacements u_i according to the constitutive law

$$t_{ij} = \lambda \delta_{ij} \, \partial u_k/\partial x_k + \mu(\partial u_i/\partial x_j + \partial u_j/\partial x_i), \tag{7}$$

where λ and μ are elastic parameters (functions of z) that are related to the elastic modulus E, and Poisson's ratio ν by

$$\lambda = E\nu/(1 + \nu)(1 - 2\nu), \qquad \mu = E/2(1 + \nu). \tag{8}$$

For the infinite inclined slab ($u_x \neq 0, u_z \neq 0, u_y = 0, \partial/\partial x = \partial/\partial y = 0, \partial/\partial z \neq 0$), the expansion of Eq. (7) gives the components $t_{xx}, t_{yy},$ and t_{zz} as

$$t_{xx} = t_{yy} = \lambda \, \partial u_z/\partial z, \tag{9}$$

$$t_{zz} = (\lambda + 2\mu) \, \partial u_z/\partial z. \tag{10}$$

From the above equations,

$$t_{yy} = t_{xx} = [\nu/(1 - \nu)]t_{zz} = -[\nu/(1 - \nu)]P \cos \theta. \tag{11}$$

In exactly analogous fashion, the identical solution is obtained if snow is assumed to be a linear, isotropic, viscous fluid, with t_{ij} a function of the deformation rate tensor, and with ν reinterpreted as the viscous analog of Poisson's ratio. In either the elastic or viscous case, the second law of thermodynamics imposes the same strict bounds on ν,

$$-1 \leq \nu \leq \tfrac{1}{2}. \tag{12}$$

Tests of the elastic response of materials invariably show that $\nu \geq 0$, although sometimes small negative ν values are reported. It is conceivable that in a slow viscous test snow could exhibit negative ν values because of internal deformation due to metamorphism; however, in the laboratory experiments to date, both elastic and viscous ν values appear to be non-negative and to increase in the range $0 \leq \nu \leq \tfrac{1}{2}$ [Mellor (1975)].

The principal stresses t_{I}, t_{II}, and t_{III} are

$$t_{\mathrm{I,III}} = -\frac{P \cos \theta}{2} \left\{ \frac{1}{1-\nu} \mp \left[\left(\frac{2\nu - 1}{1 - \nu} \right)^2 + 4 \tan^2 \theta \right]^{1/2} \right\} \tag{13}$$

and

$$t_{\mathrm{II}} = -[\nu/(1 - \nu)]P \cos \theta. \tag{14}$$

Figure 8 shows t_{I}/P and t_{III}/P as functions of ν for the range $-\tfrac{1}{2} \leq \nu \leq \tfrac{1}{2}$, and for the slope inclinations typical of starting zones $30° \leq \theta \leq 50°$. Figure 9 shows the variation of the direction of t_{I} (angle δ in Fig. 7) as a function of ν, computed from

$$\tan 2\delta = [2(1 - \nu)/(1 - 2\nu)] \tan \theta. \tag{15}$$

Figures 8 and 9 must also provide reasonable approximations for solutions based on a linear, isotropic, viscoelastic constitutive law, since the viscoelastic response falls somewhere between the purely viscous and purely elastic case. Looked at another way, the given boundary conditions and loading are not functions of time so there is really no way for viscoelastic properties such as relaxation to affect the infinite slab solutions.

There is some theoretical and experimental evidence that anisotropy and nonlinearity do not greatly increase the bounds of the above solution. Snow slabs are composed of layers and hence are macroscopically anisotropic; moreover, within each layer appears a certain degree of microscopic anisotropy as revealed by thin section analysis [Kojima (1960)]. If the slab is considered orthotropic (properties symmetrical about z axis), then Eq. (11) is replaced by

$$t_{yy} = t_{xx} = (C_{xz}/C_{zz})t_{zz}, \tag{16}$$

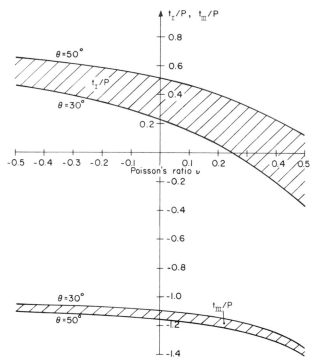

Fig. 8. Normalized maximum (t_I/P) and minimum (t_{III}/P) principal stress as a function of Poisson's ratio ν for band of typical slab inclinations $30° \leq \theta \leq 50°$. Normalizing factor is $P = g\int p \, dz$.

where C_{xz} and C_{zz} are anisotropic constants. Not much is known about the range of anisotropic constants for snow, but for simpler materials experimental measurements summarized by Ambartsumyan (1970) and the theoretical work of Pickering (1970) indicate that

$$0 \leq C_{xz}/C_{zz} \leq 1, \tag{17}$$

which corresponds to the bound on $\nu/(1 - \nu)$ in Eq. (11).

There is also some experimental evidence that the bounds apply irrespective of the linearity assumption. Suppose snow is a nonlinear, isotropic, viscous fluid. Then t_{ij} becomes a function of the deformation rate tensor

$$d_{ij} = \tfrac{1}{2}(\partial v_i/\partial x_j + \partial v_j/\partial x_i), \tag{18}$$

where v_i is the creep velocity in the ith direction. As a consequence of assumed material isotropy, the principal directions of t_{ij} and d_{ij} must coin-

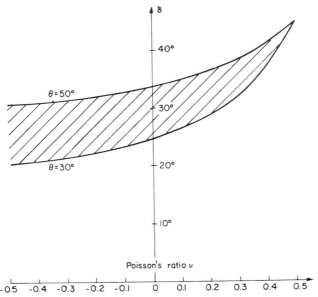

Fig. 9. Angle δ between maximum principal stress t_1 and x axis as a function of Poisson's ratio v.

cide; thus

$$t_{ik}d_{kj} = d_{ik}t_{kj}. \tag{19}$$

Carrying through the algebra of Eq. (19) for the infinite slab geometry, and substituting $P \sin \theta$ for t_{xz} and $-P \cos \theta$ for t_{zz} gives a result analogous to Eq. (11),

$$t_{xx_t} = [1 + 2[(\partial v_z/\partial z)/(\partial v_x/\partial z)] \tan \theta]t_{zz}, \tag{20}$$

which allows v to be generalized to the nonlinear case by setting

$$\frac{v}{1-v} = 1 + 2[(\partial v_z/\partial z)/(\partial v_x/\partial z)] \tan \theta. \tag{21}$$

A v value can therefore be associated with in situ measurements of $v_z(z)$ and $v_x(z)$ in a long inclined slab. Such measurements can be made by inserting lightweight tracers (Ping-Pong balls, sawdust columns) along a z reference line, and later recording the new position of each tracer. Measurements to date [Perla (1973); McClung (1975); Salm (1977)] indicate nonlinear v values are in the range of Eq. (12), although it is of interest that Perla (1973) found relatively large negative values ($v \approx -\frac{1}{2}$) for low-density snow ($\rho \approx 100$ kg/m³). For relatively high-density snow ($500 \le \rho \le 550$ kg/m³), McClung (1975) found values usually associated with engineering materials ($0.23 \le v \le 0.32$). Several investigators also have computed principal stress directly from snow deformation profiles [Yosida (1963)], and deformation ellipses [Shimizu and Huzioka (1975)], finding values within the range predicted by Figs. 8 and 9. On the other hand, the important case of shear deformations at potential slip surfaces is still open since anisotropic soils have ν values outside the linear range [Lee and Ingles (1968)]. Thus, the solutions given in Figs. 8 and 9 are considered valid only within a long, nearly homogeneous slab, but not at a potential bed surface where high shear strain rates could force ν outside the range of Eq. (12).

With the aid of Figs. 8 and 9, it is possible to examine a model of snow slab stability in the neutral zone, proposed by Haefeli (1963, 1967). According to our interpretation of Haefeli's model, if ν is below a critical value for a given slope, t_I is positive and the slab is vulnerable to tension fractures that could subsequently initiate bed-surface fracture. As the slab density increases due to natural settlement, ν increases and t_I decreases with a stabilizing effect. Although Haefeli's model provides an interesting overview of the stress transformation in the neutral zone, there are several reasons why it does not explain the essential aspects of snow slab stability. First of all, as shown in Fig. 8, t_I is not a strong function of ν, and moreover is always positive on slopes greater than 45°. Negative values (compressive t_I) are possible only for the combination of low-angle slopes and high ν values. It is difficult to believe the spectacular difference between stable and unstable slabs depends critically on the rather flat curves of Fig. 8. Also, as shown in Fig. 9, t_I does not rotate enough to explain why crown fractures are approximately perpendicular to the bed surface ($\alpha \approx 90°$ in Fig. 7). The alignment of t_I in the slope parallel direction ($\delta \rightarrow 0$) is only possible if t_{xz} decreases or t_{xx} increases outside the range allowed by the infinite slab solution. Moreover, increases in density tend to increase stability, but this is probably due to the increase in strength as a function of density rather than to a decrease in t_I. Virtually all studies of snow strength indicate that strength increases exponentially with density [e.g., Mellor (1975)].

If stress transformation in the neutral zone is important, it may be in connection with initiation of loose-snow avalanches where the slip-surface curves out of the xy plane, producing a rotational rather than slab-like failure. The loose snow may have a low (or negative) ν value, a positive t_I value, and therefore a relatively high shear stress on a potential slip-surface tilted with respect to the xy plane. A propagating shear failure would have a strong tendency to curve up through the loose snow, releasing a small localized mass (rather than an extended slab). As the layer becomes more dense, the critically stressed plane rotates towards the xy plane and the required failure stress increases, with reduced chances for a loose-snow avalanche. Jaccard (1966) studied stress metamorphism in connection with the Coulomb–Navier criterion, but did not discuss the possibility of failure propagation out of the *xy* plane.

D. Stress around Slab Boundaries

There is another important objection to Haefeli's idea that critical slab tensions t_1 depend on a relatively low ν value. This objection is based on stresses outside the neutral zone. The infinite slab solutions presented in the last section are assumed to model the stress state near the central region of the slab (the neutral zone). The solutions are less realistic near the slab boundaries where *x* and *y* gradients cannot be neglected. Terrain configurations which clamp and restrain the slab above the crown region and below the stauchwall region modify the neutral zone assumption that gradients in the *x* direction are zero; and similarly, restraints near the flanks introduce terms involving *y*-direction gradients, deformation, and velocity. This section will discuss separately the stress distribution due to the crown influence and due to the flank influence.

Figure 10 idealizes two common crown configurations: Fig. 10a, the slab which tapers against a cliff face, and Fig. 10b, the slab formed on the leeward side of rolling terrain. In either case the crown taper alters the infinite slab solutions. The stress distribution of the configuration shown in Fig. 10a was studied by Perla (1975) for two cases, $\nu = 0$ and $\nu = 0.4$. It turned out that, although the neutral zone value of t_1 was higher when $\nu = 0$, the t_1 value near the crown was higher when $\nu = 0.4$. Intuitively, it is reasonable to assume that the stiffer the slab the greater the influence of the clamping effect. It is therefore not possible to accept Haefeli's model where the critical value of t_1 is determined by relatively low ν values. In fact, it appears to be precisely opposite in nature.

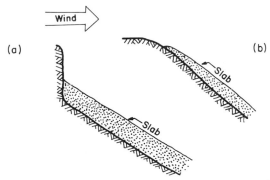

Fig. 10.   Slab tapering against cliff (a) and on lee side of slope (b).

The computations repeated by Perla (1975) indicate that even for an exaggerated case of clamping t_1 does not exceed $\sim 0.5\rho g \sin \theta$ (or about one-half the bed-surface shear stress B_{xz}). Thus, clamping effects alone are not high enough to trigger instability since most field tests to date [Perla (1969); McClung (1974)] indicate that the tensile strength of the homogeneous slab is greater than the shear strength of the bed surface. Therefore, the bed-surface conditions rather than the tensile conditions are initially critical. A case is made in the next section that shear weakening at the bed surface drives t_1 to a critical value, and that tension fracture in most cases propagates up from the bed surface. However, the crown taper causes t_1 to increase near the surface where the slab is usually unconsolidated and weaker, so the possibility of initial tensile fracture from the surface downward cannot be dismissed for all cases.

A model can also exaggerate the clamping at side boundaries. Consider an inclined rectangular gully extending to infinity in the x direction (Fig. 11), bounded by flank planes $y = 0$ and $y = L$, and bounded by planes $z = 0$ (bed surface) and $z = 1$ (snow–atmosphere interface). Displacements are held fixed along planes $y = 0$, $y = L$, and $z = 0$. The slab is assumed homogeneous. Assuming linearity, the stresses for this model can be found by superimposing the stress solution corresponding to the x-direction body force and the solution corresponding to the z-direction body force.

Considering the x-direction body force first, the new component of interest is t_{xy}, which has a maximum value of about $0.75\rho g \sin \theta$ when evaluated at the upper corner formed by the side walls ($y = 0, L$) and the

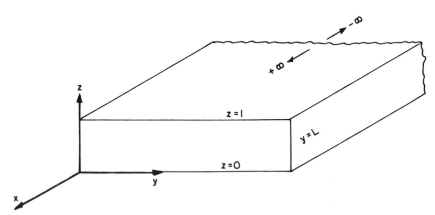

Fig. 11. Coordinates for rectangular slab in gulley. Displacement fixed on planes: $y = 0$, L and $z = 0$. Gravitational body force has components in x and z directions.

snow surface ($z = 1$). In response to the x-direction body force, the principal stresses (eigenvalues) are

$$t_{\text{I,III}} = \pm (t_{xy}^2 + t_{xz}^2)^{1/2} \tag{22}$$

with principal directions (eigenvectors)

$$(x, y, z)_{\text{I}} = (1, t_{xy}/t_{\text{I}}, t_{xz}/t_{\text{I}}), \tag{23a}$$

$$(x, y, z)_{\text{III}} = (1, t_{xy}/t_{\text{III}}, t_{xz}/t_{\text{III}}). \tag{23b}$$

At the flank sidewalls, the principal stresses (magnitude $\pm t_{xy}$) lie in the xy plane and are inclined 45° to the bed surface in the xz plane. The rotation of the principal stress would force a tensile crack to propagate along a curved trajectory into the flanks, in agreement with observations of crown fractures arcing across the top of a gully. The clamping effect of the flanks causes the stress to increase near the surface ($z = 1$), as occurred in the vicinity of the crown taper.

Solutions for the z-direction body force are conveniently found by the finite element technique, and, when superimposed on the x-body force solutions, increase t_{I} and rotate the eigenvectors around the x axis through the introduction of components t_{yy} and t_{yz}. For example, the total t_{I} value for a 45° inclined slab approaches $\sim 0.7\rho g$ at the corners ($y = 0, 1.$; $z = 1$). However, the rectangular boundary conditions grossly overexaggerate the flank clamping effect, and therefore in nature it is expected that t_{I} would be considerably less than $\sim 0.7\rho g$.

A final point concerns the curious sawtooth fracture pattern often observed at the flanks. The simplest explanation of this phenomenon is that the stress state at the flanks is a transition between tensile stress induced at the crown, and shear or compressive stress induced at the stauchwall. The sawtooth flank fracture oscillates between competing tension and shear modes. A more detailed explanation based on fracture angles predicted by the Coulomb–Navier criterion is given by Haefeli (1963). There is also an explanation in terms of slab buckling [Lang (1977)], a process that could be important in the case of a slowly creeping wet slab.

E. Failure Mechanics of Snow Slabs

Any gain in tensile stress t_{I} at the crown region due to clamping effects is accompanied by reduced shear stress at the bed surface downslope from the crown. Thus, if a tension fracture is somehow induced at the crown, it would intersect a region of lower shear stress than exists at the neutral zone, and the tensile instability would not be fully reinforced by a possible shear fracture. By contrast, in the reverse sequence, a shear failure propagating upslope along the bed surface would generate a high tensile stress immediately upslope from the advancing failure, and would therefore tend to induce tensile fracture. The opening tensile fracture would reinforce the shear failure which may not develop into a true shear fracture with complete loss of cohesion until the opening of the tensile fracture provides rapid displacement. Eventually the shear and tensile fractures reinforce one another until the entire slab fails catastrophically.

The increase in tensile stress due to a shear failure is easily demonstrated by a scale analysis of the equilibrium equations for an infinite slab of thickness Z subjected to a bed-surface perturbation across a distance X as shown in Fig. 12. Assuming linearity, the perturbation induces a stress field which can be superimposed on the infinite slab solution. The perturbation would be a combination of a shear weakness π_{zz}. A nearly pure shear weakness could be associated with a thin discontinuity such as surface hoar, radiation recrystallized grains, or meltwater lubrication at a crust. An added π_{zz} component could be induced by the weak compressive support of a thick layer of newly fallen snow of low cohesion or a thick layer of unconsolidated depth hoar on the verge of collapse. In response to shear perturbation π_{xz}, the order of magnitude of the induced stress are [Perla and LaChapelle (1970)]

$$t_{xx} \sim (X/Z)\pi_{xz}, \tag{24}$$

$$t_{zz} \sim (Z/X)\pi_{xz}, \tag{25}$$

and similarly, in response to π_{zz},

$$t_{xx} \sim (X/Z)^2\pi_{zz}, \tag{26}$$

$$t_{xz} \sim (X/Z)\pi_{zz}. \tag{27}$$

A loss of shear support of only 10% ($\pi_{xz} \sim 0.1\rho gh \sin \theta$) acting over a distance of ten times the slab thickness would approximately double the

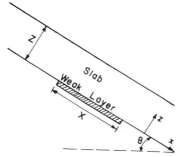

Fig. 12.   Increase of stress due to weak layer depends on ratio X/Z.

neutral zone value of t_1. A shear loss of 50% ($\pi_{xz} \sim 0.5\rho gh \sin \theta$) over the same distance would increase t_1 by a factor of 5 and would rotate t_1 to within 10° of the x axis ($\delta \to 10°$), consistent with the observed crown-bed surface fracture angle. A compressive perturbation (π_{zz}) is even more influential on t_1 since the weakness is amplified by (X/Z).[2] The physical interpretation of $(X/Z)^2$ is that the bending stress develops in the slab analogous to a beam supported at both ends and free to sag in the middle.

The influence of a bed-surface weakness has been studied in considerably more detail with analytical solutions [Brown *et al.* (1972)], and with finite-element solutions [Smith and Curtis (1975)]. Viscoelastic models have also been developed that try to explain the growth of the weak region due to a shear perturbation [Lang and Brown (1975)] and due to slab buckling [Lang *et al.* (1973)]. The above models are consistent with the possibility that shear failure somehow spreads and induces critically high tensile and shear stress around the boundaries. Although further details of the failure process are presently speculative, theoretical computations to date support the following very general model:

(1)   Due to stress increase (new snow load, skier weight) or strength loss (metamorphism, bond-melting), shear deformation increases at a potential bed surface somewhere in the neutral zone.

(2)   Supporting stress at the bed surface falls below the neutral zone value ($\rho gZ \sin \theta$) and stress increases at the slab boundaries. At the crown region, t_1 increases and rotates toward the slope parallel direction. Shear stress increases at the stauchwall region. Tensions and shears both increase at the flanks.

(3)   Elastic strain energy increases around the slab boundaries.

(4)   An increased load or shock, induced naturally or artificially (explosive blast, earthquake, sonic boom, skier, cornice fall), may initiate fractures at the boundary regions (crown, flank, stauchwall) where stress has increased. These fractures convert stored elastic energy into surface energy associated with the displacement of propagating fractures. Tension fractures are always observed to propagate in dry slabs at speeds suggestive of brittle fracture, about one-half the speed of the elastic shear wave in snow, ~ 100 m/sec, depending on snow density.

(5)   Crack displacement provides the necessary jolt to convert a portion of the bed surface into a state of kinetic friction with a corresponding large drop in the angle of repose. The slab is then undercut by rapid propagation of a shear fracture across a large bed-surface area.

(6)   Propagating tension and shear fractures reinforce one another until the entire slab (and often a series of neighbouring slabs) is completely detached.

(7)   If catastrophic failure does not occur, then cohesion will eventually recover through sintering, stress will relax at the slab boundaries, and the instability will disappear.

Although wet slab release may also follow closely the above steps, tension fractures in wet snow often open gradually as the slab slowly glides ($\sim$10 mm/day) over wet snow, soil, or rock [In der Gand and Zupancic (1965); Akitaya (1974); McClung (1975); Salm (1977)]. In this slow failure process, stored elastic energy and gravitational energy are fully dissipated at the slowly advancing cracks, and kinetic energy only enters the energy balance if the process suddenly accelerates, as is sometimes observed just prior to wet slab release. Kinetic jolt (Step 5) is not needed since the wet interface is characterized by sufficiently low friction (or low angle of repose).

For both wet and dry slabs, the crux of the stability problem is the development and spread of the shear failure at the bed surface. In the simplest model, bed-surface failure occurs when shear stress ($\sim B_{rz}$) exceeds shear strength τ, which, analogous to well-known methods used universally in slope stability analysis, can be expressed in terms of a Coulomb–Navier criterion

$$\tau = C + \sigma \tan \phi, \tag{28}$$

where C is the cohesion, ϕ the angle of friction, and σ is the stress normal to the slip surface ($\sim t_{zz}$). The relative importance of C compared to $\sigma \tan \phi$ depends on development of grain to grain bonds by sintering (diffusion transfer of water molecules to the concavities formed at grain contact points as described in Chapter 6). One complication in the application of Eq. (28) to snow in comparison to soil slope application is that, due to sintering, C and $\tan \phi$ are strong functions of time and temperature. For example, Haefeli [Bader *et al.* (1939)] found that C values for disaggregated depth hoar grains increased by an order of magnitude in a period of 6 days because the grains resintered. Moreover, depending on grain geometry and stress levels, C may either increase if sintering is dominant or possibly decrease if viscous (or plastic) deformation of intergranular bonds results in strain softening. Another problem is low-density snow (100–400 kg/m³) either becomes more dense or collapses when normal stress σ is applied, and thus exhibits rather drastic structural changes even before the onset of brittle shear failure.

The prominent nonlinearity in the C–ϕ envelope that appears in the test results of Haefeli [Bader *et al.* (1939); Roch (1966a)] is also a problem. The latter two problems are not completely unique to snow. Soil investigators [e.g., Lee and Ingles (1968)] are careful to point out that the parameters C and ϕ do not have strict physical meaning, and that, in the final analysis, Eq. (28) is chosen as a failure criterion for its simplicity, to insure solutions are tractable, and because it opens up a large backlog of experience.

The τ values reported by Haefeli were of the same order ($10^2 - 10^4$ Pa) as shear stress conditions in natural snowpacks. He also found that the τ value of newly fallen snow initially dropped with increase in normal load σ, and noted these results confirmed the universally accepted observation that new snow is highly susceptible to shear failure if the rate of increase of new snow load outpaces the gain in cohesion. In practice, a critical limit is assumed surpassed when the load of the new precipitation exceeds about 100 Pa at a sustained rate of about 10 Pa/hr [Atwater (1952); Perla (1970)].

Coulomb–Navier parameters reported by Roch (1966a) and Voitkovsky (1977) also are in approximate agreement with expected stress levels in natural snowslabs and with Haefeli's values. Moreover, there is general agreement that ϕ has a relatively low value for unconsolidated new snow ($\phi < 30°$), even initially negative in Haefeli's study, but increases ($\phi > 40°$) for more cohesive snow. Cohesionless depth hoar has an intermediate value, $30° < \phi < 40°$, that may persist for extended periods (~ 1 month) as long as the depth hoar remains unconsolidated. To summarize, reported ϕ values are at least in qualitative agreement with the observational evidence that new snow is highly unstable with respect to additional load, and the depth hoar persists in a state of marginal stability at an angle of repose approximately equal to the slope angles of interest which produces the most dangerous state in terms of unpredictability.

With regard to the details of progressive failure at a potential bed surface, there is recent interest [McClung (1979)] in Haefeli's (1938, 1950, 1966a) concept of residual strength, the strength remaining along a failure surface after shear displacement removes a large portion of the cohesion. From field studies of overconsolidated clay slopes it is known that in a period of about 10 yr there is a relative slip of about 1 m along the main failure surface [Skempton (1964)]. This produces a definite alignment of clay particles [Morgenstern and Tchalenko (1967)], a drop in shear strength ($\sim 50\%$ due to loss of cohesion) to a residual value, and a lowering of ϕ from 2° to as much as 10° (Fig. 13). Laboratory studies indicate the drop to residual occurs after displacement of about 10 mm across the slip surface; natural slopes remain in this residual state through far greater displacements.

It is reasonable to expect that slip displacement across a snow failure surface could proceed far more rapidly than in the analogous case of clay slope failure; in hours or days instead of months or years. McClung (1977) has found evidence of strain softening in laboratory shear tests of a wide

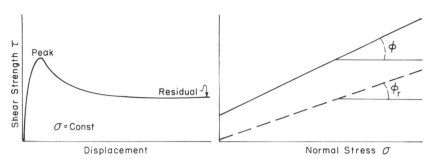

Fig. 13. Typical strain-softening characteristics of clay. The residual fraction angle ϕ_r is usually less than the peak friction angle ϕ. According to McClung (1977), the shear strength of snow sometimes exhibits similar behavior.

variety of snow samples subjected to shear displacement of 1–10 mm at rates of 1–10 mm/hr. He suggests that strain softening of some critical area of bed surface to a residual value about equal to $\sigma \tan \phi$ is a precursor to complete shear failure as predicted by models of progressive failure in clays by Bjerrum (1967) and Palmer and Rice (1973).

There remains some speculation as to what extent a shear fracture can propagate along the bed surface prior to the opening of boundary fractures (crown, stauchwall, flanks). The seven-step model presented above assumes that the high angle of static friction ($\phi > 30°$) due to continual sintering between grains opposed rapid and total shear fracture until the kinetic jolt of the opening of a boundary crack reduces ϕ to a kinetic value. Alternately, if a weakened shear band develops at the bed surface, it is conceivable, in principle, for a shear fracture to propagate over an extensive area, perhaps connecting en échelon tension cracks that form within the band [Skempton (1966)] before the opening of a boundary fracture. Future experiments are needed to clarify the shear fracture process.

F. Stability Evaluation

The stability problem, one of the more challenging problems in avalanche technology, has been approached by comparing in situ measurements of strength and load. Drawing on analogous techniques in soil mechanics, Roch (1966a,b) determined in situ the C–ϕ parameters of snow samples using a device known as the shear frame which is loaded with weights to stimulate the variation in normal stress σ. Over a period of 15 years he toured 35 avalanche fracture areas and measured C values at the bed surface in pits excavated into the slab crown. For each slab he computed t_{xz} and t_{zz} from thickness, density, and slope angle, as $\rho g Z \sin \theta$ and $\rho g Z \cos \theta$, respectively. He was then able to compute a strength load ratio τ/t_{xz} for the 35 cases. Perla (1977) performed a similar study in the U.S.A. and Canada, and summarized a total of 80 cases combining Swiss and North American data. The mean and standard deviations of τ/t_{xz} were 1.66 and 0.98 over a range $6.4 \geq \tau/t_{xz} \geq 0.19$. Although the range and variance are large, the shear frame criterion (τ/t_{xz}) is employed routinely as part of the stability evaluation program at Rogers Pass, British Columbia [Gardner and Judson (1970); Schleiss and Schleiss (1970)].

Sommerfeld *et al.* (1976) suggested that a larger shear frame would decrease the variance of τ/t_{xz} and shift the mean closer to 1.0. He bases his argument on Daniels' (1945) statistics, which predict that a model of parallel threads will have a strength decrease with increasing cross-sectional area. Perla (1977), in support of Sommerfeld's idea, found the shear frame index was sensitive to frame area; a 0.1-m² frame indicated 64% of the strength values obtained with traditionally sized frame (0.01 m²). Size effects have long been recognized in tests on overconsolidated clays [Bishop and Little (1967) found a 55% decrease in the shear strength of London clay as the sample size increased from 1.45×10^{-3} to 0.37 m²] but a fundamental explanation of statistical effects in snow testing is still controversial [Gubler (1978b)].

The size effect is only one of many factors that complicates the in situ measurement of τ. Operator variability in aligning the frame on a weak layer and variability of load rate are also problems. The rapid pull that is normally applied in the shear-frame test does not measure strain-softening properties which could be more meaningful than the brittle (or quasi-brittle) behavior induced by the rapid pull. Besides instrument problems, τ values vary strongly due to spatial inhomogeneity. Measurements at a level study plot and at the bed surface along the crown of a released slab suggest that about 10 shear frame measurements are required to obtain an average τ value within 15% of the population mean at 90% confidence [Perla (1977)]. Size effects, rate effects, and variance due to spatial inhomogeneity are problems shared by all in situ tests.

In view of these uncertainties, most field workers do not believe the in situ measurement of τ/t_{xz} is a practial index for prediction of instability, and cannot justify the time required to locate potential bed surfaces and to perform the shear tests. As the critical layers become thinner the operation becomes more difficult and time consuming.

In practice, one has a far better chance to find relatively thick (>10 mm), obvious layers such as depth hoar. The most popular approach is to test the snow profile from the snow surface to the ground with a device known as the ram penetrometer [Bader *et al.* (1939)]. The profile is interpreted qualitatively along with input data from snowpits and meteorological instruments. A sharp reduction of ram hardness with depth is reason to suspect slab instability; however, there are few quantitative guidelines. R. L. Armstrong (1977) suggests the integrated ram profile is a useful index for comparing snowpack changes.

As an alternative to the ram penetrometer, Bradley (1966, 1970) and Bradley and Bowles (1967) developed a convenient strength–load ratio based on an in situ device known as a resistograph which gives a continuous strength profile with a resolution of at least 10 mm. The resistograph index is based on a complex combination of shear and hardness strength but can be obtained quickly. In Bradley's system, the load is also quickly obtained with a Mt. Rose sampling tube. Bradley (1970) mapped strength–load values at 27 sites in a study area ($\sim 2.5 \times 10^4$ m²) and was able to demonstrate systematic correlations between his strength–load index and deep slab instability over the period January–May.

In principle, it also should be possible to evaluate slope stability by measuring slab deformation in the starting zone. One straightforward technique is the use of glide shoes to measure the displacement over a wet ground surface [In der Gand and Zupancic (1965)]. Although wet slab instability due to glide is usually a slowly developing process accompanied by ample visual warning (slowly opening tensile cracks), there are nevertheless practical advantages in learning the times of increasing glide, since these are the most probable times of critical wet slab instability.

A more difficult problem is to establish correlations between slab deformations and failure on bed surfaces which are not necessarily at the snow–ground interface. The experimental problems are formidable in comparison to the glide problem because (1) the failure surface is not known in advance, and (2) the precursory deformations for dry slab release are considerably smaller than the precursory glide, perhaps one or two orders of magnitude smaller (1–10 mm compared to 10 or 100 mm). Slab deformation experiments are reported by Lang and Sommerfeld (1977) who were able to correlate two-dimensional deformations (x, z directions) with values predicted by orthotropic viscoelasticity, but failed to detect deformational patterns prior to slab failure.

Part of the experimental problem is the critical slip deformation, possibly no greater than 1–10 mm, is obscured by the overall viscous creep of the slab. Creep rates in excess of 10 mm/hr have been recorded [In der Gand (1956)]. However, it is possible the initiation of failure in a shear band at the bed surface involves more subtle deformation that radiates elasticlike pulses. One means of detecting these pulses is with seismic geophones tuned to frequencies up to about 100 Hz. Seismic investigations have been performed near avalanche paths in the Bridger Range of Montana [St. Lawrence and Williams (1976); St. Lawrence and Bradley

(1977)]. Three types of seismic signals were correlated with visual observations of starting zone events:

(1) *Slab release.* An initial spike associated with slab failure, followed
by quiescent period of 0.5 sec, followed by a gradual signal increase as the
slab accelerated down slope (Fig. 14a).

(2) *Loose snow avalanches.* A gradual increase of signal but no initial
spike (Fig. 14b).

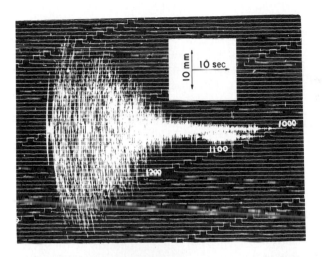

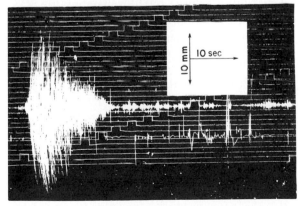

Fig. 14.   (a) Seismic recorder trace of a naturally released hard-slab avalanche; (b)
trace recorded by the release of a loose snow avalanche. [From St. Lawrence and Williams
(1976), reproduced from the *Journal of Glaciology* by permission of the International Glaciological Society.]

(3) *Cornice fall.* A spike associated with the impact of the cornice on the slope, and additional spikes as the cornice rolls down slope.

St. Lawrence and colleagues also noticed a fourth cluster of seismic signals which consisted of a series of spikes of low amplitude. These signals occurred during a period of slope instability, when there was no visual evidence of slab movement or other disturbances. They suggested these signals were radiated by internal fractures which could have propagated further and induced slab release under slightly more unstable conditions. These signals were obtained with a 28-Hz geophone cemented into a rock outcrop. Peak sensitivity of the geophone and seismic drum recorder was approximately 50 Hz. However, they concluded that seismic signals associated with slabs could easily be distinguished from extraneous noise (trucks, earth tremors), and therefore a lower frequency system could be used to extend the range of surveillance.

While recognition of analog signal patterns was successful, no definitive correlations were found between digital counts of seismic events prior to slab failure. If anything, instability appeared to correlate with a reduction in seismic counts below a preset trigger level [W. F. St. Lawrence (personal communication)]. However, Sommerfeld (1977) noted an increase in seismic signal events above a preset level correlated with slab instability. Considering his results controversial, he emphasized the preliminary nature of his study.

Ultrasonic emission (30–200 kHz) generated by microscopic and submicroscopic crystal deformations in connection with variation of slab stress (bed-surface failure, t_1 variation) also may provide prior warning for avalanches. Comprehensive laboratory and field tests have shown a correlation between ultrasonic emission from snow and the history of deformation and stress [St. Lawrence and Bradley (1973, 1975); Bradley and St. Lawrence (1975); St. Lawrence (1977)]. Increased attenuation at high frequencies places a practical limitation on ultrasonic applications in field stability evaluation. In retrospect, St. Lawrence and Bradley (1977) conclude that, although ultrasonics provide fundamental information about snow stress, deformation, and failure, the seismic range will probably prove more useful for routine stability evaluation.

G. Ice Avalanches

Enormous tragedies have been caused by the sudden failure of glacier ice. [Haefeli (1966b) and Scheidegger (1975) list some recent ice avalanche catastrophes and give additional literature sources.] What little information has been published on ice avalanches is also reviewed by

Mellor (1968) and Röthlisberger (1977). Ice avalanche phenomena are intimately related to mechanisms of glacier flow; hence, the material developed in Chapter 2 provides the essential framework for understanding the problem. Differences between snow and ice avalanches are discussed by Oulianoff (1954) and Lliboutry (1971).

Viewing the problem in the context of long-term glacier flow, ice avalanches generally originate where a portion of the glacier terminates at a steepened slope or cliff band. The glacier may reach this position either by advancing downslope or, as is more commonly observed, by retreating upslope away from a zone of compressive support. The impact energy of a large ice block dropping down a steeply inclined slope or over a cliff band is occasionally strong enough to trigger a massive landslide that can involve, by an order of magnitude, more earth material than ice. There is also the possibility of falling ice triggering a massive snow avalanche.

As perennial snow becomes dense enough to form glacier ice, stratigraphic weaknesses are obscured, and the most probable surface of slip instability becomes the ice–bedrock interface. Englacial failure (ice-to-ice slip surface) is rare, but not impossible [Haefeli (1966b)]. In terms of strength versus load, it is unlikely that gravitational shear stress ($\rho g Z \sin \theta$) can reach the magnitude necessary to induce massive englacial shear fracture. However, the basal stress state is high enough ($> 10^5$ Pa) to induce active stages of slip at the ice–bedrock interface according to mechanisms discussed in Chapter 2. Most investigators believe active stages of slip and glacial surges are only possible if there is a "cushion" of liquid water to reduce friction at the ice–bedrock interface. In terms of the Coulomb–Navier model, this essentially means $C = 0$ and $\sigma \tan \phi$ is lowered through a reduced ϕ or [Haefeli (1966b)] by a water pressure correction P_w, as is commonly introduced in soil mechanics in the form $(\sigma - P_w) \tan \phi$. Röthlisberger (1977) feels that, although meltwater played an important role in the release of the Allalin ice avalanche, a pressure effect $(\sigma - P_w)$ was unlikely. Emphasizing the diversity of ice avalanche failure mechanics, Röthlisberger (1977) notes that slip instability in cold glaciers precludes a large meltwater effect, and that for many cases a fracture or collapse hypothesis is necessary. He gives an empirical rule that cold glaciers tend toward instability on slopes of 45°, while temperate glaciers are unstable on 30° slopes.

It is possible to differentiate three broad categories of ice avalanches on the basis of initiation mechanics (see Fig. 15):

(1) *Frontal block failure.* Frontal blocks are continually calved from the glacier toe as the ice flows at rates varying from 0.1 to 10 m/day, depending on ice thickness and temperature. For example, during August

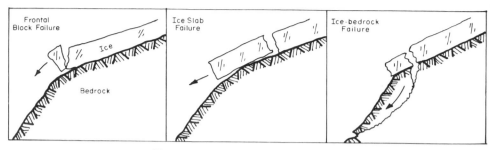

Fig. 15.   Initiation of ice avalanches.

1973 the flow rate of the Weishorn glacier (Randa, Switzerland) increased in 10 days from 0.3 to a failure rate of 1.2 m/day [B. Salm (personal communication)]. Shinoda (1963) photographed an ice block failure in the Nepal Himalayas where the movement was 6 m/day prior to failure. Haefeli (1966b) subdivides frontal block failure into two subcases: in the first, there is no slip at the ice–bedrock interface, and therefore the ice flow is essentially due to viscous creep; in the second, ice flow is due to slip as well as creep. In both cases, a slowly opening transverse crevasse usually appears before the block detaches. Iken (1977) has studied a model of progressive slow tensile fracturing in a viscously deforming frontal block to test empirical relationships for the time of failure. She also provides references describing some successful attempts at real time forecasting.

   (2)   *Ice slab failure.* The failure mechanics of large ($\sim 10^6$ m^3) ice slabs may be quite similar to the mechanics of snow slab release as presented in Section II.E. For some reason a slip discontinuity develops at the ice–bedrock interface. High tension stress t_l develops upslope and eventually causes tensile fracture, which in turn triggers catastrophic shear fractures. Alternatively, viscous creep could first induce slow tensile fracture. The resulting transverse crevasses could then free the ice slab which would slowly accelerate downslope on the bedrock, in the same manner as a frontal failure, but on a larger scale. When the slab encountered a steepened slope, its velocity would increase and it would detach. Haefeli (1966b) assumed a frictional resistance $(\sigma - P_w) \tan \phi$ over an interface length X. Analogous to Eq. (24), the t_{xx} stress required for equilibrium across a slab thickness Z is on the order of

$$t_{xx} \sim \rho g X(\sin \theta - \cos \theta \tan \phi) + (X/Z)P_w \tan \phi. \qquad (29)$$

As a numerical example, Haefeli considers data from the Altels ice slab disaster (11 September 1895): $X = 300$ m, $Z = 30$ m, $\phi = 32°$, and $\rho = 900$ kg/m^3. He concludes that t_{xx} was in the range of 5×10^5 to 14×10^5 Pa for various combinations of ϕ and P_w. This range corresponds closely to strength values of ice determined in laboratory tensile tests ($\sim 10 \times 10^5$ Pa).

(3) *Ice–bedrock failure*. The third category occurs when the weight and shear of the flowing ice induces failure of the underlying bedrock. Seepage of glacier meltwater into bedrock pores or joints also may play an important role. In any case, the glacial factors (ice weight, shear induced by ice flow, and meltwater pore pressure) combine to generate very unstable terrain. Mokievsky–Zubok (1978) briefly describes how ice weight and melt caused bedrock failure and led to an enormous avalanche ($\sim 3 \times 10^7$ m³) from the Devastation Glacier in British Columbia on 22 July 1975.

The difficulty of evaluating stability of glacial bedrock systems is compounded if there is a possibility of an earthquake. The 1970 Huascaran disaster is a particularly well-documented earthquake-induced avalanche, and there are other examples from the Caucasus [Heybrook (1935)]. Seismic shocks have triggered snow and debris avalanches in Alaska [Field (1966); LaChapelle (1968)], but there is no firm evidence that seismic energy played a role in the European glacial disasters.

III. AVALANCHE MOTION AND IMPACT

A. Observations

The study of avalanche dynamics involves quantifying such parameters as avalanche velocity, mass, runout distance, and impact force. Estimates of these parameters are prerequisite to the design of avalanche defense structures, and for the preparation of avalanche zoning plans.

Almost immediately after a snow slab breaks away from the starting zone and begins accelerating down the track, the moving mass disaggregates into blocks, and then into round chunks and various-sized particles. If the snow is dry, the smaller dustlike particles will diffuse turbulently into a cloud which envelopes the descending mass. Wet avalanche snow invariably flows close to the surface without forming a dust cloud, but even very wet snow impacting after freefall over a cliff band can pulverize and form dust clouds. An example of a moving avalanche is shown in Fig. 16.

The density profile of a moving avalanche is presently a topic of speculation and no doubt varies significantly from case to case. Examination of damaged buildings and trees provides at least qualitative evidence that density decreases rapidly with height above the snow surface [de Quervain (1966a)]. Irrespective of the height of the dust cloud, most of the

Fig. 16.   Large powder avalanche in the Canadian Rockies (photo courtesy of Jim Davies).

momentum is probably concentrated within a flowing layer 1–10 m thick. Although the dust cloud may rise much higher (10–100 m), most of the cloud is probably of relatively low density (<10 kg/m^3), and thus carries only a small proportion of the total momentum.

There is also speculation concerning the range of avalanche velocities. Voellmy's (1955) summary of Oeschlin's 1938 measurements is quoted repeatedly in the literature. According to Oeschlin, snow avalanches have attained speeds of 125 m/sec on extended 35° slopes. It may well be that some of the largest avalanches in the world reach speeds in excess of 100 m/sec. On the basis of indirect evidence, Körner (1976) computes that the 1962 and 1970 Huascaran avalanches attained maximum speeds in the ranges 60–80 and 110–125 m/sec, respectively. Körner's figures are consistent with average velocities of approximately 50 and 80 m/sec computed by Plafker and Ericksen (1978). Plafker and Ericksen (1978) note that a historic avalanche, which was even larger than the 1970 event, ran the 16-km track from Mt. Huascarán with an estimated average velocity of 90 m/sec.

Several investigators have measured avalanche velocities using photo-grammetry, mechanical switches mounted in the track, or simple hand timing. Results are summarized in Table 4. Measured velocities are clearly less than the above-mentioned values, although the investigators did not study extremely large avalanche paths. On the other hand, until detailed measurements similar to Briukhanov's (1968) photogrammetric data are avaliable for large avalanche paths, one should view some of the high-velocity measurements of Oeschlin with at least mild skepticism.

From a continuum mechanics viewpoint, de Quervain (1966a) and Salm (1966) point out that mass elements within the moving avalanche

TABLE 4

Avalanche Speeds and Volume

Investigator	Location	Maximum speed (m/sec)	Vertical drop (m)	Path slope angle (deg)	Volume (m³)
Cunningham (1887)	Rogers Pass, British Columbia	54	500–1000	—	2×10^5
Schaerer (1967, 1972, 1973)	Rogers Pass, British Columbia	53	500–1000	—	$\sim 10^5$
Shoda (1966) 11 experiments	Tsuchitaru, Japan	20 (midtrack)	115–290	35	$\sim 10^2 - 10^3$
Van Wijk (1967)	Banff, Alberta	17.7	250	45–33	10^4
Briukhanov (1968)	Central Asia, USSR	27.9	500	45–20	1.9×10^3
	Central Asia	20.8	300	31–22	3.5×10^3
	Central Asia	13.1	300	34–24	4.6×10^3
	Khibiny	21.2	300	37–21	$\sim 10^4$
	Khibiny	16.4	250	47–22	5×10^3
	Khibiny	12.2	125	30–15	3×10^2
	Tien-shen	9.6	150	20–15	—
Tsomaia (1968)	Krestovy Pass, Caucasus, USSR	63	—	—	2×10^5
Bon Mardion et al. (1974)	Col du Lautaret, France	16.5	400	29	—
Bon Mardion et al. (1975)	Col du Lautaret, France	28	400	29	—
Körner (1976)	Zugspitze, Germany	17	400	27	—
Nakajima (1976)	Mitsumata, Japan	24	235	38	10^4
Kotlyakov et al. (1977)	Khibiny, USSR	50	450	24	5×10^4

may be moving at higher speeds than the observed avalanche front (or head). High-speed transients should exist as a consequence of turbulence since a moving avalanche must contain mass moving with a statistical distribution of velocities. Shimizu *et al.* (1977) have installed pressure sensors at several positions in a major avalanche track and recorded internal transient velocities up to 150 m/sec.

As the avalanche descends the track, it incorporates additional mass into the flow. At the same time, a certain amount of mass with zero momentum will drop out of the flow onto the track. In most cases, a snow avalanche can only reach its full potential of velocity and runout distance if the track is covered with loose snow to appreciable depth (>0.3 m), either newly fallen snow or wet snow, so the surface layer is easily entrained into the flow. Thus, the avalanche mass is rarely constant during the flow, tending to increase during the initial movement in the starting zone and on relatively steep portions of the track, and to decrease as the avalanche decelerates on less steep terrain. As will be discussed, motion is partly controlled by the inertial effects of entrainment.

Present computational models of avalanche flow include an assumption that the avalanche path consists of three sections: the starting zone, the track, and the runout-deposition zone. The starting zone is assumed to supply a quantity of snow that feeds into the track where the avalanche attains its maximum velocity. In the models presently used in Switzerland [reviewed by Leaf and Martinelli (1977)], entrainment and mass loss are assumed to balance within the track, and the initial starting zone mass is therefore assumed to reach the runout-deposition zone which is normally a low-angle slope where the avalanche decelerates rapidly. The division of an avalanche path into three sections is arbitrary, and in practice is based on boundaries drawn subjectively on topographic maps. However, with this approach it is possible to estimate maximum velocities and runout distances by overestimating the starting zone area and by assuming thick slab layers that could be deposited during exceptionally severe winters. Moreover, a large number of avalanche paths have well-defined starting zones in the form of catchment bowls and well-defined gully shaped tracks that open out to low-angle valley bottoms.

Table 5 gives examples of geometric parameters measured on large avalanche paths. The mean starting zone inclination (34°), measured by Chitadze and Zalikhanov (1967), appears reasonable since the overall inclination of a catchment area should be somewhat less than the bed-surface angle ($\bar{\theta} = 38°$) at the location of fracture initiation (see Fig. 3). Frutiger (1964) has prepared maps for 80 large avalanche paths in Colorado and found that the track portion had an average inclination of approximately 29°, but a large variation was noted from case to case.

TABLE 5

Avalanche Path Parameters

Investigator	Location	Number of paths	Parameters
Frutiger (1964)	Colorado (highways)	80	Starting zone area up to 6×10^5 m²; maximum vertical drop 1.2 km; maximum path length 2.4 km; average slope of path 25°; average slope of track 29°
Chitadze and Zalikhanov (1967)	Mestia Rd., Caucasus, USSR	86	Maximum path length 2.5 km; starting zone volume up to 7×10^5 m³; starting zone inclination 34° average
Novikov (1971)	Carpathians, USSR	17	Maximum vertical drop 0.5 km; maximum path length 1.2 km; total path inclination 29° average
LaChapelle and Leonard (1971)	Cascades, Washington (highways)	70	Maximum vertical drop 1.65 km; maximum path length 2.4 km; inclination starting zone to road 31° average
Schaerer (1973)	Rogers Pass, B.C.	41	Inclination starting zone and track 37° average
Bovis and Mears (1976)	Colorado (highways)	67	Inclination track 26°; inclination runout zone 12°; mean length of runout zone 380 m; range 10^2–10^3 m
Lied (1978)	Norway	423	Maximum vertical drop 1.7 km; maximum path length 3.5 km; average inclination of total path 33°; range of inclination 50–18°

Bovis and Mears (1976) found a mean runout inclination of 12° for 67 large Colorado paths. Avalanche paths in the Cascades in Washington and in the Selkirks in British Columbia are steeper according to the measurements of LaChapelle and Leonard (1971) and Schaerer (1973).

Frutiger (1964) estimated starting zone areas up to 6×10^5 m². However, the majority of his 80 samples had starting zone areas less than 2×10^5 m². LaChapelle and Leonard (1971) have mapped path lengths up to 3.5 km and vertical drops up to 1.65 km. These distances are probably typical upper bounds in populated areas of North America (ignoring remotely located paths in the Yukon and Alaska), and are comparable to some of the larger paths in the Swiss Alps. For example, one of the larger paths that ran during the severe 1950–1951 winter in the Lötschental Valley, Switzerland, had a vertical drop of 1.5 km and a path length of approximately 3.8 km [EIDG (1951)]. Avalanche paths in the Himalayas and Andes are considerably larger, but are presently unmapped with the exception of notorious paths such as Huascaran (4-km vertical, 16-km path length).

There is continuing interest in the development of an adequate scale for comparing avalanche events, perhaps analogous to earthquake magnitude scales. Atwater and Koziol (1953) introduced a subjective estimate of potential capacity for destruction according to a 1–5 scale similar to the levels presented in Table 1. Perla (1970) correlated avalanche events measured on the Atwater–Koziol scale with precipitation, wind, temperature, and snowpack parameters. More refined scales (1–10) based on the logarithm of energy $(mgh, \frac{1}{2}mv^2)$ have been proposed by Shoda (1966) and Shimizu (1967) and could gain in popularity as observations improve.

B. Acceleration of a Unified Snow Mass

We first assume that an avalanche accelerates as a single, unified snow mass. For many cases this is a simplification since an avalanche typically spreads a large volume $(10^3–10^5$ m³) over a large distance $(10^2$ m), and therefore certain aspects of the phenomenon are more akin to fluid flow in a channel than to the motion of a unified mass. Furthermore, the moving snow consists of an assortment of particles which differ widely in mass and velocity. Thus, the intrinsic behavior of the system is determined by particle statistics. Nonetheless, the unified mass model provides some results that do not differ drastically from present studies based on fluid models. Incorporation of fluid and particle assumptions are deferred until later sections.

On a slope of inclination θ, an avalanche mass m experiences a gravitational component $mg \sin \theta$. As shown in Fig. 17, the forces resisting

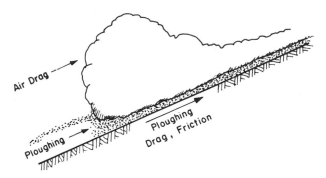

Fig. 17. Forces resisting avalanche acceleration.

acceleration can be lumped into three categories: friction underneath the avalanche, air drag at the front and top boundary, and plowing at the advancing tip and underneath surface. These resistive forces are undoubtedly complex functions of velocity. However, the obvious fact that avalanches accelerate on steep slopes and decelerate on less steep slopes can be explained by simply assuming friction is proportional to the component of force normal to the bed ($mg \cos \theta$). A coefficient of friction μ is introduced, and the difference between the accelerating and frictional force is $m\alpha$, where α is

$$\alpha = g \sin \theta - \mu g \cos \theta. \tag{30}$$

From hydrodynamics, the air drag force along the front and upper boundary is assumed proportional to v^2 according to the form

$$\text{air drag} = \tfrac{1}{2} C_{\mathrm{D}} A \rho_{\mathrm{a}} v^2, \tag{31}$$

where C_{D} is a drag coefficient dependent on the Reynolds number and avalanche shape, A the area over which the air drag operates, and ρ_{a} is the air density which is about 1 kg/m³ at higher elevations.

Plowing resistance should in principle have a very strong v^2 dependence. For simplicity, the model will be assumed proportional to only v^2 through a conventional expression for plastic impact derived from the conservation of mass and momentum [Dolov (1967)], or

$$\text{plowing force} = A_{\mathrm{s}} \rho [(\rho + \Delta\rho)/\Delta\rho] v^2, \tag{32}$$

where A_{s} is the avalanche area over which the plowing force operates, ρ the density of the undisturbed top layer of the snow, and $\Delta\rho$ is the change in density due to the compressional force of plowing.

The combined force of plowing and air drag can be expressed as βv^2, where

$$\beta = \tfrac{1}{2}C_D A \rho_a + A_s \rho[(\rho + \Delta\rho)/\Delta\rho]. \tag{33}$$

These forces control the change of momentum of the avalanche. The balance is

$$d(mv)/dt = m\alpha - \beta v^2. \tag{34}$$

Equation (34) models the essential features of the acceleration problem, and provides a control against unlimited velocity because of the βv^2 term. A wide range of motions can be modeled, depending on the values of α, β, and m. It is possible to write a more complex equation including a v dependence as well as a v^2 term [Salm (1966)] but, considering the presently available data, even Eq. (34) may be too flexible.

An equation similar to Eq. (34) has been derived and integrated for $v(t)$ by Voellmy (1955) and Mellor (1978). Rather than solve for $v(t)$, it is preferable to expand Eq. (34) using the operation

$$\frac{d}{dt}(\ ) = v\frac{d}{dx}(\ ), \tag{35}$$

where x is the distance measured along the path. Equation (34) may be then rearranged as

$$mv\, dv/dx = m\alpha - (\beta + dm/dx)v^2. \tag{36}$$

The term dm/dx represents the entrainment per unit path length (with units kg/m), and appears as an inertial factor resisting acceleration. Entrainment is not as easily introduced into a $v(t)$ formulation. Moreover, the important practical advantage of the $v(x)$ formulation is that it can be used to compute runout distance as well as velocity.

Equation (36) is a fairly good approximation for a general slope profile (provided $d\theta/dx$ is small) and could be solved for a given profile $\theta(x)$. However (as will be explained shortly), with present data it is only possible to estimate the order of magnitude of the controlling parameters β, m, and dm/dx; thus, detailed solution of Eq. (35) for an exact $\theta(x)$ is often an inappropriate complication. The problem is greatly simplified if $\theta(x)$ is approximated by two slopes of constant inclination: an upper slope consisting of the starting zone and track where $\tan\theta > \mu$ and the mass accelerates to its maximum velocity; and a lower slope consisting of the runout zone, where $\tan\theta < \mu$ and the mass with its initial velocity (determined by solving the upper slope problem) decelerates to rest. With this formulation, we must first find the velocity $v(x)$ on the upper slope; and second, the runout distance x_r on the lower slope.

Assuming initial conditions $v = 0$, $x = 0$, the velocity solutions to Eq.

(36) for the upper slope can be expressed as

$$v(x)/\sqrt{2\alpha} = f(x, \beta, m, dm/dx), \tag{37}$$

where $\alpha > 0$. If entrainment and v^2 forces are neglected ($\beta = 0$, $dm/dx = 0$), then the solution is simply

$$v(x)/\sqrt{2\alpha} = \sqrt{x}. \tag{38}$$

If $\beta \neq 0$, but entrainment is neglected ($dm/dx = 0$), then

$$v(x)/\sqrt{2\alpha} = \{(m/2\beta)[1 - \exp(-2\beta x/m)]\}^{1/2}. \tag{39}$$

For this case velocity increases asymptotically toward a mathematical maximum

$$v_{\max} = (\alpha m/\beta)^{1/2}. \tag{40}$$

The most general case includes both a β term and an entrainment term dm/dx. If entrainment occurs at a constant rate R along the path, then

$$m(x) = M_0 + Rx, \tag{41}$$

where M_0 is the initial mass of the avalanche at $x = 0$. The corresponding solution, which does not contain a mathematical maximum, is

$$\frac{v(x)}{\sqrt{2\alpha}} = \left[\frac{(M_0 + Rx)^{\gamma+1} - M_0^{\gamma+1}}{(2\beta + 3R)(M_0 + Rx)^{\gamma}}\right]^{1/2}, \tag{42}$$

where

$$\gamma = (2/R)(\beta + R). \tag{43}$$

Evaluation and comparison of velocity solutions depends on establishing a range of numerical values for α, β, γ, and $m(x)$. For the more restrictive purpose of making a qualitative comparison of solutions an evaluation of $v(x)/\sqrt{2\alpha}$ suffices. Unfortunately, there are few hard data available to help narrow the range of values. For example, air drag ($C_D A \rho_a/2$) could have a very wide range. An avalanche may move with a spherically shaped front or with a streamlined shape resembling an air foil. For an avalanche of characteristic dimension L (height, width, or length), the Reynold's number is computed as

$$\mathrm{Re} = \rho v L/\mu. \tag{44}$$

Substituting 1.47×10^{-5} m²/sec for the kinematic viscosity of air gives a Reynolds number equal to $7vL \times 10^4$. After a short period of initial acceleration, Re exceeds 10^5, and standard tables indicate that C_D decreases to less than one, probably to within the range $0.1 \leq C_D \leq 1$, depending on the shape of the front. The drag area could conceivably have a range

$10^2 < A < 10^4$ m² or greater, depending on the avalanche size. Thus, the corresponding range of $C_D A \rho_a / 2$ is from less than 10 to perhaps 10^4 kg/m.

The plowing contribution to β may vary over a comparable range. The top snow would have a typical density $50 < \rho < 300$ kg/m³ which would perhaps double or triple after avalanche impact [Bon Mardion *et al.* (1975)]. The plowing depth beneath the snow would have some inverse relationship with density, becoming as deep as one meter for the lightest snow. The combination of a one meter plowing depth and dense snow ($\rho \geq 300$ kg/m³) may be possible for wet cohesionless snow. Assuming an avalanche width on the order of 100 m, the plowing contribution to β could range up to about 10^4, and in some special cases (such as for wet avalanche) up to 10^5.

Figure 18 shows the variation of $v(x)/\sqrt{2\alpha}$ as a function of x for a relatively large avalanche, $M_0 = 10^7$ kg, and for path lengths up to 10^4 m. As

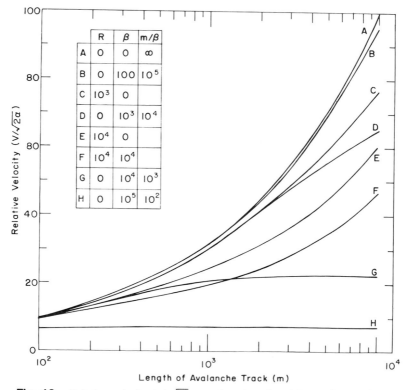

	R	β	m/β
A	0	0	∞
B	0	100	10^5
C	10^3	0	
D	0	10^3	10^4
E	10^4	0	
F	10^4	10^4	
G	0	10^4	10^3
H	0	10^5	10^2

Fig. 18. Relative velocity ($v/\sqrt{2\alpha}$) as a function of path length for possible values of drag β and entrainment R. Computations based on initial mass m of 10^7 kg; $\alpha = g \sin \theta - \mu g \cos \theta$. All units in mks.

seen from Curve H, if $\beta = 10^5$, then an avalanche would reach 90% maximum speed before traveling 100 m. This may possibly occur for a slow-moving wet avalanche ($v < 15$ m/sec).

The following qualitative deductions can be made from Fig. 18. For the conditions $\beta < 10^3$, $R < 10^3$, $v(x)/\sqrt{2\alpha}$ is reasonably approximated as $\sqrt{x}$, except for very long path lengths (>2 km). The $\sqrt{x}$ approximation is considerably worse over medium path lengths (5×10^2 to 10^3 m) for combinations of $\beta = 10^4$ and $R = 10^4$; and appears completely invalid for $\beta = 10^5$ (a wet avalanche).

The validity of the $\sqrt{x}$ approximation improves with the increase in the ratio m/β. Roughly speaking, mass tends to vary as the third power of the characteristic dimensions, whereas the β drag varies as the square of the dimensions. Initially entrainment and drag cooperate to oppose acceleration, but if entrainment continues the mass increase will eventually work against the β drag, and velocity will increase. However, a comparison of curves F and G reveals that the velocity boost is only significant given continual entrainment on very long paths.

C. Computation of Velocity and Runout Distance

Velocity depends on the parameter $\sqrt{2\alpha}$, a function of slope angle θ and the coefficient of friction (shown in Fig. 19). The choice of μ can be

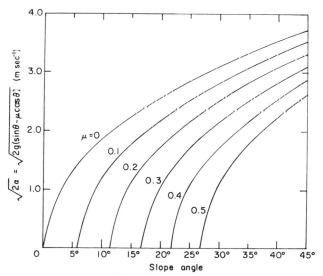

Fig. 19. Variation of factor $\sqrt{2\alpha}$ as a function of slope angle θ and coefficient of friction μ. All units mks system.

narrowed by examining a few observed facts. Based on Tables 4 and 5, it appears that maximum avalanche velocities range from about 20 to 30 m/sec for path lengths ranging up to about 500 m and tracks with slope angles in the range 25–35°. From Fig. 18, it is apparent these velocities can be obtained by multiplying the factor $v/\sqrt{2\alpha}$ in the range $0 < \beta < 10^4$ by a $\sqrt{2\alpha}$ factor in the range $1 < \sqrt{2\alpha} < 2$. Figure 19 indicates that for this range of values of $\sqrt{2\alpha}$ with θ between 25 and 35° $\mu \geq 0.25$. However, the $\mu = 0.25$ curve intersects the abcissa at approximately $\theta = 14°$, implying that an avalanche would not begin to decelerate unless the slope decreased to less than 14°. This value is too low, at least for dry avalanches which often come to rest on steeper slopes (~20°). At the other extreme, μ cannot exceed 0.45, since dry avalanches initiate on slopes steeper than 25° ($\tan^{-1} 0.45 \approx 24°$). It appears that a reasonable range of values for μ for dry avalanches is $0.25 \leq \mu \leq 0.45$. From analogous theoretical considerations, Kozik (1962) reaches a similar conclusion concerning μ. Heimgartner (1977) has measured μ on small test slopes near Davos, Switzerland, and reports values between 0.22 and 0.34 for dry snow. He reports lower values ($0.14 \leq \mu \leq 0.25$) for snow near the melting point, but notes the test slide was covered with a thin ice glaze.

It has been suggested [Voellmy (1955)] that μ is a strong function of snow fluidity. According to his model, criticized in detail by Eglit (1968), a dry powder avalanche with a well-developed dust cloud is characterized by a low μ value, perhaps 0.1 or lower. Such low values are difficult to justify in terms of the observed evidence; for example, $\mu = 0.1$ implies an avalanche would not decelerate on a 6° slope. Similarly, Schaerer (1975) has suggested that μ decreases with speed to a value of 0.1 or less. He attempted to fit the relatively high avalanche velocities observed at Rogers Pass, British Columbia, to Voellmy's equations (discussed in the next section), and found it necessary to reduce μ. However, his published result $\mu = v/5$, now quoted in the literature, is based on an erroneous correlation which confuses maximum and instantaneous velocity. His data are better explained by assuming a relatively high value for m/β.

A relatively low value of μ may be characteristic of an avalanche mass containing a significant quantity of liquid water, for example, an avalanche of slush or of a slurry of melted snow, ice, rock, and soil. Plafker and Ericksen (1978) computed values of μ about equal to 0.22 for the Huascaran avalanches, which were classic examples of slurrylike flow. Slush avalanches and mud flows probably have values of μ of 0.1 or less since these movements continue at very low speeds on slopes of 5° or less.

In order to compute velocities, it is also necessary to restrict β, m, and

R. First, it should be noted from Fig. 18 that the entrainment curves ($R \neq 0$) cross and lie between the nonentrainment cases ($R = 0$). Thus, in order to estimate the upper bound on velocity in terms of the simplest expression, it is preferable to set $R = 0$, and choose a value of β small enough to include any possible entrainment effects. Furthermore if R is equal to zero it is possible to model the motion in terms of the combined parameter m/β in Eq. (39), and thus replace two variables (m and β) by a single parameter.

Recall that for the case of a dry avalanche on slopes up to 500 m long, the range of $\sqrt{2\alpha}$ was $1 < \sqrt{2\alpha} < 2$. If this range also applies to slopes 10 km long, then according to Fig. 18, for $x = 10^4$ m, $m/\beta = 10^2$ implies $7 \leq v(10^4 \text{ m}) \leq 14$ m/sec, $m/\beta = 10^3$ implies $22 \leq v(10^4 \text{ m}) \leq 44$ m/sec, $m/\beta = 10^4$ implies $65 \leq v(10^4 \text{ m}) \leq 130$ m/sec, and $m/\beta = 10^5$ implies $95 \leq v(10^4 \text{ m}) \leq 190$ m/sec. According to all observations, on very long slopes dry avalanches may reach speeds in the range $50 < v < 100$ m/sec, so $m/\beta = 10^4$ provides a fairly good upper bound. For wet avalanches, it is more appropriate to reduce m/β to 10^3 or 10^2.

Given the preceding discussion, $v(x)$ on the upper slopes may be estimated by

$$v(x) = \{(mg/\beta)(\sin \theta - \mu \cos \theta)[1 - \exp(-2\beta x/m)]\}^{1/2} \qquad (45)$$

Dry-avalanche velocities are slightly overestimated by setting $m/\beta = 10^4$ and $\mu = 0.25$. Slow-moving wet avalanches are more properly modeled by reducing m/β and possibly μ.

The second portion of the analysis is to find the runout distance on the lower slope or runout zone (where $\alpha < 0$) when the avalanche enters the top of the zone with a velocity V_0 [determined by Eq. (45) or an equivalent expression]. The solutions to Eq. (36) are easily extended for a new reference system $x = 0$, $v = V_0$, and the resulting expressions for the distance x_r required for an avalanche to decelerate to rest are as follows:

Case I $m/\beta = \infty$ and $R = 0$

$$x_r = V_0^2/2|\alpha|; \qquad (46)$$

Case II $R = 0$

$$x_r = (m/2\beta) \ln(1 + BV_0^2/m|\alpha|); \qquad (47)$$

Case III

$$x_r = \frac{M^{\gamma/(\gamma+1)}}{R} \left[M + \frac{V_0^2(2\beta + 3R)}{2|\alpha|} \right]^{1/(\gamma+1)} - \frac{M}{R}; \qquad (48)$$

where α, β, and γ are as previously defined in Eqs. (1), (4), and (12), M is the mass entering the runout zone, and m is $(m + Rx)$.

Figure 20 shows the variation of x_r and V_0, computed from Eqs. (46), (47), and (48) with $\mu = 0.35$, $\theta = 12°$, and $M = 10^7$ kg. According to Bovis and Mears (1976), the mean length of the runout zone for 67 Colorado avalanche paths was 380 m with a mean inclination $\bar\theta = 12°$. In Fig. 20, a runout path of 380 m corresponds to the approximate intervals $10^3 \le m/\beta \le \infty$ and $30 \le v \le 45$ m/sec, allowing for entrainment. These are reasonable values and indicate a certain self-consistency with the velocity model. Analogous to the velocity case, the assumption $m/\beta = 10^2$ underestimates the Bovis–Mears $\bar x_r$ by a wide margin. Again, unless only slow-moving wet avalanches are being modeled, it is necessary to set m/β to 10^3 or greater. For consistency with the velocity information, the suggested value is $m/\beta = 10^4$, which will overestimate x_r. The overestimate will be increased by using $\mu = 0.25$ instead of 0.35.

The decrease in avalanche impact pressure with runout distance depends on the velocity decay in the runout zone. Since dynamic pressures

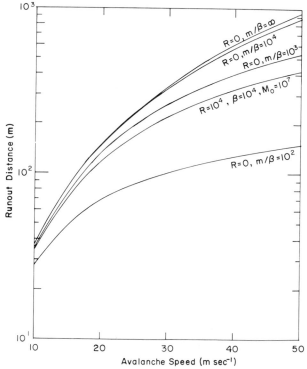

Fig. 20.   Runout distance as a function of velocity at beginning of runout zone. Computations according to Eqs. (46)–(48) for the case $\theta = 12°$, $\mu = 0.35$.

are proportional to ρv^2, the crucial parameter is the decay of v^2. From Eq. (36) for the case m/β equal to infinity, it is easily shown that v^2 decays linearly, according to

$$v^2 = V_0^2 - 2|\alpha|x, \tag{49}$$

when V_0 is the velocity at the top of the runout zone. For the case of finite m/β and R equal to 0, v^2 decays as

$$v^2 = (m|\alpha|/\beta + V_0^2)\exp(-2\beta x/m) - m|\alpha|/\beta. \tag{50}$$

In the numerical example given in Fig. 21, it appears a linear decay is a good approximation for the range $10^3 \le m/\beta \le \infty$. Thus, for engineering design within the boundary $0 \le x \le x_r$, one may use

$$v^2 = V_0^2(1 - x/x_r). \tag{51}$$

Similar models derived by Salm (1966) and Voellmy (1955) also predict a linear decay of v^2.

Finally, it should be noted that the runout zone may in some cases extend across a level valley floor and then continue upslope. This added complication is easily handled by using Eq. (50) to find v at the initiation of the upslope portion, and then once again using Eq. (50) with $\alpha = g \sin\theta + \mu g \cos\theta$.

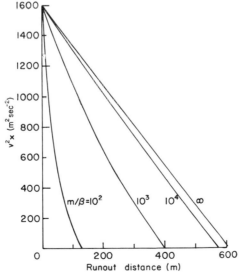

Fig. 21. Velocity decay in the runout zone for the case $\theta = 12°$ and $\mu = 0.35$.

D. Voellmy's Fluid Model

Voellmy (1955) was the first to draw an analogy with open-channel flow and regard avalanches as fluid. In his and other models, terminal velocity is assumed to occur very quickly and entrainment is ignored. If the endless fluid is considered in a state of nonaccelerating flow ($\partial/\partial x = \partial/\partial t = 0$) through Eulerian coordinates fixed to the channel (Fig. 22), the equation of motion reduces the equilibrium equation. At the center of the channel, where there is no variation in the y direction, the equilibrium equation is

$$dt_{xz}/dz + g(\rho - \rho_a) \sin \theta = 0, \tag{52}$$

where ρ is the density of the avalanche fluid (a mixture of snow and air), with the density of air ρ_a subtracted to account for buoyancy. Assuming an average $\overline{\rho - \rho_a}$ across a flow height h, the integration of Eq. (52) gives

$$t_{xz}(0) - t_{xz}(h) = gh(\overline{\rho - \rho_a}) \sin \theta. \tag{53}$$

The shear stress at the channel interface ($z = 0$) consists of a dynamic drag and a friction component due to the normal stress t_{zz}; hence,

$$t_{xz}(0) = k_1 V_{max}^2 + \mu gh(\overline{\rho - \rho_a}) \cos \theta. \tag{54}$$

Shear stress at the avalanche–atmosphere interface ($z = h$) is assumed to consist only of dynamic drag,

$$t_{xz}(h) = k_2 V_{max}^2. \tag{55}$$

Substituting and rearranging terms gives a result analogous to Eq. (40),

$$v_{max} = [\xi h(\sin \theta - \mu \cos \theta)]^{1/2}, \tag{56}$$

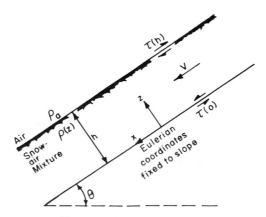

Fig. 22.   Endless fluid model.

where density and drag coefficients are lumped together into a single constant ξ. Voellmy (1955) suggests from experience that ξ is typically 500 m/sec², ranging from 400 to 600 m/sec² with decreasing roughness of the path. However, this narrow range is disputed. For example, Shen and Roper (1970) use the experimental results of Michon *et al.* (1955) on density currents and find $\xi = 750$ m/sec². From his Rogers Pass data, Schaerer (1975) computes $1000 < \xi < 1800$ m/sec² values well outside of Voellmy's range. Leaf and Martinelli (1977) allow ξ to range from 500 m/sec² for very rough slopes (boulders, trees) to 1800 m/sec² for very smooth slopes, and compute reasonable velocities for their sample of Colorado avalanche paths. Comparing ξ to constants used to describe the roughness of natural streams in Chézy's equation of open channel flow, Leaf and Martinelli (1977) demonstrate that ξ could range from 150 to 3000 m/sec².

It is interesting to compare these suggested values of ξ with values derived from the unified mass model. Comparing Eqs. (40) and (56) gives

$$\xi h = (m/\beta)g. \tag{57}$$

It was argued in the preceding sections that fitting known data requires that m/β range from 10^9 for slow moving wet avalanches to 10^4 for fast-moving, dry-powder avalanches. For consistency, the corresponding range of ξh is $10^9 < \xi h < 10^5$ m⁹/sec⁹. From field observations, it appears that the flow height h can vary between $1 \le h \le 10^2$ m, with the high limit attained in massive powder avalanches and the low limit typical of slow-moving, wet avalanches. Thus, if ξ did have a narrow range as suggested by Voellmy, it would be about 10^3 m/sec². However, since it is not obvious that ξ is restricted to a narrow range, we can only conclude that strong variations in both ξ and h contribute to the range of ξh.

Voellmy's theory and subsequent Swiss guidelines include techniques for estimating h beginning with the continuity assumption

$$h\rho = h_0\rho_0, \tag{58}$$

where h_0 and ρ_0 are the slab thickness and snow density in the starting zone. In present Swiss engineering practice [see Sommerhalder (1966)], ρ is assumed to range from $2 \le \rho \le 30$ kg/m³ for well-developed powder avalanches, and to be equal to ρ_0 for wet avalanches. For some powder avalanches on long tracks (fully developed dust clouds), Swiss engineers will sometimes apply Voellmy's approximation based on minimum potential energy,

$$\rho = (\rho_a/2g)\xi \sin\theta, \tag{59}$$

which, if substituted into Eq. (56) with the further assumption that μ is equal to zero, yields the curious result that V_{max} is independent of θ and depends only on $\rho_o h_o$,

$$V_{max} = [(2g/\rho_a)\rho_o h_o]^{1/2} \approx 4\sqrt{\rho_o h_o}. \qquad (60)$$

To justify Eq. (60), Voellmy notes that Oeschlin's data do not show a θ dependence, but instead correlate with $\sqrt{h_0}$ as computed from historical meteorological records. In a related study, Bovis and Mears (1976) did not find a significant correlation between track angle θ and runout distance x_r, inferring that track angle θ does not correlate significantly with V. However, both the Oeschlin and the Bovis and Mears studies are unconvincing in the sense that values of θ were restricted to small standard deviations about the mean.

A correlation between $\rho_o h_o$ and V_{max} is also implicit in the unified mass model since V_{max} increases with $\sqrt{m}$ which should increase with $\sqrt{\rho_o h_o}$; m should also increase with starting zone area A_s. Although a correlation between V_{max} and A_s is not reported in the present literature, Bovis and Mears (1976) find that x_r correlated significantly with A_s, implying possible $V_{max} - A_s$ correlation. In any case, it appears the unified mass and fluid models lead to similar results.

Acceleration does not appear in the above equations as they emphasize only the terminal or equilibrium flow. Voellmy's original equations allow for acceleration, but he computed that $0.8V_{max}$ is reached as $x \approx 25h_0$, a rather far-fetched result if one accepts that V_{max} can exceed 50 m/sec, while $h_0 \approx 1$ m. Mellor (1978) also strongly questions such a short acceleration period, and Shen and Roper (1970) caution against a routine assumption of terminal speed. Moreover, deceleration must be introduced to treat the problem of runout distance. Voellmy's equations for x_r are in fact identical to the unified mass equations, except Voellmy includes a potential energy correction for the buildup of debris. According to Voellmy,

$$x_r = V_0^2/[2|\alpha| + (V_0^2/2\xi h_d)], \qquad (61)$$

where h_d is the mean deposition depth.

Perhaps the "endless" fluid model is a suitable approach when wet snow from a massive starting zone discharges into a gully. According to Swiss guidelines [Salm (1966); Sommerhalder (1966)], the maximum velocity of a wet avalanche through a gully is computed by replacing h in Eq. (56) with the hydraulic radius R, defined as the cross-sectional flow area divided by the "wetted" perimeter of the gully. The guidelines begin with a series of assumptions concerning discharge rates from a starting zone of length L_s and width W_s, namely,

$$\left(\begin{array}{c}\text{discharge time}\\\text{from starting zone}\end{array}\right) = \Delta t = \frac{L_s}{(\xi h_0 \alpha_s/g)^{1/2}}, \tag{62}$$

$$\left(\begin{array}{c}\text{discharge rate}\\\text{from starting zone}\end{array}\right) = Q_s = \frac{W_s L_s h_0}{\Delta t}, \tag{63}$$

$$\left(\begin{array}{c}\text{discharge rate}\\\text{through gully}\end{array}\right) = Q_s = RP(\xi R\alpha_G/g)^{1/2}, \tag{64}$$

where α_s and α_G are computed from Eq. (31) for the respective cases of starting zone and gully. From Eqs. (62)–(64), R is determined uniquely from a given gully profile using

$$R^3 P^2 = h_0{}^3 W_s{}^2 \alpha_s/\alpha_G. \tag{65}$$

The R determined from Eq. (65) is then used in Eq. (56) to compute V_{max}. According to Eq. (65), R approaches $h_0\alpha_s/\alpha_G$ on a flat, open slope where $P \approx W_s$. This suggests that the above theory predicts h values far too small for the case of a well-developed dust cloud in a wide gully, but could be a reasonable approximation for a wet flowing avalanche.

In terms of a unified mass model, the increase in the speed of an avalanche confined to a gully in comparison to an avalanche on an open slope is explained by the decrease in the dynamic drag term β that results when an avalanche of given mass enters a gully, decreases its frontal area, and assumes a more projectilelike shape. Conversely, when an avalanche leaves a gully and fans out over an unconfined runout zone, β increases, and x_r decreases.

Pressure due to air mass displacement by the advancing avalanche, the so-called air blast, is known to cause damage [Tøndel (1977)], and Voellmy and many others argue that a compressional shock wave sometimes propagates out ahead of an avalanche. This possibility is not presently accepted in Switzerland [Salm (1964); de Quervain (1966a)] or in North America [Mellor (1978)], and seems unlikely since avalanche speeds are far lower than the sonic speed (~ 300 m/sec). However, the possibility of subsonic surges of mass within the avalanche moving at a higher speed than the observable front cannot be excluded. The front sustains the greatest drag pressures and smooths a path, so material releasing from a high position in the starting zone could overtake and collide with slower-moving mass near the front, producing momentary surges.

Swiss engineers recognize that many of Voellmy's equations are of marginal validity but insist that an approximate formalism is better than none, and have acquired considerable experience fitting his equations to a large number of avalanche paths.

E. Laboratory Modeling

Since field measurements of large avalanche motion are difficult, avalanche motion is modeled at a smaller scale in the laboratory. In these models, geometric and dynamic similarity between laboratory and natural scales should be maintained and the boundary conditions should be preserved. Dynamic similarity is maintained by insuring that certain characteristic numbers of the nondimensionalized equations of motion are the same in laboratory and field situations.

For example, suppose a moving avalanche is to be modeled using the Navier–Stokes equation in conjunction with the continuity equation. Dynamic similarity depends on the Euler, Reynolds, Froude, and Strouhal numbers, which involve pressure P, density ρ, speed V, characteristic length L, characteristic time T, and viscosity μ [Malvern (1969)]. It is virtually impossible to devise a general avalanche model that simultaneously satisfies all characteristic numbers. However, it is possible to restrict the model and thereby eliminate one or more of the characteristic numbers.

If the model is limited to regions where the frontal velocity is nearly steady, then the requirement of matching the Strouhal number can be omitted with the understanding that all portions of the motion which depend intrinsically on temporal variations (e.g., vortex shedding) are not necessarily modeled. It is also possible to omit the Reynold's number requirement, since avalanches appear fully turbulent after a short period of initial acceleration and the Reynolds number defined according to Eq. (44) reaches well into the turbulent regime at the terminal speeds of interest (10 − 50 m/sec). Hence, the requirement of matching the Reynolds number can be replaced by the less restrictive requirement that the Reynolds number in the model be sufficiently high to guarantee turbulence. Finally, the Euler number is not of intrinsic importance since the pressure gradient force due to the decrease of atmospheric pressure with elevation (about 10 Pa) is negligible compared to the gravitational and inertial forces. Thus, for the case of steady turbulent flow, dynamic similarity is approximated by satisfying only the Froude number requirement with α in place of g,

$$(V^2/\alpha L)_{\text{model}} = (V^2/\alpha L)_{\text{nature}}. \tag{66}$$

For a steady flow problem, the characteristic length L can be set equal to the flow height (or hydraulic radius). A feasible flow height in a laboratory model would be about 0.1 m, or about $\frac{1}{100}$ the scale of nature. Assuming the speeds of interest in nature are 10–50 m/sec, then, according to Eq. (66), the model would have to allow for speeds in the range 1–5 m/sec.

Although these speeds can be reached in inclined open channels 10 m long, such experiments have not been performed.

The relevant modeling to date involved inclined water tanks (about 3 m long) where the avalanche was modeled as a salt solution wedging under the ambient water fluid [Tochon-Danguy and Hopfinger (1975); Hopfinger and Tochon-Danguy (1977)]. In this type of experiment it is necessary to redefine the Froude number since the lighter ambient fluid of density ρ_{amb} provides buoyancy for the heavier "avalanche" fluid ρ_{aval}, so the gravitational force is about $\alpha L(\rho_{aval}-\rho_{amb})$. Assuming the inertial force is about $\rho_{aval} V^2$, the Froude number requirement is replaced by the analogous requirement

$$\left[\frac{V^2}{\alpha L}\left(\frac{\rho_{aval}}{\rho_{aval} - \rho_{amb}}\right)\right]_{model} = \left[\frac{V^2}{\alpha L}\left(\frac{\rho_{aval}}{\rho_{aval} - \rho_{amb}}\right)\right]_{nature} \qquad (67)$$

In nature, the density of the ambient fluid air is about 1 kg/m³. Since ρ_{aval} averaged over observed height rarely falls below 10 kg/m³, the density ratio

$$[\rho_{aval}/(\rho_{aval} - \rho_{amb})]_{nature} \qquad (68)$$

rarely exceeds unity. However, in the laboratory model of Hopfinger and Tochon-Danguy, $[\rho_{aval} - \rho_{amb}]_{model}$ was about 0.01 with a corresponding decrease in V_{model} by an additional factor of 10; that is, V was about 0.1 m/sec. Hopfinger and Tochon-Danguy were able to produce avalanche flow patterns in their water tank that strongly resembled powder avalanches in nature and reached conclusions about the structure and growth of the head and the flow interaction of avalanches with structures. They were also able to simulate snow entrainment by distributing a thin layer of brine along the floor of the tank. Unfortunately, their experimental results are not on firm theoretical grounds, because a scale analysis of the Navier–Stokes equation for the case of stratified fluids indicates that Eq. (67) is not justified physically, independent of approximate agreement in the density ratios [Turner (1973)]. Hopfinger and Tochon-Danguy emphasize that their experiments do not satisfy this requirement, but feel this does not necessarily invalidate their results and propose future experiments to test the effect of varying the density ratios.

In support of their conjecture, it is interesting that many atmospheric and oceanic flows that come under the common heading of gravity and density currents (e.g., turbidity currents, haboobs, catabatic winds) bear a striking resemblance to snow avalanches, even though the density ratios of these atmospheric and oceanic flows range from 10 to 10^2, compared to unity for snow avalanches. Hopfinger and Tochon-Danguy borrow heavily from the extensive theoretical and experimental model data already developed to describe gravity currents. It should be understood that mdoeling of atmospheric and oceanic gravity currents presents a much smaller scale problem compared to snow avalanche modeling. The characteristic flow height of the gravity currents in nature may be 10^2 m or greater and, since they travel at speeds less than 50 m/sec, it is possible to reach feasible water tank speeds with a scale of about 10^{-3}.

One interesting idea that applies to avalanche models [Hopfinger and Tochon-Danguy (1977)] from gravity current experiments is that mass velocities in the steady layer upstream of the front are in excess of the frontal velocities [Middleton (1966)]. This produces mass transfer and consequent head growth along with a possible sorting whereby heavier and larger snow clumps are concentrated in the head. Middleton (1966) also found a strong attenuation of the velocity of the ambient fluid with distance downstream d from the front of the gravity current (Fig. 23). According to Middleton,

$$\log_{10} V_{amb} \approx \log_{10} V_{aval} - d/H, \tag{69}$$

where H is the height of the head of the gravity current. This expression contradicts the traditional belief of field observers [beginning in North America with Cunningham (1887)] that avalanche winds are important over long ranges. Unfortunately, no quantitative data about avalanche winds exist.

This section has emphasized the problem of dynamic similarity and avoided the problem of finding even an approximate replication of the complex boundary conditions involving entrainment, ploughing drag, and frictional drag that exists at the interface between the moving avalanche, the fixed slope, and the ambient fluid. Putting aside the unresolved question of scaling according to Eq. (67) without satisfying the density ratio, the water tank experiments of Hopfinger and Tochon-Danguy at least pro-

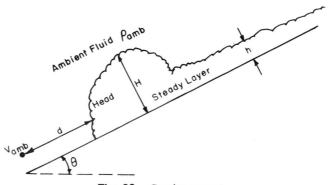

Fig. 23.   Gravity current.

vided for entrainment from the ambient fluid as well as from a heavier brine layer at the lower interface. Their studies appear consistent with the gravity current model by Ellison and Turner (1959) in which the thickness h of the steady gravity layer increases with distance x in response to entrainment according to

$$dh/dx \approx (5 + \theta)/1000 \qquad (70)$$

and where the increase in head thickness H is larger by a factor of 2 or 3.

F. Impact Pressure

Table 6 provides a partial summary of attempts to measure or compute the horizontal impact pressure of moving avalanches; that is, the impact pressure P_x against a plane perpendicular to the direction of motion. The proposed values scatter wildly from Voellmy's (1955) assertion that P_x rarely exceeds 2×10^5 Pa to the measued value 22.6×10^5 Pa found by Shimizu *et al.* (1977). Some of the higher values are questionable because most are based on indentation measurements which are usually calibrated by measuring penetration in response to a static load. The load cell and penetrometer measurements could be an order of magnitude higher than Voellmy's computations of pressures required for actual destruction and survival of structures since Voellmy is referring to pressures averaged over sides of large objects. The load cell and penetrometer measurements involve relatively small target plates which would register comparatively high P_x values when hit by a snow particle with dimensions comparable to the target area. Although target size effects have not yet been quantified, the $P_x(t)$ recordings show quite clearly that avalanche impact consists of a series of discrete collisions.

Another complication is the variation of P_x with distance z above the snow surface. Shoda (1966) indicates that P_x usually has a maximum in the range $0 < z < 3\,m$, and P_x decreases on both sides of the maximum according to a function that resembles a Gaussian distribution (see Fig. 24a). Kotlyakov *et al.* (1977) find a P_x maximum in the range $0 < z < 2$ m, but without further justification extrapolate their data to show that the maximum is sustained at $z = 0$ (see Fig. 24b). Both the Shoda and Kotlyakov studies utilized indentation penetrometers. Bon Mardion *et al.* (1974) have obtained load cell recordings at heights of 0.2 and 0.4 m but do not find any major differences in P_x values at these two levels. The above studies therefore suggest that P_x reaches a relatively broad maximum 1 or 2 m above the surface. Evidence that the maximum may shift to a much higher level is presented in Roch's (1962) study of the Val Buera avalanche (see Fig. 24c).

TABLE 6

Avalance Impact Experiments and Computations

Investigators	Location	Impact pressure (10^5 N/m²)	Type of experiment or computation	Avalanche characteristics
Goff and Otten (1939)	Caucasus and Apatit Mine, USSR	>4.0 peaks	Force transducers target 0.1 m²	Volume ~10^5 m³
Puzanov (1943) Shinoda (1953)	Apatit Mine, USSR Japan	>1.18 sustained 0.33	Force transducers target 0.1 m² —	Volume ~2.5 × 10^4 m³ —
Voellmy (1955)	Blons, Sonntag, and Dalaas, Switzerland	Typically less than 0.5; max. 2.0; air blast 0.05	Calculated from damage and survival of structures; tipping and movement of massive objects	$V \leq 60$ m/sec Vertical ~1 km $\theta \sim 30°$
de Spindler (1957)	Val Mila, Switzerland	2.7	Calculated damage to transmission tower	Vertical ~800 m
Killer (1957)	Val Mila, Switzerland	1.47 peak	Indentation penetrometer 0.2-m² target area	Vertical ~800 m
	British Columbia, Canada	1.75	Computed damage to Kemano-Kitimat H.V. transmission line (by M. de Quervain)	—
Roch (1962), Salm (1966)	Val Buera, Switzerland	10.8 peak	Indentation penetrometer 0.2-m² target area	$V \sim 16$ m/sec Vertical ~700 m $22° < \theta < 29°$ Volume ~3 × 10^5 m³

Reference	Location		Instrument	Parameters
Shoda (1966)	Tsuchitaru, Japan	3.0 peak	Indentation penetrometer 0.1-m² target area	$v \sim 20$ m/sec Vertical ~290 m $\theta \sim 35°$ Volume ~10^3 m³
Schaerer (1973)	Rogers Pass, British Columbia	2.56 peak 1 05 ave	Force transducers 25-mm-diam target	$v \lesssim 53$ m/sec Vertical ~500–1000 m
Bon Mardion et al. (1974, 1975)	Col du Lautaret, France	3.0 peak 1.5 sustained	Force transducers 100-mm-diam target	$v \lesssim 28$ m/sec Vertical ~400 m $\theta \sim 29°$
Kotlyakov et al. (1977)	Khibiny, USSR	11.0 peak	Indentation penetrometer 300-mm-diam target	$v \gtrsim 50$ m/sec Vertical 450 m $\theta \sim 24°$ Volume ~5×10^4 m³
Shimizu et al. (1972, 1973, 1974, 1975, 1977)	Kurobe Canyon, Japan	22.6 peak 13.4 peak 8.7 peak	Indentation penetrometer Load cell Stress gage	Transient speeds up to 150 m/sec Vertical ~800 m $\theta \sim 30°$

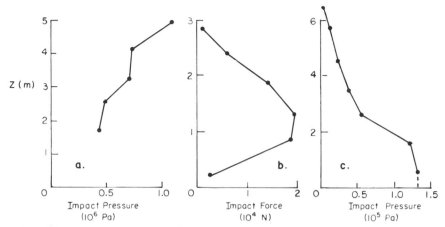

Fig. 24. Example of avalanche pressure (and force) as a function of height Z above the snow surface. (a) Val Buera avalanche from Roch (1962). (b) "Surface flow" avalanche from Shoda (1966). (c) Mean data from Kotlyakov *et al.* (1977).

From case histories of damage, Voellmy computed that the vertical component of pressure P_z should range between $P_x/4$ to $P_x/2$. Mellor (1968) and Sommerhalder (1972) summarize measurements made on avalanche sheds in Switzerland, indicating that P_z values reach 0.7×10^5 Pa. Mellor (1968) also notes that the successful designs of two sheds at Rogers Pass, British Columbia, were based on the lower values: $P_z = 0.15 \times 10^5$ Pa, and 0.34×10^5 Pa. Nakajima (1976) attached strain gages to snowsheds (see Table 4) and recorded a lower P_z value of 0.11×10^5 Pa. With respect to shear load P_{xz}, Sommerhalder (1972) reports that measurements on top of Swiss avalanche sheds indicate $P_{xz} < 0.2 \times 10^5$ Pa, and further suggests it is reasonable to assume that μ is about equal to $\frac{1}{3}$; however, in Switzerland and Rogers Pass, shed designers have assumed μ about equal to $\frac{1}{2}$ as a safety allowance [Mellor (1978)]. The shed studies reported by Sommerhalder, Nakajima, and Mellor involve large-scale measurements and seem to confirm indirectly Voellmy's contention that loads over 2×10^5 Pa are found in only extreme instances.

According to Swiss guidelines [Sommerhalder (1966)], the total avalanche load can be divided into static and dynamic components. The static component, estimated as $\rho g L$, where L is the thickness of the snow load, generally will not reach a maximum at the same time as the dynamic component, but will peak after the avalanche comes to rest and deposition builds up on top of or against the structure. For a severe combination of factors, for example L greater than 3 m and ρ greater than 300 kg/m³, the purely static component of P_z may be in the range 10^4–10^5 Pa, which may be a significant or perhaps dominant portion of the design load.

An estimate of the dynamic component is not as straightforward. Although conventional theory suggests that

$$P_x = k\rho V^2, \tag{71}$$

the value of k is unknown. If the avalanche is approximated by an incompressible fluid, then k is equal to $\frac{1}{2}$, analogous to the stagnation pressure. However, if the avalanche is assumed to transport momentum through collision and increased density, then in accordance with impact models discussed by Dolov (1967)

$$P_x = [\rho_2/(\rho_2 - \rho_1)]\rho_1 V^2, \tag{72}$$

where ρ_1 and ρ_2 are the densities before and after impact. Thus, k could assume almost any value greater than unity, depending on the density increase. The density of avalanche-deposited snow is typically two or three times the density of the undisturbed surface snow [Schaerer (1967); Bon Mardion et al. (1975)], but it is unclear how much of this increase is due to impact, and how much is due to compaction and ploughing during the motion prior to impact.

It could be argued that k approaches $\frac{1}{2}$ as avalanche speed and fluidity increase, with the net result that P_r is no longer proportional to V^2, but to some lesser power of V, perhaps according to empirical functions suggested by Furukawa (1957) and Isaenko (1975). However, in keeping with Hopfinger's comments on Isaenko's paper, it seems preferable to retain the V^2 dependence and compile as much empirical data as possible on the variation of the combination k_ρ. For rough estimates, Voellmy's large-scale pressures are reasonably predicted by setting k equal to one, and estimating ρ from expected flow heights consistent with the formalism presented in Section III.D. In practice, Swiss engineers routinely apply a safety factor of about two, thus allowing for contingencies such as vibrational loads [B. Salm (personal communication)].

As the impact area decreases, discrete collisions become important, and, as suggested by the experimental evidence summarized in Table 6, there is probably a substantial increase in P_x over small areas for short time intervals (~ 10 m sec). The practical ramification is that it may be risky to design within the above limits if the structural members are relatively narrow, as may be the case in the design of bridges and transmission towers.

One simple analytical ploy to account for small area peak pressure is to increase kP by an appropriate factor. Schaerer (1973) suggests that

$(P_x)_{\text{ave}} = 0.3(P_x)_{\text{peak}}$, and Bon Mardion *et al.* (1974) obtain a similar result: $(P_x)_{\text{ave}} = 0.35(P_x)_{\text{peak}}$, although it is not clear how these investigators averaged their spiked analog traces. Schaerer (1973) further suggests that $(P_x)_{\text{peak}}$ can be approximated by setting kP equal to $\frac{1}{2}$ the density of the deposited snow ρ_D, which in his 1973 study was in the range $210 \leq \rho_D \leq 350$ kg/m³, but in an earlier study [Schaerer (1967)] was considerably higher, $350 \leq \rho_D \leq 450$ kg/m³ for dry snow, and up to 650 kg/m³ for wet snow. In the study of Bon Mardion *et al.* (1975) the deposited snow was lighter, $165 \leq \rho_D \leq 259$ kg/m³, yet their values of both $(P_x)_{\text{ave}}$ and $(P_x)_{\text{peak}}$ were higher than Schaerer's, and they were therefore unable to substantiate Schaerer's approximation ($k_\rho = \frac{1}{2}\rho_D$).

Perla *et al.* (1979) studied peak impacts by dropping blocks of snow from a 10-m tower onto a circular target supported by load cells. Analysis of results indicate that peak impact of a discrete snow mass of density ρ_m is approximately

$$(P)_{\text{peak}} \approx 1.5\rho_m V^2. \tag{73}$$

If an avalanche consists of particles with $\rho_m = 300$ kg/m³, moving at $V = 50$ m/sec, the peak pressures attain 10^6 Pa, which is consistent with the Table 6 values recorded in small-scale experiments. The impact force, which is proportional to the particle mass, attains a peak in a rise time about equal to $r/2v$, where r is the particle radius.

IV. CONCLUDING REMARKS

Some of the questions raised in this chapter were answered at the recent conference Snow in Motion (*Journal of Glaciology,* in press) held at Fort Collins, Colorado, 12–17 August 1979. However, at the end of this conference it was clear that even the most fundamental measures of a snow avalanche remain uncertain. Avalanche research continues today as a fascinating and active discipline.

ACKNOWLEDGMENTS

The following have provided valuable input: D. Bachman, R. Brown, K. Lied, E. La-Chapelle, M. Martinelli, D. MacKay, D. McClung, H. Röthlisberger, B. Salm, R. Sommerfeld, and W. St. Lawrence. The author especially wishes to thank Diane Schoemperlen for her help.

REFERENCES

Akitaya, E. (1974). *Low Temp. Sci., Ser. A* **32**, 77–104.
Allix, A. (1925). *Mater. Etude Calam.* **2**, 37–57.
Ambach, W., and Howorka, F. (1966). *Int. Assoc. Sci. Hydrol., Publ.* **69**, 65–72.
Ambartsumyan, S. A. (1970). ''Theory of Anisotropic Plates.'' Technomic Publ. Co., Stamford, Connecticut.

Armstrong, B. R. (1976). "Century of Struggle Against Snow: A History of Avalanche Hazard in San Juan Country, Colorado." INSTAAR, University of Colorado, Boulder.

Armstrong, B. R. (1977). "Avalanche Hazard in Ouray Country, Colorado, 1877–1976." INSTAAR, University of Colorado, Boulder.

Armstrong, B. R. (1978). In "Avalanche Control, Forecasting, and Safety" (R..Perla, ed.), Tech. Memo. No. 120, pp. 199–218. Assoc. Comm. Geotech. Res., Natl. Res. Counc., Ottawa, Ontario, Canada.

Armstrong, R. L. (1977). J. Glaciol. **19**, 325–334.

Armstrong, R. L., La Chapelle, E. R., Bovis, M. J., and Ives, J. D. (1974). Development of methodology for evaluation and prediction of avalanche hazard in the San Juan Mountain area of southwestern Colorado. Occasional Paper No. 13, INSTARR, Univ. of Colorado, Boulder, Colorado.

Atwater, M. M. (1952). Proc. 20th Annu. Meet. West. Snow Conf., 1952 pp. 11–19.

Atwater, M. M. (1970). In "Ice Engineering and Avalanche Forecasting and Control," Tech. Memo. No. 98, pp. 99–105. Natl. Res. Counc., Ottawa, Ontario, Canada.

Atwater, M. M. (1968). "The Avalanche Hunters." Macrae Smith Co., Philadelphia, Pennsylvania.

Atwater, M. M., and Koziol, F. C. (1953). "Avalanche Handbook." U.S. Dept. of Agriculture, For. Serv., Washington, D.C.

Aulitzky, H. (1974). "Endangered Alpine Regions and Disaster Prevention Measures," Nature Environ. Ser. No. 6 Council of Europe, Strasbourg, France.

Bader, H., Haefeli, R., Bucher, E., Neher, J., Eckel, O., Thams, C., and Niggli, P. (1939). Beitr. Geol. Schweiz, Geotech. Ser., Hydrol. No. 3.

Bishop, A W , and Little, A, L. (1967). Proc. Geotech. Conf., Oslo Vol. 1, pp. 89–96.

Bjerrum, L. (1967). J. Soil Mech. Found. Div. **93**, 1 49.

Björnsson, H. (1977). In "Snow Avalanche Studies in Iceland," Glaciol. Data, Rep. GD-1, pp. 39–41. World Data Center A for Glaciology, University Colorado, Boulder,

Bon Mardion, G., Coche, G., Eybert-Berard, A., Jourdan, P., Perroud, P., and Rey, L. (1974). "Measures dynamiques dans l'avalanche. Premiers résultats expérimentaux," CENG/ASP No. 74 01. Centre d'Etude de la Neige, Grenoble, France.

Bon Mardion, G. Eybert-Berard, A., Guelff, C., Perroud, P., and Rey, L. (1975). "Measures dynamiques dans l'avalanche. Resultats expérimentaux de la saison de neige 1973–1974," CENG/ASP No. 75-01. Centre d'Etude de la Neige, Grenoble, France.

Bovis, M. J., and Mears, A. I. (1976). Arct. Alp. Res. **8**, 115–120.

Bradley, C. C. (1966). Int. Assoc. Sci. Hydrol., Publ. **69**, 251–260.

Bradley, C. C. (1970). J. Glaciol. **9**, 253–261.

Bradley, C. C., and Bowles, D. (1967). Phys. Snow Ice, Conf., Proc., 1966 Vol. 1, Part 2, pp. 1243–1253.

Bradley, C. C., and St. Lawrence, W. (1975). IAHS–AISH Publ. **114**, 145–154.

Bradley, C. C., Brown, R. L., and Williams, T. (1977). J. Glaciol. **78**, 145–147.

Briukhanov, A. V. (1968). Bull. Soc. Fr. Photogr. **29**, 9–28.

Brown, C. B., Evans, R. J., and LaChapelle, E. R. (1972). Geophys. Res. **77**, 4570–4580.

Brown, R. L., Lang, T. E., St. Lawrence, W. F., and Bradley, C. C. (1973). J. Geophys. Res. **78**, 4950–4958.

Burrows, C. J., and Burrows, V. L. (1976). "Procedures for the Study of Snow Avalanche Chronology Using Growth Layers of Wood Plants," Occas. Pap. No. 23. INSTAAR, University of Colorado, Boulder.

Carrigy, M. A. (1970). Sedimentology **14**, 147–158.

Castelberg, F., In der Gand, H. R., Pfister, F., Rageth, B., and Bavier, G., eds. (1972). "Lawinenschutz in der Schweiz," Beih. No. 9.

Chitadze, V. C., and Zalikhanov, M. C. (1967). *Tr., Vysokogorn. Geofiz. Inst.* **6**, 47–65.
Cunningham, G. C. (1887). *Trans. Can. Civ. Eng.* **1**, 18–31.
Daniels, E. H. (1945). *Proc. R. Soc. London, Ser. A* **183**, 405–435.
de Quervain, M. R. (1946). *Prisma* **1**, 36–38.
de Quervain, M. R. (1966a). *Int. Assoc. Sci. Hodrol., Publ.* **69**, 15–22.
de Quervain, M. R. (1966b). *Int. Assoc. Sci. Hydrol., Publ.* **69**, 410–417.
de Spindler, A. (1957). *Schweiz. Bauztg.* **75**, 675–678.
Dolov, M. A. (1967). *Tr., Vysokogorn. Geofiz. Inst.* **9**, 3–12.
Eglit, M. E. (1968). *Itogi Nauki, Gidrol. Sushi* pp. 69–97.
EIDG. (1951). "The Avalanche Winter 1950/51 (Der lawinen Winter 1950/51)." *Veröffentlichung über Verbauungen,* Department des Innern
EIDG. (1968). "Lawinen verbau in Anbruchgebiet. Richtlinien des GIDG. Oberfort inspektorates für del stutzverbau." Institut für Schnee und Lauwinen Forschung.
Ellison, T. H., and Turner, J. S. (1959). *J. Fluid Mech.* Vol. 6. 423–448.
Field, W. O. (1966). *Int. Assoc. Sci. Hydrol., Publ.* **69**, 326–331.
Flaig, W. (1955). "Lawinen." F. A. Brockhaus, Weisbaden.
Föhn, P., Good, W., Bois, P., and Obled, C. (1977). *J. Glaciol.* **19**, 375–387.
Fondation Internationale "Vannie Eigenmann" (1963). "Sauvetage des victimes d'avalanches." Fond. Int. "Vannie Eigenmann." Davos, Switzerland.
Fondation Internationale "Vannie Eigenmann" (1975). "Lawinenunfalle-Prophylaxe, Ortung, Rettung." Fond. Int. "Vannie Eigenmann." Sulden, Italy.
Fraser, C. (1966). "The Avalanche Enigma." Murray, London.
Frutiger, H. (1964). *U.S., For. Serv., Rocky Mount. For. Range Exp. Stn., Res. Pap. RM* **7.**
Frutiger, H., and Martinelli, M., Jr. (1966). *U.S., For. Serv., Rocky Mount. For. Range Exp. Stn. Res. Pap. RM* **19.**
Furukawa, I. (1957). *Seppyo* **19**, 140–141.
Gallagher, D. (1967). "The Snowy Torrents; Avalanche Accidents in the United States 1910–1966." U.S. For. Serv., Alta Avalanche Study Cent., Utah.
Gardner, N. C., and Judson, A. (1970). *U.S., For. Serv., Rocky Mount. For. Range Exp. Stn., Res. Pap. RM* **61.**
Goff, A. G., and Otten, G. F. (1939). *Geogr. Geofiz.* **3**, 303–308.
Good, W. (1972). *Winterber. Eidg. Inst. Schnee-Lawinenforsch.* **35**, 154–162.
Gubler, H. (1977). *J. Glaciol.* **19**, 419–429.
Gubler, H. (1978a). *J. Glaciol.* **20**, 329–341.
Gubler, H. (1978b). *J. Glaciol.* **20**, 343–357.
Haefeli, R. (1938). *Schweiz. Bauztg.* **111**, 321–325.
Haefeli, R. (1950). *Geotechnique* **2**, 186–208.
Haefeli, R. (1963). *In* "Ice and Snow" (W. D. Kingery, ed.), pp. 560–575. MIT Press, Cambridge, Massachusetts.
Haefeli, R. (1966a). *Proc. Int. Conf. Soil Mech. Founda. Eng., 6th, 1965* Vol. III, pp. 134–148.
Haefeli, R. (1966b). *Int. Assoc. Sci. Hydrol., Publ.* **69**, 316–325.
Haefeli, R. (1967). *Phys. Snow Ice, Conf., Proc.,* Vol. 1, Part 2, pp. 1199–1213.
Heimgartner, M. (1977). *J. Glaciol.* **19**, 357–363.
Heybrook, W. (1935). *Geogr. Rev.* **25**, 423–429.
Hopfinger, E. J., and Tochon-Danguy, J. C. (1977). *J. Glaciol.* **19**, 343–356.
Iken, A. (1977). *J. Glaciol.* **19**, 595–605.
In der Gand, H. R. (1956). *Winterber. Eidg. Inst. Schnee- Lawinenforsch.* **19**, 85–101.
In der Gand, H. R., and Zupancic, M. (1965). *Int. Assoc. Sci. Hydrol., Publ.* **69**, 230–243.
Isaenko, E. P. (1975). *IAHS-AISH Publ.* **114**, 433–440.
Ives, J. D., and Bovis, M. J. (1978). *Arct. Alp. Res.* **10**, 185–212.

Jaccard, C. (1966). *Int. Assoc. Sci. Hydrol., Publ.* **69**, 170–81.
Jónsson, O. (1957). "Skriduföll og snjoflod." Annad Bindi, Bokautgafan Nordri.
Judson, A. (1977). *Proc. 45th Annu. Meet. West. Snow Conf., 1977* pp. 1–11.
Killer, J. (1957). *Schweiz. Bauztg* **75**, 561–568.
Kojima, K. (1960). *Low Temp. Sci. Ser. A* **16**, 47–59.
Körner, H. J. (1976). *Rock Mech.* **8**, 225–256.
Kotlayakov, V. M., Rzhevskiy, B. N., and Samoylov, V. A. (1977). *J. Glaciol.* **19**, 431–439.
Kozik, S. M. (1962). "Calculation of the Movement of Snow Avalanches." Gidiome-
    teoizdat, Leningrad.
Kuroiwa, D., Mizuno, Y., and Takeuchi, M. (1967). *Phys. Snow Ice, Conf., Proc.,* Vol. 1,
    Part 2, pp. 751–772.
LaChapelle, E. R. (1967). *Phys. Snow Ice, Conf., Proc., 1966* Vol. 1, Part 2, pp. 1169–1175.
LaChapelle, E. R. (1968). *In* "The Great Alaska Earthquake of 1964," Hydrology. Publ.
    1603 pp. 355–361. Natl. Acad. Sci., Washington, D.C.
LaChapelle, E. R. (1970). *In* "Ice Engineering and Avalanche Forecasting and Control,"
    Tech. Memo. No. 98, pp. 106–113. Assoc. Comm. Geotech. Res., Natl. Res. Counc.,
    Ottawa, Ontario. Canada.
LaChapelle, E. R. (1977). *J. Glaciol.* **19**, 313–324.
LaChapelle, E. R., and Leonard, R. (1971). "North Cascades Highway SR-20 Avalanche
    Atlas." University of Washington, Seattle.
Lang, T. E., and Brown, R. L. (1975). *IAHS-AISH Publ.* **114**, 311–320.
Lang, T. E. (1977) *J. Glaciol.* **19**, 365–373.
Lang, T. E., and Sommerfeld, R. A. (1977). *J. Glaciol.* **19**, 153–163.
Lang, T. E., Brown, R. L., St. Lawrence, W. F., and Bradley, C. C. (1973). *Geophys. Res.*
    78, 339–351.
Leaf, C. F., and Martinelli, M. Jr. (1977). *U.S. For. Serv., Rocky Mount. For. Range Exp.
    Stn., Res. Pap. RM 183*
Lee, I. K., and Ingles, O. G. (1968). *In* "Soil Mechanics: Selected Topics" (I. K. Lee, ed.),
    pp. 195–294. Elsevier, New York.
Leon, R. E. (1978). *In* "Avalanche Control, Forecasting and Safety" (R. Perla, ed.), Tech.
    Memo. No. 120, pp. 219–228. Assoc. Comm. Geotech. Res., Natl. Res. Counc., Ottawa,
    Ontario, Canada.
Lied, K. (1978). Personal communications. Norw. Geotech. Inst., Oslo.
Lliboutry, L. (1971). *Recherche* **2**, 417–425.
Luckman, B. H. (1977). *Geogr. Ann. Ser. A.* **1-2**, 31–48.
McClung, D. M. (1974). Ph.D. Thesis, University of Washington, Seattle.
McClung, D. M. (1975). *IAHS-AISH Publ.* **114**, 236–248.
McClung, D. M. (1977). *J. Glaciol.* **19**, 101–109.
McClung, D. M. (1979). *J. Geophys. Res.* **84**, no. B7, 3519–3526.
Malvern, L. E. (1969). "Introduction to the Mechanics of a Continuous Medium."
    Prentice-Hall, Englewood Cliffs, New Jersey.
Martinelli, M., Jr. (1974). *U.S., Dep. Agric., Agric. Inf. Bull.* **360**.
Matznetter, J. (1955). *Geogr. Rundsch.* **7**, 47–52.
Mellor, M. (1968). "Avalanches," Monogr. III A3d. U.S. Army Cold Reg. Res. Eng. Lab.,
    Hanover, New Hampshire.
Mellor, M. (1975). *IAHS-AISH Publ.* **114**, 251–291.
Mellor, M. (1978). *In* "Rockslides and Avalanches" (B. Voight, ed.), Vol. 1, pp. 753–792.
    Elsevier, Amsterdam.
Michon, X., Goddet, J., and Bonnefille, R. (1955). "Etude théorique et expérimentale des
    courants de densité," Vols. I and II. Natl. Hydraul., Chatau, France.
Middleton, G. V. (1966). *Can. J. Earth Sci.* **3**, 523–546.

Mokievsky-Zubok, O. (1978). *J. Glaciol.* **20**, 215–217.
Morales, B. (1966). *Int. Assoc. Sci. Hydrol., Publ.* **69**, 304–315.
Morgenstern, N. R., and Tchalenko, J. S. (1967). *Geotechnique* **17**, 309–328.
Nakajima, H. (1976). "Measurement of Snowshed Stress in Artifical Avalanche," Intern. Rep. Nippon Kokan K. K., Yokohama, Japan.
Nobles, L. H. (1966). *Int. Assoc. Sci. Hydrol., Publ.* **69**, 267–272.
Novikov, B. I. (1971). *Sov. Hydrol., Sel. Pap.* **2**, pp. 124–129.
Ommanney, C. S. L. (1978). *In* "Avalanche Control, Forecasting and Safety" (R. Perla, ed.), Tech. Memo. No. 120, pp. 195–198. Assoc. Comm. Geotech. Res., Natl. Res. Counc., Ottawa, Ontario, Canada.
Oulianoff, N. (1954). *Bull. Soc. Vaudoise Sci. Nat.* **66**, 19–25.
Pahaut, E. (1975). "Les cristaux de neige et leurs métamorphoses." Météorologie National, Paris.
Palmer, A. C., and Rice, J. R. (1973). *Proc. R. Soc. London, Ser. A* **332**, 527–548.
Perla, R. (1969). *J. Glaciol.* **8**, 427–440.
Perla, R. (1970). *Can. Geotech. J.* **7**, 414–419.
Perla, R. I. (1971). "Slab Avalanche," Alta Avalanche Study Cent., Rep. No. 100. U.S. For. Serv., Wasatch Natl. For., Salt Lake City, Utah.
Perla, R. I. (1972). *In* "Environmental Geology of the Wasatch Front, 1971," pp. 1–25. Utah Geol. Assoc., Salt Lake City.
Perla, R. I. (1973). *Int. J. Non-Linear Mech.* **8**, 253–259.
Perla, R. I. (1975). *IAHS-AISH Publ.* **114**, 208–221.
Perla, R. I. (1977). *Can. Geotech. J.* **14**, 206–213.
Perla, R. I. (1978a). *Arct. Alp. Res.* **10**, 235–240.
Perla, R. I., ed. (1978b). "Avalanche Control, Forecasting and Safety," Tech. Memo. No. 120. Assoc. Comm. Geotech. Res., Natl. Res. Counc., Ottawa, Ontario, Canada.
Perla, R. I. (1978c). *In* "Rockslides and Avalanches" (B. Voight, ed.), Vol. 1, pp. 731–752. Elsevier, Amsterdam.
Perla, R. I., and LaChapelle, E. R. (1970). *J. Geophys. Res.* **75**, 7619–7627.
Perla, R. I., and Martinelli, M. (1976). *U.S. Dep. Agric., Agric. Handb.* **489.**
Perla, R. I., Beck, T., and Banner, J. (1979). "Impact Force of Snow," NHRI Paper No. 2, IWD Sci. Ser. No. 97. Natl. Hydrology Research Institute, Ottawa, Canada.
Pickering, D. J. (1970). *Geotechnique* **20**, 271–276.
Plafker, G., and Ericksen, G. E. (1978). *In* "Rockslides and Avalanches" (B. Voight, ed.), Vol. 1, pp. 277–314. Elsevier, Amsterdam.
Puzanov, V. P. (1943). *Ser. Geogr. Geofis.* **2**, 86–88.
Ramsli, G. (1974). *In* "Natural Hazards: Local, National, Global" (G. F. White, ed.), pp. 175–180. Oxford Univ. Press, London and New York.
Rapp, A. (1963). *Proc. Permafrost Int. Conf., 1963* pp. 150–154.
Roch, A. (1955). *Alpen* **31**, 94–104.
Roch, A. (1961). *J. Glaciol.* **3**, 979–983.
Roch, A. (1962). *Winterber. Eidg. Inst. Schnee-Lawinenforsch.* **25**, 124–136.
Roch, A. (1966a). *Int. Assoc. Sci. Hydrol., Publ.* **69**, 86–99.
Roch, A. (1966b). *Int. Assoc. Sci. Hydrol., Publ.* **69**, 182–195.
Röthlisberger, H. (1977). *J. Glaciol.* **81**, 669–671.
St. Lawrence, W. F. (1977). Ph.D. Thesis, Montana State University, Bozeman.
St. Lawrence, W. F., and Bradley, C. C. (1973). *Proc. Symp. Adv. North Am. Avalanche Technol., 1972* pp. 1–6.
St. Lawrence, W. F., and Bradley, C. C. (1975). *IAHS-AISH Publ.* **114**, pp. 155–170.

St. Lawrence, W. F., and Bradley, C. C. (1977). *J. Glaciol.* **19**, 411–417.
St. Lawrence, W. F., and Williams, T. R. (1976). *J. Glaciol.* **17**, 521–526.
Salm, B. (1964). "Verdictungswellen bei Staublawinen," Interne Not. No. 47. Eidg. Institutes für Schnee-und Lawinenforschung, Davos-Weissfluhjoch, Switzerland.
Salm, B. (1966). *Int. Assoc. Sci. Hydrol., Publ.* **69**, 199–214.
Salm, B. (1975). *IAHS-AISH Publ.* **114**, 222–235.
Salm, B. (1977). *J. Glaciol.* **19**, 67–100.
Salway, A. A. (1976). Ph.D. Thesis, University of British Columbia.
Schaerer, P. A. (1967). *Phys. Snow Ice, Conf., Proc.*, Vol. 1, Part 2, pp. 1255–1260.
Schaerer, P. A. (1972). *Geogr. Ser.* **14**, 215–222.
Schaerer, P. A. (1973). *Proc. Symp., Adv. North Am. Avalanche Technol., 1972* pp. 51–54.
Scheidegger, A. E. (1975). "Physical Aspects of Natural Catastrophies." Elsevier, Amsterdam.
Schild, M. (1952). *Winterber. Eidg. Inst. Schnee-Lawinenforsch.* **15**, 86–202.
Schleiss, V. G., and Schleiss, W. E. (1970). *In* "Ice Engineering and Avalanche Forecasting and Control," Tech. Memo. No. 98, pp. 115–122. Natl. Res. Counc., Ottawa, Ontario, Canada.
Seligman, G. (1936). "Snow Structure and Ski fields." Macmillan, New York.
Shen, H. W., and Roper, A. T. (1970). *Bull. Int. Assoc. Sci. Hydrol.* **15**, 7–26.
Shimizu, H. (1967). *Phys. Snow Ice, Conf., Proc.*, Vol. 1, Part 2, pp. 1269–1276.
Shimizu, H., and Huzioka, T. (1975). *IAHS-AISH Publ.* **114**, 321–331.
Shimizu, H., Akitaya, G., Nakagawa, M., and Toshio, O. (1972). *Low Temp. Sci., Ser. A* **30**, 103–114.
Shimizu, H., Akitaya, E. Huzioka, T., Nakagawa, M., and Kawada, K. (1973). *Low Temp. Sci., Ser. A* **31**, 178–189.
Shimizu, H., Huzioka, T., Akitaya, E., Narita, H., Nakagawa, M., and Kawada, K. (1975). *Low Temp. Sci., Ser. A* **33**, 109–116.
Shimizu, H., Huzioka, T., Akitaya, E., Narita, H., Nakagawa, M., and Kawada, K. (1977). *Low Temp. Sci., Ser. A* **35**, 117–132.
Shinoda, G. (1963). *Nature (London)* **199**, 165–166.
Shinoda, N. (1953). *Res. Snow Ice* **1**, 215–217.
Shoda, M. (1966). *Int. Assoc. Sci. Hydrol., Publ.* **69**, 215–229.
Skempton, A. W. (1964). *Geotechnique* **14**, 77–101.
Skempton, A. W. (1966). *Congr. Int. Soc. Rock Mech., 1st, 1966* pp. 329–335.
Smith, F. W., and Curtis, J. O. (1975). *IAHS-AISH Publ.* **114**, 332–340.
Sommerfeld, R. A. (1977). *J. Glaciol.* **19**, 399–409.
Sommerfeld, R. A., King, R. M., and Budding, F. (1976). *J. Glaciol.* **17**, 145–147.
Sommerhalder, E. (1966). *Winterber. Eidg. Inst. Schnee-Lawinenforsch.* **29**, 134–141.
Sommerhalder, E. (1972). Deflecting structures. *In* "Lawinenschultz in der Schweiz" (F. Castelberg, H. R. in der Gand, F. Pfister, B. Rageth, and G. Bavier, eds.). Beih. No. 9. Genossenschaft der bündnerischen Holzproduzenten. [*English transl.:* "Avalanche protection in Switzerland." U.S. Dept. of Agriculture, Forest Service, Gen. Tech. Rep. RM-9 (1975).]
Southern, G. A. (1976). *J. Meteorol.* **1**, No. 6, 182–183.
Stethem, C. J., and Schaerer, P. A. (1979). "Avalanche Accidents in Canada. I. A Selection of Case Histories 1955 to 1976." DBR Paper No. 834. Natl. Res. Counc. Canada, Ottawa, Canada.
Swanston, D. N. (1970). *U.S., For. Serv., Res. Pap. PNW* **103**, 1–17.
Swiss Institute for Snow and Avalanche Research (1947–1973). "Schnee und Lawinen in

den Schweizerlapen,'' Nos. 1–37. Weissfluhjoch-Davos, Switzerland.

Tochon-Danguy, J. C., and Hopfinger, E. J. (1975). *IAHS-AISH Publ.* **114**, 369–380.

Tøndel, I. (1977). ''Sikring av veger mot Snøskred,'' Medd. No. 17. Institutt for Veg-og Jennbanebygging, Trondheim.

Tsomaia, V. Sh. (1968). *Tr. Zakavk. Nauchno-Issled. Gidrometeorol. Inst.* **27**, 53–66.

Turner, J. S. (1973). ''Buoyancy Effects in Fluids.'' Cambridge Univ. Press, London and New York.

Van Wijk, M. C. (1967). *J. Glaciol.* **6**, 917–933.

Varnes, D. J. (1958). *In* ''Landslides and Engineering Practices'' (E. G. Eckel, ed.), Spec. Rep. No. 29, pp. 20–47. Natl. Res. Counc., Highway Res. Board, Washington, D.C.

Voellmy, A. (1955). *Schweiz. Bauztg.* **73**, 159–165, 212–217, 246–249, and 280–285.

Voitkovsky, K. F. (1977). ''Mechanical Properties of Snow.'' Acad. Sci. USSR, Nauka Publ., Moscow.

Wakabayashi, R., and Yamamura, M. (1968). *Jpn. Soc. Snow Ice* **30**, 75–80.

Wechsberg, J. (1958). ''Avalanche.'' Knopf, New York.

Williams, K. (1975a). *U.S., For. Serv. Rocky Mount. For. Range Exp. Stn., Gen. Tech. Rep. RM* **8**, 1–190.

Williams, K. (1975b). *U.S., For. Serv., Rocky Mount. For. Range Exp. Stn., Res. Note RM* **300**, 1–4.

Yosida, Z. (1963). *Low Temp. Sci., Ser. A* **21**, 1–12.

Yosida, Z., Huzioka, T., and Kinosita, S. (1963). *Low Temp. Sci., Ser. A* **21**, 75–94.

INDEX